AUTOMATION AND SOCIETY

AUTOMATION
AND SOCIETY

Edited by

Howard Boone Jacobson

Chairman, Departments of Industrial Journalism & Journalism
University of Bridgeport

AND

Joseph S. Roucek

Chairman, Departments of Sociology & Political Science
University of Bridgeport

GREENWOOD PRESS, PUBLISHERS
NEW YORK

TO THORSTEIN VEBLEN,

who foresaw the modern social implications
of technological progress.

PREFACE

Archeology tells us that man's discovery of fire, its preservation and its utilization, was a revolutionary departure, placing him firmly apart in behavior from all other animals. This was man at his best, "asserting his humanity and making himself." Fire gave Stone Age man a mastery over nature, the New Stone Age found man controlling his own food supply, the Bronze Age evolved specialized industries and organized trade, the Iron Age permitted universal application of implements for cultivation and production, and blossomed into the great Industrial Age.

Throughout each of these "archeological ages," man's biological evolution has become almost imperceptible and his cultural change more recognizable and definable. Each of these "ages" has revealed evidence of being ushered in by an economic revolution, particularly improvements in implements which men use, and from a biological standpoint each of these "revolutions" was a "success" because the species survived and multiplied.

Is Automation a new economic revolution in the making? Some Jeremiahs have called it economic revolution, but the realists have labeled it no more than an evolution in culture, a mere extension of what has been known for some two centuries as the Industrial Revolution. What the contributors to this book, who represent the combined resources of government, industry and education, have demonstrated most importantly is that Automation is really a twentieth century symbol for progress and change.

Each economic revolution reacted upon man's attitude towards nature, affected his values, changed his institutions and determined his social relations—and civilization as we know it. Automation has produced such a reaction, a clamor for re-educating for change, for a better educated man generally, for reconsideration of our fundamental concepts about work and how we work with one another, and for new values of human conduct and intellectuality. Even survival itself is an issue.

One thing is certain. In the centuries-long process of substituting the machine for muscle and human drudgery, man is slowly giving up the idea that he has to be a beast of burden. Automation is man asserting his humanity once more, making fuller use of his brain capacity for social cooperation and welfare, making himself more human. This book means to show that Automation is more than just an act or method of doing something differently, that it is an intellectual and social challenge, "heavy with promise and with problems."

The Editors

CONTENTS

SECTION I—THE CONCEPT OF AUTOMATION

SECTION II—AUTOMATION APPLIED

SECTION IV—AUTOMATION AND SOCIETY

AUTOMATION AND SOCIETY

Section I

The Concept of Automation

CHAPTER 1

THE RISE OF THE CONCEPT OF AUTOMATION

PAUL T. VEILLETTE

*Acting Director, Division of Operator Control
Connecticut Department of Motor Vehicles*

Albertus Magnus, the German Dominican monk, is said to have spent 30 years in building a robot which advanced to the door when someone knocked, opened it, and greeted the visitor. Although this seems a pleasant fable to us it gained Magnus a reputation as a sorcerer among his contemporaries in 13th-century Europe. A learned scientist—second only to Roger Bacon in his time—Magnus might today be known as the "Father of Automation" if fact and fancy had been joined. Instead, the world had to wait over four centuries for the first known inventor of an automatic device.

The term "automation" has two claimants to authorship. John Diebold, the well-known management consultant, and Delmar S. Harder, a Ford executive, created the term independently of each other and gave it different meanings.[1] Diebold used it to describe "both automatic operation and the process of making things automatic." To Harder the term denoted automatic transfer of work-pieces from one machine to another in the production process without human aid. His definition, more limited and specific than Diebold's, grew out of his experience

---

1. Harder coined the word at a Ford staff meeting in late 1946; Diebold, who gave it its popularity, simplified and shortened "automatization" into "automation" while writing a student report at the Harvard Graduate School of Business in May, 1951.

at Ford in the middle forties. At that time, Ford was replacing the assembly belt with transfer machines, which, as their name indicates, transfer work-pieces from one station to the next. Diebold and Harder both concede that their original definitions do not adequately describe automation today. Diebold, speaking before a CIO-sponsored conference in 1955, found the word in its current loose use "very difficult to define." Automation, he said, had acquired two meanings: (1) self-regulation through feedback and (2) the integration of machines with one another. The reader should be warned that, although the writer accepts this particular concept of Diebold's, there is no generally-held definition of automation. Its meaning is nebulous; its use varies with the user.

Feedback, the key to Diebold's first meaning of automation, entails the use of sensing media that continuously compare the actual condition of a controlled quantity or mechanism with a desired condition. The difference between these conditions then creates a corrective force which brings them into harmony. Because the sequence of operations is such that the results of a mechanism's actions are fed back to it and modify its further behavior, feed-back is sometimes called closed-loop control.

Illustrations of this type of control are not limited to automatic devices but can be found in abundance in nature. The human body constantly employs feedback control. For example, the maintenance of body temperature within narrow limits is accomplished through feedback. Acting in response to a temperature-sensitive area at the base of the brain, the body decreases its temperature through evaporation (sweating) and through dilation of the blood vessels near the skin. Conversely, contraction of the surface vessels helps to prevent loss of heat when body temperatures are low.

An example of feedback in the social sciences is found in the Keynesian school of economic thought.[2] John Maynard Keynes,

---

2. In theorizing on the impact of feedback on another social science, government, Andre Marie Ampere (1775-1836), the renowned mathematician-physicist, created the word *cybernetique* to denote feedback. Its English cognate, "cybernetics" (the science of communication and control mechanisms), has been popularized in recent years by Dr. Norbert Wiener, M.I.T.'s social-conscious mathematical theorist. When Wiener revived the word he was unaware of its use by Ampere, his source being the Greek word for "governor" or "steersman." Wiener views cybernetics as the nexus of many disciplines, ranging from servomechanism theory to Gestalt psychology.

its founder, postulated that an industrial economy contains two interacting feedback loops—one represented by consumer activity, the other by capital activity. Both affect the general level of the economy, and their proper manipulation, according to Keynes, tends to decrease the swings in the business cycle.

EVOLUTION IN COMMERCE AND INDUSTRY

The first devices deliberately contrived to control a process automatically, Denis Papin's pressure cooker, did not employ the closed-loop, but the open-loop system. This is a relatively simple means of automatic control, wherein a mechanism is capable of acting automatically but is incapable of self-correction. Papin's pressure cooker (heralded as a new method of cooking meat and of "softening bones") was invented around 1680, a few years after Papin, a physicist and a convinced Calvinist, fled from his native France to England to avoid religious persecution. It was followed, in 1713, by another open-loop device, invented by Humphrey Potter, an English teen-ager. Potter (whose existence is questioned by some historians) developed a method of controlling the flow of steam in steam engines by using a slide valve mechanism linking piston and valves. Through this linkage, steam was admitted to or exhausted from an engine cylinder automatically. Potter's mechanism has been used in every subsequent steam engine to control cylinder steam pressure.

Closed-loop control has a lengthy but sporadic history. The float control valves in the plumbing systems of ancient Rome utilized feedback, as did Dutch windmills some centuries later. However, the most significant event in the early history of feedback control occurred in 1788. In that year the Scottish instrument-maker and inventor, James Watt, invented the flyball, or centrifugal, governor to control the speed of steam engines. Previously speed had been regulated manually by a throttle valve. But by linking a flyball governor with the output shaft of the engine and also with the valve that controlled steam input, Watt was able to automatically maintain constant engine speed. As the engine's speed increased the spinning governor impelled the flyballs outward proportionately, gradually decreasing the supply of steam to the engine and consequently slowing it down. Conversely, if it went too slowly, the balls collapsed inward, gradually opening the input valve. Thus, a closed loop—with input inducing output, output modifying input—enabled the engine to regulate itself.

The next noteworthy advance, in 1868, applied feedback to the steering of a large steamship, the *Great Eastern*. This was accomplished mechanically by (1) cable connections between the helmsman's wheel and the throttle of a steam-operated steering engine in the stern, (2) gearing this engine to the ship's rudder post, and (3) linking the rudder post and the throttle of the steering engine by means of a "follow-up," or feedback, linkage. As the follow-up mechanism responded to the movement of the ship's rudder, it gradually closed the engine's valves, thereby prohibiting additional steam from entering the cylinder once the rudder had moved to the position fixed by the helmsman's wheel. Hence, by feedback control of rudder position, zig-zagging was reduced, and the ship was kept on an even course.

Four years after this achievement, Joseph Farcot, in Paris, coined the word "servo-motor" in naming a similar, though more advanced, ship's steering engine. Servo-motor, or servomechanism (a more general term),[3] is now used to denote devices for automatic position control. The *Great Eastern's* follow-up mechanism represents the maiden application of closed-loop position control.

Until early in this century progress in automatic control centered in the hydraulic and mechanical fields. However, since that time electronics has played an increasingly prominent role, and, in recent years, electro-mechanical and electronic computers, in particular, have been in the forefront of automatic control. During the 1930's, Dr. Vannevar Bush at M.I.T. developed the modern analogue computer. It was not electronic but electro-mechanical. One of the two basic computer types—the other is the digital computer[4]—an analogue computer functions by setting up a physical model, or analogue, of the problem to be solved, operating on the same principle as a slide rule. Like many modern developments in automation, the Bush computer had predecessors, although they were less complex or even un-

---

3. A servomechanism is defined by the Feedback Control Committee of the American Institute of Electrical Engineers as "a feedback control system in which the controlled variable is mechanical position."

4. Simply stated, the modern digital computer is a counter that operates by adding and subtracting digits electronically. It performs multiplication through a series of additions, and divides through subtractions. Since it has been highly publicized for use in the clerical field its evolution will be discussed in that context later in this chapter.

successful. One of these, designed by Lord Kelvin (William Thomson) *circa* 1875, was called the "tide engine," and, through the analogue principle, mechanically predicted the rise and fall of tides.

Essentially measuring devices, analogue computers are used today in some of the liquid-products industries to control processing. This is done by continuously measuring, through process controllers (simple analogue computers), the deviation of the actual processing condition from the desired. The amount of error serves then to adjust the controllers, which automatically produce a compensating force via electronic feedback. Analogue computers were also used in World War II to direct naval and anti-aircraft guns automatically. Their design and that of the electronic digital computer resulted from the pressure of military demands—demands which gave impetus to basic and applied research in feedback control systems generally.

The added knowledge of control systems gained during World War II stimulated the application of automatic controls to a variety of tasks during the post-war period. For instance, computers, both digital and analogue, had originally been built to help solve the mathematical problems facing scientists, but during the past decade have been adapted to industrial and business uses. Not only are analogue computers being employed as industrial process controllers, but plans are under way to combine them systemwise with digital computers to effect overall control of automatic factories. In other words, numerous process controllers—each dealing with only one variable, such as temperature —are to be supervised by digital computers. In this way a continuous process, from beginning to end, can be analyzed and adjusted.

To this point, we have traced the evolution of industrial automation in relation to one of the main processes in manufacturing —control. Another process is material handling, which includes assembly as well as the transfer of material from one machine to another. It is common for students of automation to say that automating control (and its auxiliary, inspection) means replacing Man's sensory organs and brains by machines, while automating material handling means supplanting his muscles. The latter transition—which relates to Diebold's second meaning of automation, i.e., the integration of machines with one another —has as its object continuous production with a minimum of human labor.

Automatic material handling, like automatic control, has a long lineage, dating back at least to 1784. About that time Philadelphia's Oliver Evans established a fully automatic flour mill, producing cleaner and more uniform flour than conventional mills of its day. Using three types of conveyor mechanisms in a continuous production line, as well as controls to regulate grinding, Evans mechanized the entire process from raw grain to bagged flour. This achievement marked the beginning of continuous transfer and assembly, with the product untouched by human hands during processing.

As Evans demonstrated, the continuous method of processing is well-suited to liquid and fluid products. In fact, material handling in two liquid-process industries, oil refining and chemical production, was largely automatic a generation ago. And today several Atomic Energy Commission plants are almost completely automated. But the hard, discrete products—metal goods, for example—pose a more difficult problem, since they are unyielding and often awkward masses. The initial impetus toward automatic assembly of this type of product occurred in 1789 when Eli Whitney first manufactured one of its prerequisites, the interchangeable part. Although Whitney was concerned with gun manufacture the principle he discovered had general applicability. A century and a quarter later, the interchangeable part coupled with the assembly line permitted Henry Ford to inaugurate mass production in the automobile industry.

As every school child learns, mass production had a revolutionary effect on American industry. Yet Ford, in reviving the assembly line—it had been used in Illinois for dismembering hogs in 1869—did not employ automatic assembly or transfer. The assembly line, important though it was, only partially mechanized the flow of parts and subassemblies throughout the Ford plant; manual material handling was still very much in evidence. Instances of automatic material handling were soon to come, however. In 1915, the year after Ford initiated his assembly line, one of the earliest illustrations of automatic assembly had its genesis, fathered by a decision of the A. O. Smith Company of Milwaukee to build a plant for manufacturing automobile frames automatically. Some years later the plant, complete with automatic handling and assembly devices, was able to produce a frame every eight seconds. One authority describes it thus: "There were nine separate units linked by automatic handling devices; these units inspected the incoming strips of steel, rejecting all that were unsatisfactory; they then cleaned, bent, and drilled the remainder, assembled and riveted them, washed and painted them, and finally dispatched them to store."[5] Another advance in material handling, often cited in automation literature, is the automatic transfer machine. A British contribution of the early 1920's, it was first used in the Coventry plant of Morris Motors to machine flywheels for automobiles, but due to then existing gaps in technical knowledge was soon discontinued. Ford successfully adopted it twenty years later to machine cylinder blocks, and was followed by the French government which, at that time, was modernizing the newly-nationalized Renault Company. An ingenious invention, the transfer machine unites varied operations in a single device and can move a work-piece from one station to another without human aid. In some cases, although these are exceptional, the assembly of parts is also performed automatically.

Today, transfer machines are not uncommon in the automotive and electronics industries. But despite their use, material handling in the discrete-products industries remains a frontier to be conquered.

EVOLUTION IN THE OFFICE

The *raison d'être* of the office is data-processing, just as that of the factory is product-processing. Data-processing, or information-handling as it is sometimes called, has within it three functions, broadly stated; these are classifying, computing, and recording. The data-processing functions themselves, taken as a totality, comprise only one of several major office processes, the others being data-inspection, -storage, and -transportation. As one would expect, data-processing, as the core office process and the most complex, has been historically the object of attention and of attempts at improvement. The history of these attempts is the history of office automation.

The rise of office automation can be thought of as occurring in stages, even though these overlapped historically and co-exist today. Each of the stages—manual, mechanical, punched card, punched paper tape, and electronic—has had its distinctive impact on the data-processing functions and on administration.[6]

---

5. R. H. MacMillan, *Automation: Friend or Foe?* (London: Cambridge University Press, 1956), 31.

In the manual stage, the origins of which are lost in antiquity, the data-processing functions are usually separate. For example, sorting (classifying) office documents into some type of order is operationally independent of computing or recording. The mechanical stage, in contrast, introduces the combination of two or more functions within basically the single operation. Adding machines, for instance, both compute (summarize) and record; bookkeeping machines also compute (calculate, summarize) and record.

The adding machine—and with it office machine methods—had its beginning in 1642 when Blaise Pascal invented his *Machine Arithmetique,* a device for automatic addition and subtraction. Twenty-five years earlier a simple multiplying device, called Napier's Rods, had appeared in France. But this was not a machine, as commonly thought of; it was a calculating tool based upon the Pythagorean tables, and can be classified with the Chinese abacus. A genuine multiplying machine did not make an appearance until 1670. An adaptation of Pascal's adding machine, it was invented by Gottfried Wilhelm von Leibniz, the Rationalist, and multiplied by repeated additions, just as the electronic computer does today. But in spite of the contributions of Pascal and Leibniz, calculators were unknown in the offices of the 17th century. The first practical commercial calculator was not marketed until 1820 when Thomas de Colmar of France introduced his Arithometer.

Intercommunication among similar machines, through the medium of punched cards, is the identifying feature of the punched card stage. The "native language" through which punched card machines communicate consists of patterns of holes in cards, which can only be understood, or "read," by similar machines. A given set of punched cards can activate a machine that will compute and record, another that will sort and add, and still another that will interfile the punched cards into a file of similar cards. The originator of the first punched card machines used in the United States was an employee of the United States Census Bureau, Dr. Herman Hollerith. In 1897, Hollerith devised a punched card adding unit to process census statistics, and it was used for this purpose in the 1910 census. This occasion was not the dawn of punched card equipment,

6. For a discussion of this subject see Paul T. Veillette, "The Impact of Mechanization on Administration," *Public Administration Review,* XVII (Autumn, 1957), 231-237.

however. About a century before, in 1801, Joseph Marie Jacquard utilized punched cards to activate an automatic loom. The holes in the cards contained the weaving pattern. Jacquard's cards permitted diverse and intricate patterns to be produced cheaply and accurately, and within a dozen years of their introduction more than 11,000 card-activated looms were operating in France alone.

The fourth evolutionary stage, distinguished by the use of punched paper tape, brought intercommunication among machines that were formerly unable to communicate with one another. Prior to this development raw data from a typewriter, for example, could not be transferred directly to punched card equipment. The human factor had to be interjected into the process, i.e., a key punch operator was needed to create punched cards from the data produced on the typewriter. The operator served, in effect, as an interpreter, since typewriters and punched card equipment have different native languages. The birth of punched paper tape brought with it a common machine language.

In 1870, Jean Baudot, a civil servant in the French Ministry of Posts and Telegraphs, perfected a punched tape containing a series of holes in vertical columns.[7] Each column had up to five holes. The tape served as the common language for various models of telegraph machines, being emitted by some and read by others. As the office equipment industry advanced, the five-hole, or five-channel, tape became the common language link among a wide variety of office machines. Typewriters, calculators, bookkeeping machines, telegraph equipment, and graphotypes, as well as punched card equipment, were now able to be functionally integrated into a single data-processing system. For this reason, the common language concept of paperwork management is often referred to as integrated data processing (IDP).

The latest stage in the evolution of office automation, the electronic, introduced intercommunication among data-processing functions in a single machine at fantastically high speeds.

---

7. The common progenitor of Baudot's tape and Jacquard's punched cards was an invention by Basile Bouchon in 1725, a loom controlled by a roll of perforated paper. Twelve years later Bouchon substituted punched cards for the paper roll. The significance of Baudot's and Jacquard's creations lies in their practical application on a large scale.

Through such internal intercommunication a complex series of data-processing operations could be performed automatically once the machine had been programmed. A brief description of the electronic digital computer will illustrate the point. The digital computer has five components: input, memory, arithmetic, control, and output. The input media include punched cards, punched paper tape, and magnetic tape. The memory unit—which stores information until it is ready for use—consists of a magnetic core, a magnetic drum, or a magnetic tape, among other possibilities. Capable of intercommunication with the memory, the arithmetic unit adds, subtracts, multiplies, divides, compares, and makes logical choices between alternatives. Instructions for the arithmetic unit come from the control unit which is also capable of communicating with the memory, where the instructions are stored. Output media are usually documents or reports, but can consist of punched cards, punched paper tape, or magnetic tape.

A product of World War II, the digital computer was created, in part, to satisfy military needs for trajectory and ballistics tables, but today it is used largely in the research and clerical fields. Conceptually, it is not new. Over 100 years ago Charles Babbage, an English mathematician, designed a general-purpose digital computer which he called an "analytical engine." This was to be a mechanical computer, of course, since electronics was unknown. But Babbage was a victim of his time, for the technical limitations of the 19th century thwarted his plans. The computer was never built.

The first general-purpose digital computer actually constructed, christened the Mark I, was completed at Harvard in late 1945. It was not electronic but operated through electromagnetic relays. ENIAC, completed in early 1946 at the University of Pennsylvania, was the pioneer electronic digital computer. It was conceived by Dr. John Mauchley, then of the United States Bureau of Standards, and developed jointly with J. P. Eckert, Jr. for the Ballistic Research Laboratory at Aberdeen, Maryland. The post-ENIAC decade saw the rise of numerous computers, some designed to solve intricate scientific and mathematical problems, others to serve as filing systems. It is not within the scope of this chapter to discuss current applications for digital computers; those will be treated elsewhere in this book. Suffice it to say that the computer is constantly being modified and refined to adapt it to the varied and changing requirements of business, government, and industry. Although the thread running through this short history of office automation has been the data-processing functions, the writer does not mean to imply that the other major office processes have not felt the impact of technological progress. Indeed, a moment of reflection will reveal that the obverse is true, for the processes are intertwined and cannot be separated completely except for purposes of analysis. In varying degrees changes have occurred in all four processes. Data-storage, for one, has been literally revolutionized. Data are no longer stored almost exclusively on paper, but are captured in punched cards and paper tapes, in magnetic tapes and discs. Further, in regard to data-inspection, modern business machines have the ability, within limits, to make logical checks of data for accuracy. And Baudot's teletype tape—to recall but one illustration—has had an obvious impact on the data-transportation process.

The march toward fully-automatic offices and factories has steadily increased in tempo over the past 300 years, but the ground to be covered is still immense, and progress, relative to the magnitude of the task, is not as great as commonly assumed. Whether most organizations—large or small, regardless of type—can be automated is a moot point among students of automation. The technological obstacles confronting us are formidable and the problems perplexing. But equalling or exceeding these are the social and management problems inextricably accompanying them. To master them, Man's capacity to guide social change must keep pace with his technical progress; otherwise, the dignity of the individual, that cherished cornerstone of our Judaeo-Christian and democratic society, will be jeopardized, and automation's promise, "the human use of human beings," will be an illusory one to thousands of our fellow men.

Selected Bibliography

Esther R. Becker and Eugene F. Murphy, *The Office in Transition* (New York: Harper and Brothers, 1957). The opening chapter deals briefly with the developmental stages of office automation, while the second predicts the impact of automation on the office in the next decade. The remainder of the book discusses automation from the standpoint of the systems and procedures analyst, with the exception of the closing chapters which cover automation's effect on office unionization and management. Thornton F. Bradshaw and Maurice S. Newman, "The Evolution of Data-Processing," *Electronic Data-Processing in Industry* (New York: American Management Association, 1955), 17-38. This paper traces the evolution of data-processing in the office as a series of technical stages, without regard to the personalities involved in its development or the chronology of events. Only moderately technical, it is perhaps the best succinct treatment of the "building block" approach to data-processing.

James R. Bright, *Automation and Management* (Boston: Harvard University Graduate School of Business Administration, 1958). A newly-published volume, it competently discusses the evolutionary nature of automation in manufacturing, using the electric lamp and shoe industries as examples. In addition, case studies of automation experience are presented, and critical areas of automation impact—such as sales, personnel, work force, maintenance, and management—are examined.

H. R. Hunt Brown, *Office Automation* (New York: Automation Consultants, Inc., 1956). A loose-leaf book periodically brought up to date by the author, it gives concise summaries of the many office machines on the market today. The author includes a short history of office automation, complete with illustrations.

P. E. Cleator, *The Robot Era* (New York: Thomas Y. Crowell, Inc., 1955). On a popular level, with some good illustrations, this wide-ranging book gives numerous examples of automation, historical and current, in a variety of fields. In expanding on his theme, i.e., that organizations are evolving steadily toward complete automation, the author describes a myriad of machines at work.

John Diebold, *Automation: The Advent of the Automatic Factory* (New York: D. Van Nostrand Company, 1952). Including within it a history of closed-loop control and computers, this informative book approaches automation from several standpoints, including the technical, sociological, economic, and educational. It is easily understandable to the intelligent lay reader, and has the standing in automation literature of an early popular classic.

Carl Dreher, *Automation: What It Is, How It Works, Who Can Use It* (New York: W. W. Norton & Company, 1957). A popularized work interestingly written, it is broad in scope.

A chapter is devoted to the evolution of automation, with emphasis on industry and commerce. Also included: the use of automatic machines in war, sports, the arts, medicine, and interpreting languages.

E. M. Hugh-Jones (ed.), *The Push-Button World: Automation Today* (Norman: University of Oklahoma Press, 1956). The implications of automation for industry, labor, administration, and society are analyzed in this work by seven British engineers and scientists. Among other historical notes, it contains an authoritative statement on the origin of transfer machines by an automation pioneer, F. W. Woollard, who was directly connected with their introduction into the English motor car industry.

Howard S. Levin, *Office Work and Automation* (New York: John Wiley and Sons, 1956). Dealing in the main with computers and common language equipment for the office, the author also comments on the technical origins of electronic computers. The effect of science on business resulting from the rise of the automated office is examined with perception.

R. H. MacMillan, *Automation: Friend or Foe?* (London: Cambridge University, 1956). The author traces the development of automatic control and production, discusses control system design problems and the social and economic aspects of automation. Emphasis is on automatic control in industry, with a brief discussion of computer use in the office and the future of automation. A logically organized and clearly written book.

Editors of "Scientific American," *Automatic Control* (New York: Simon and Schuster, 1955). Paperbound, this book presents reprints of important articles from past issues of "Scientific American." It stresses the importance of feedback control and distinguishes between automation and mechanization. Historical comments are few but where given are valuable, since they are accompanied by penetrating analyses of pertinent concepts.

David O. Woodbury, *Let Erma Do It* (New York: Harcourt, Brace & Company, 1956). A readable survey of the current status of automation and its development historically. The text—devoted to automation in industry, the office, and in commerce—was written for the general reader.

PAUL T. VEILLETTE is acting director, Division of Operator Control, Connecticut Department of Motor Vehicles. A graduate, with a *cum laude,* from the University of Connecticut, he received his M.A. degree in public administration from the Maxwell Graduate School of Public Affairs, Syracuse University (where he was a Maxwell Fellow). A former Fellow of the National Records Management Council, he has been conducting management surveys in Connecticut state agencies and institutions for the past six years; these projects have included feasibility studies for punched card and electronic equipment. He is the author of several articles on mechanization, and has lectured at the International Busines Machines School, Endicott, New York, at the University of Connecticut, and before professional societies. He is a member of the Management Institute Advisory Committee of the American Society for Public Administration, and is currently president of the Connecticut chapter of that society.

Section II

Automation Applied

CHAPTER 2

AUTOMATION IN THE MANUFACTURING INDUSTRIES

RALPH J. CORDINER, *Chairman of the Board, General Electric Company*

It is important that the public have an informed picture of the effects of automation and technological change on our national life because everyone has a personal stake in this progress. Frequently one reads statements which give lip service to the obvious benefits of automation, but go on to exaggerate the problems of adjustment until they make automation appear to be a public menace. Automation is *not* a menacing development. It is, in fact, a concept which has already raised the nation's standard of living and has had a stimulating and stabilizing effect on the economy. It will continue to have this welcome effect in the future.

For practical purposes in planning manufacturing facilities, automation can be defined as "continuous automatic production," largely in the sense of *linking together* already highly mechanized individual operations. Automation is a way of work based on the concept of production as a continuous flow, rather than processing by intermittent batches of work.

AUTOMATION ONLY ONE PHASE IN TECHNOLOGICAL PROGRESS

There are those who try to make "automation" a catch-all term to apply to every improvement, whether new or thoroughly familiar, that occurs in a factory or office. It is important to recognize that "automation" is only one phase in the process of technological progress, a natural evolutionary step in man's continuing effort to use the discoveries of science in getting the world's work done.

The "flow" concept of automation is most easily applied in the processing of liquids, gases, and energy such as electric energy. Thus, in the chemical, petroleum, and electric utility industries you will find a high degree of automation already in existence.

For several decades, our engineers have also developed electric drives with feedback controls for the continuous, increasingly automatic production of materials that can be produced in a continuous strip or web, such as paper, cloth, steel, rubber, and plastics. In parts of these processes "automation" is quite well advanced.

In recent years we have seen increased efforts to apply the principles of continuous flow and automaticity to the production of separate pieces or products. Here, you are likely to find the greatest progress in the manufacture of simple, standardized items, usually small ones. Among our more than 45,000 suppliers, we find many small companies that have done a superb job of mechanization and automation. They thrive on the fact that simple, standard items are easier to produce by automatic techniques than complex and variable products. Nevertheless, we may expect a step-by-step progress toward greater automation even in the industries producing fairly complex products.

MORE AUTOMATIC PROCESSING OF DATA TO COME

In the office and laboratory too, we may expect to see more speedy and automatic processing of data—which might be considered under the heading of automation. In this area the giant computers have caught the popular imagination. Their potential value in helping man to extend the scope of his knowledge, and the things he can dare to do, is greater than we can comprehend at this time. General Electric has been pioneering in the use of computers in scientific, engineering, accounting and marketing work.

Thus, in the factory and in the office, you could say that progress toward greater automation is nothing new; only the expression "automation" is new.

Technological change in industry is a gradual process. Most products are first made by hand, or with hand tools. Then industry mechanizes; it introduces machines for some parts of the process, although many hand operations usually remain. As the economics of the situation warrant, the machines are made more and more automatic. Finally, where we can do it technically and where economic considerations warrant the investment, we link together parts of the process to achieve more continuous and automatic operation. As industry moves its operations up this scale toward automation, there is a greater demand for more highly trained people to handle the larger responsibilities.

Most of industry is still in the *lower mechanization* stage. There are literally millions of hand operations in manufacturing today. Highly skilled machinists, for example, usually spend much of their time and effort placing material in machines, and removing the material. Only part of their time is used as skilled machinists. There are many other wasteful and burdensome hand operations that add to the ultimate cost of goods going to the consumer. This situation makes two comments appropriate:

1. American industry has a challenging opportunity to increase its productivity, reduce the cost of goods, and increase the real wealth and purchasing power of all Americans by using every feasible application of the automation concept all along the line from factory to customer. We will at the same time be taking the danger and dullness and drudgery out of industry, and giving people more satisfying work to do, at higher pay for higher skills.

2. This will be a step-by-step process. In the years ahead, our guess is that 90 per cent of the technological changes in American industry, including General Electric, will be the familiar process of increased mechanization and electrification, along with improvements in manufacturing layout, product design, material selection, data processing, marketing, management methods, and a host of other techniques to make human work more productive—which has been a major preoccupation of man ever since the invention of the wheel.

With this background on the nature of automation and technological change, let us evaluate their future effect on the economy.

GENERAL ELECTRIC'S VIEWS ON AUTOMATION

Let me first state General Electric's views on automation and technological change, and then I will present the evidence supporting these views.

1. Technological progress is necessary in order to maintain our national security.

2. It is necessary if we are to continue to raise the American standard of living even at the same rate as in the last decade. It is even more urgent if we are to *accelerate* the rate of progress.

3. Technological progress is also necessary from the point of view of individual companies; those companies, large or small, which continually modernize in order to serve their customers better, will prosper in our competitive economic system.

4. Progress toward greater mechanization and automation is in the best interests of all the groups that business must serve—customers, share owners, employees, suppliers, and the public.

5. Technological change is a gradual, evolutionary process which creates employment and exerts a stimulating and stabilizing effect on the economy. This can and will continue as long as business has the incentives and freedom to grow, and to create new products and industries.

6. The benefits of mechanization and automation are so profound—and so urgently needed—that we must encourage those companies that push the advances which make industry more productive.

Now let me present the evidence supporting these views.

AUTOMATION AND THE NATIONAL DEFENSE

Many of the key items of military equipment today, such as radar, gunfire control systems, guided missiles, and atomic weapons, are themselves products of automation principles. These military developments have spurred industrial technology; but even more important, continuing progress in industrial technology is essential as a source of knowledge for further improving the nation's military equipment.

For example, during World War II our industrial engineers helped design radical new systems of aircraft armament in which the aerial gunner had push-button control over his guns, and even had computers to help him aim the guns. In coming up with the gunfire control systems for the B-29, the A-26, the B-36, and other airplanes, the engineers used circuits and equipments originally developed for industrial use. The amplidyne generator in the B-29 was originally developed for the steel industry. Thyratron tubes, very important in the B-36 armament, were originally industrial electronic tubes developed for the paper and resistance welding industries. The totally new approach in electric generating systems developed for the B-29 was directly based on our industrial experience.

Thus continuing advances in industrial technology are vital to military technology.

My other point is that our margin of safety in modern arms depends on continuously increasing the productivity of American industry which produces them.

For example, the United States could not even consider a radar defense system if it did not have highly productive electronics and communications industries to design and produce the necessary elements in the huge quantities required.

Another example is jet engines. When the Korean War broke out, the Air Force needed thousands of jet engines as fast as possible. By applying the principles of mechanization and automation in General Electric's Evendale, Ohio, plant, we were able to boost our monthly production of engines 1200 per cent. These J47 jet engines powered the F-86 Sabres which scored a 14 to 1 edge over the Communist jet aircraft in the Korean War.

As an illustration of technical progress, the thrust of a J47 engine has been increased 23 per cent through design improvements and one J47 model has an allowable flying time of 1700 hours, compared to 150 hours in 1950.

In spite of the cost of making more than 20,000 design improvements, we were able progressively to reduce the price by about $15,000 per engine. This resulted in a total saving to the taxpayers, from cost reductions on jet engines, of millions of dollars.

This illustrates how technological progress is essential to our national defense, and effects important savings for the taxpayer.

AUTOMATION AND THE STANDARD OF LIVING

But we are a peaceful nation, and far more important is the necessity to improve our technology in order to increase our standard of living. The new abundance which is possible through greater automation is one of our major weapons in waging peace.

According to our Company's estimates, the United States will require about 40 per cent more goods and services by 1967 with only 14 per cent more people in the labor force. To produce 40 per cent more goods and services with only 14 per cent more people, either everyone must work harder and longer—which is neither a realistic nor a good solution—or industry must be encouraged to invest in more productive machinery and methods. Faster progress in the newer field of automation seems to be the only available solution to this problem, particularly in situations where we have exhausted the known economic possibilities in the more familiar field of simple mechanization.

From all that we can foresee, it appears that there will be a shortage of men and women to fill the work opportunities in the coming decade. After 1965, when the proportion of labor force to the total population increases, some feel that there may be a trend toward the shorter work week. But our feeling is that the demand for goods may rise so fast in the 1970's that we will still be hard put to produce enough goods to satisfy the market on a 40-hour basis, and the American public will choose more goods in preference to a shorter work week.

AUTOMATION IS IN BEST INTERESTS OF ALL

As Plutarch said, "He who would divine the future must study the past." With that in mind, let us look at a case history of an industry where the so-called "Automation Revolution" has already taken place—namely, the production and distribution of electric power.

When you flip the light switch on the wall, you start up one of the most completely automated processes in the world. Electricity cannot be made and stored in advance. It is made and delivered immediately, to your order, at the speed of light. To make such an incredible process safe, reliable, and low priced, it has been made increasingly automatic. From your light switch, back through wires, meters, transformers, substations, switchgear, generators, turbines—right back to the fuel sources such as the coal pile, gas, oil, or waterhead—there are hundreds of self-supervising and self-regulating devices, many of them developed by General Electric engineers and scientists over the past 75 years.

For example, the first completely automatic hydroelectric station, with *no* attendants, was installed in 1917. The first completely automatic substation was installed in 1914, by the Elgin and Belvidere Electric Company. Prior to this, there were usually three attendants at the substation to read meters and turn switches, and obviously they could not deal with interruptions or load changes with the same speed as automatic equipment.

The granddaddy of modern computers was a d-c network analyzer built by our engineers in 1915 to reduce the amount of pencil-and-paper calculation required to design power systems. The principles now referred to as "automation" have been familiar in the electrical industry for many years.

Now, has this progressively automated process of producing electricity injured the consumer and put people out of work? To the contrary, our highly automatic network of power has created a huge electrical industry employing 2,740,000 people and it underlies America's productive capacity and her standard of living.

As the electric utilities have increased their efficiency, the price of electricity to the residential consumer has gone down steadily. At the turn of the century, the price of residential electricity on a national average was 17 cents per kilowatt-hour. By 1957 the price was down to 2.56 cents per kilowatt-hour on the average.

Because electricity costs so little, the *average* home uses more than 3150 kilowatt-hours a year—the energy equivalent of 47 servants. A *fully* electrified home uses 30,000 kilowatt-hours a year—equivalent to 450 servants.

Since 1939 the cost of living has gone up 105 per cent, but the cost of residential electricity has stayed the same, as measured by the Consumer's Price Index. This is one way that the consumer benefits from an industry that has progressively employed the techniques of mechanization and automation.

As the price dropped, volume rose—the classic formula in industries which aggressively improve their methods of production and distribution. Sales of electricity have doubled every decade since the turn of the century. In the past ten years, sales of electricity have gone up 2½ times, with tremendous effect on industry and the home.

In 1957 the average worker in manufacturing had 22,000 kilowatt-hours of electrical energy to help him get his work done —the equivalent, in human energy, of 330 assistants. This is one of the important reasons why our country is so much more productive than the rest of the world.

To design, build, sell, and service the equipment to produce and use this low-cost power, there has come into being a major industry, today employing 2,740,000 people. This includes 400,000 in the utilities, 1,400,000 in electrical contracting, 550,000 in the electrical wholesale and retail trade, and 160,000 in electrical service and repair. Employment in the total industry has more than doubled since 1939, with increases in every segment of the industry.

It is necessary to look at a *whole* industry to assess the full effect of mechanization and automation. The highly automatic production and distribution of power requires "only" 400,000 people, but the process itself creates employment opportunities for *six and a half times* as many people. And the really incalculable effect is on users of electricity—the industries that are powered, the cities that are lighted, and the homes that are made more livable by low-cost electric power.

The increasing use of electricity provides one of the important stabilizers in our expanding economy.

This seems, in fact, to be a characteristic phenomenon. The manufacturing industries which have gone the farthest in automation, and the industries which supply equipment for automation, are the ones in which employment has been rising the fastest in the past twenty years. This would include the communications, electrical, machinery, chemical, rubber, automobile, and petroleum industries. The static, unchanging industries are not the sources of growth in employment in our economy.

In the General Electric Company we have been mechanizing, improving methods, and automating as fast as we can economically develop and apply the required technology. It may be of interest to consider, as a second case study, the results of 19 years of unremitting efforts to introduce greater mechanization and automation in General Electric so that we can judge the effect on all groups that our Company is set up to serve—namely, customers, share owners, employees, suppliers, and the public. We chose the year 1939 as our base because it provides a long enough time span for both good and bad effects to show up; because it is considered by economists to be something like a "normal" year; and because it was a year when the new techniques of industrial electronics were just beginning to emerge.

Results show that consumers have shared in the benefits of increasing automation in General Electric. On a weighted average, General Electric prices went up 75 per cent, while the price of all commodities except farm and food rose 117 per cent. That difference represents money saved by consumers. In the same period we have had an increase of 143 per cent in the cost of our basic raw materials and a 200 per cent increase in average earnings and benefits paid to our employees.

The housewife buying many of our consumer products found a lower price tag in 1957 than she did in 1950, even though the products had improved. For example, the 1950 refrigerator, with eight cubic feet capacity, cost $329.95. In 1957, a comparable refrigerator with 10 cubic feet capacity sold for $219. A 12-inch television set in 1950 cost $230.90. In 1957 you could get the comparable 21-inch set for $209. Much of this price reduction can be attributed to our investment in more automatic production.

For the nearly half a million owners of General Electric, their equity in the Company—their money in the business—has tripled, mostly through retained earnings. But their percentage return on the equity has stayed about the same.

Employment at General Electric has been increasing since 1939 at a rate six times as fast as in the country as a whole. The number of employees at General Electric has tripled from 79,500 in 1939 to 282,000 in 1957, including those employed on atomic projects. Thus 202,500 General Electric employees hold positions that did not exist in 1939—certainly evidence that technological progress, together with skillful marketing, creates *new* employment opportunities.

Employee compensation and other benefits paid by the Company have grown more than eight and a half times, from $153,400,000 in 1939 to $1,715,300,000 in 1957.

Translating that into individual terms, in 1939 the average General Electric employee earned $2,026 a year, including the value of benefit programs. In 1957, a General Electric position was worth $6,082 a year, on the average, and that included a splendid package of pension, insurance, vacation, holiday, and other benefits providing better economic security. Our pension plan began in 1912, and we've had group life insurance since 1920. When you take out the effect of inflation, since 1939 the average employee has had a 49 per cent increase in real purchasing power, except for taxes. This increase also reflects the general upgrading of jobs as we advance toward greater automation. Work is cleaner, safer, and more pleasant in modern factories and offices.

General Electric's progress from technological advance has

been shared by other businesses, large and small. Since 1939, our payments for materials, supplies, and services have gone up more than 10 times. At the present time we have more than 45,000 suppliers, most of them small businesses and many of them aggressively mechanizing and automating their operations. This flow of business to other companies of course creates growing employment opportunities. In addition, roughly 400,000 small companies gain all or part of their income from selling and servicing our products.

The public and its representative, government, have also shared in the benefits of automation in two principal ways. First, there are the improved product values and new products to which I have already referred. Second, there is the effect on the national defense, and the tax savings that automation makes possible such as the multimillion-dollar savings in jet engine costs.

Results such as these, in General Electric and in the electrical industry, would not have been possible without mechanization and automation, although other factors have obviously contributed. It is in the setting of these immense benefits to consumers, share owners, employees, suppliers, and the public that we must consider the human problems which go along with technological advance.

ADJUSTING TO TECHNOLOGICAL CHANGE

Automation usually involves labor-saving machinery and methods, just as simple mechanization does. And, theoretically, in individual instances it would appear that people would be laid off because they are no longer needed on a particular job. But in the General Electric Company, because of the gradual nature of the improvements and the small ratio of the improvements to the continuing base of operations in the plant by unchanged methods, it is seldom that a person is put out of work by an improvement. Other factors—particularly national economic conditions and their effect on General Electric sales—are the significant determinants of employment levels.

Naturally, General Electric management puts extra thought and effort into minimizing lay-offs by having other work ready for the employees involved whenever possible. Good planning for automation includes planning for the all-important human problems as well as the mechanical and financial problems.

The fear that automation will move too swiftly for orderly adjustment overlooks the powerful factors which govern the pace of technological advance.

First of all, there is the difficulty of actually thinking through and designing workable automation developments.

Second, the financial risks involved must be evaluated—and they are serious enough to make a businessman weigh carefully each investment in automation.

And third, management must work out some way to assure the wider, steadier, market which will justify the investment in new machinery and methods.

At General Electric, we try to plan any substantial technological changes in such a way that normal attrition of our work force—the people who quit, retire, or die—will absorb the shift in employment. The factor of normal employee turnover has not been adequately appreciated in most discussions of automation. This, combined with job changes within the Company, is how General Electric—and probably industry generally—takes care of short-term adjustments related to technological change.

Naturally, the Company provides any training required to enable employees to handle new assignments. Most of this training is informal in nature and is done by individual supervisors on the job or through vestibule training schools that run from one to two weeks, in preparation for a specific kind of work. In addition, our Company conducts *more than a thousand courses* in factory skills and at least 500 courses at its various locations for professional, technical, and semi-technical personnel in the areas of finance, manufacturing, engineering, supervision and management, and marketing. We estimate that in an average year *one out of every eight* General Electric people at all levels of the organization takes advantage of Company-conducted courses.

We estimate that we spend on the order of 35 to 40 million dollars a year to train or retrain our employees. This figure includes idle time and scrap and waste resulting from inexperience and the training needed.

Such statistics give you some measure of General Electric's willingness and ability to handle whatever retraining activities are required by technological advance. In addition, we have an expanding program of support for education generally, which in 1957 included about $1,500,000 in scholarships, grants, and fellowships.

If, in spite of the best planning we can do, some people are temporarily unemployed because of technological change, both industry and government have recognized a responsibility to help families through any such periods of transition as they seek new employment. The states provide unemployment compensation, and in all but two states, the entire cost is borne by employers.

But even more important is the role of industry, and of automation itself, in making new employment available. Generally speaking, there are four sources of new employment that arise from automation and technological change.

HOW TECHNOLOGICAL PROGRESS CREATES EMPLOYMENT

1. *Technological progress sets off a sort of chain reaction of economic growth:* more productive machines reduce costs and prices; this increases volume of business, creating a need for more workers. The period between the installation of new machines and the build-up of business is generally very short. It has to be, or the company could not afford to invest in the machinery.

The reverse is also true: if a company fails to modernize, it will lose business, and *fewer* workers can be employed. A company *owes* it to its customers, share owners, *and* employees to modernize and thus remain competitive.

2. *The service industries provide new employment.* Our economy, as its progresses toward greater automation, spends less of its effort (proportionately) in making things and more in selling, servicing, and using things. In 1947, purchase of services accounted for 31 cents of the consumer dollar. In 1957, the figure was 38 cents—up 23 per cent in 10 years. Technological progress creates more leisure and wealth for cultural and educational activities. Hobbies, sports, travel, entertainment, and retail trade are increasingly important sources of employment.

3. *The industries supplying automation and technological advance also create new employment opportunities.* We have what might be described as a "bow-wave" theory of technological employment. When a boat moves at high speed, the water it displaces piles up in *front* of the boat, in what is called the "bow-wave." In an analogous manner, there is a wave of new

employment opportunities that runs in front of automation and technological change—the employment involved in designing, selling, building, and installing the new machinery and controls, along with the new buildings required. In addition, there is additional employment required to maintain and service the equipment after it is installed—and to sell and service its increased output.

This "bow-wave" of technological employment has not yet been adequately studied, statistically, and deserves the attention of interested economists.

4. *Entire new industries, employing thousands, are created by the new automation technologies.* The great chemical, petroleum, and electrical industries, among the fastest growing industries in America, simply could not exist without mechanization and automation. You cannot make chemicals, gasoline, and electricity by hand.

On the horizon we see an atomic energy industry, a transistor and semi-conductor industry, an industry for the production of the supermetals like titanium and zirconium, and even the man-made diamonds that came out of the General Electric Research Laboratory. These and many others will grow into sizable areas of employment. Advanced techniques make such difficult products possible, and of course create new employment opportunities.

In General Electric nearly 100,000 of our employees work on new types of products we did not make in 1939, such as television, jet engines, chemical products, and atomic energy. Not only research, but advanced manufacturing methods, makes such new products possible.

The computer, extending man's mental capacities beyond anything we can imagine, will create fantastic increases in human knowledge and thus vastly increase the number of things we can make and enjoy. Based on General Electric's experience with these machines, it may well be that the computer-derived technologies will be a major source of new employment in the 1960's and 1970's and that they will keep us perpetually short of manpower to take advantage of our opportunities.

Thus, we have four factors at work to provide new and increased employment opportunities:

1. The chain reaction of lower costs, higher volume, and higher employment.
2. The expanding service industries.
3. The automation-supply industries.

4. The new products and industries growing out of the new technologies.

There are two additional factors which indicate that automation will have a *stabilizing* effect on the national economy and employment.

Automation programs require long-range, detailed planning of capital investment, and the pursuit of these plans regardless of temporary ups and downs in annual sales. Thus, investment in automation will increasingly serve as a general stabilizer in our economy.

As automation and mechanization are introduced in a company's operations, fixed costs go up. With high investments in machinery, industry has one more incentive to keep those machines running as steadily as possible. This provides a great stimulus for better planning, more professional marketing, and all the other techniques for maintaining steady demand and employment.

Now, when you couple these factors with the simple fact that the nation's appetite for goods in the next decade will rise faster than the number of people available to produce them, you can see why we feel that automation and other technological progress are necessary and beneficial, and that they exert a stimulating and stabilizing effect on the economy.

On balance, it appears that automation is part of the general picture of research and progress which has characterized our strikingly successful American competitive business system. It is in the public interest to have tax policies and other economic policies which will encourage business to invest in research and greater productivity because these are the sources of new employment and national wealth. They are the real and substantial sources of increased purchasing power throughout the economy, not only among the 25 per cent of the labor force engaged in manufacturing, but among all the families in America.

Concepts like automation are at once an expression and an instrument of the vitality of the American people. They serve us well in our continuing search for better ways to work and live.

RALPH J. CORDINER was president of the General Electric Company with headquarters in New York City, until April, 1958, when he became Chairman of the Board. Born in Walla Walla, Washington, on March 20, 1900, Mr. Cordiner attended

Whitman College there, majoring in economics. In 1923 he joined a General Electric affiliate, the Edison General Electric Appliance Company, and five years later became its Northwest manager. He was appointed manager of the General Electric Heating Device Division in Bridgeport, Conn., in 1932. In December, 1942, Mr. Cordiner went to Washington as director general of war production scheduling for the War Production Board, and three months later was appointed vice chairman. In July, 1943, he returned to General Electric as assistant to the president, and in February, 1945, he became a vice president and assistant to the president, Charles E. Wilson. On April 21, 1949, he was elected executive vice president and a director of the Company. When Mr. Wilson retired from General Electric on December 15, 1950, to become director of the Office of Defense Mobilization, Mr. Cordiner was elected to succeed him as president of the General Electric Company. Under Mr. Cordiner's leadership the Company has experienced rapid growth, and in 1956 its sales reached the $4 billion figure for the first time.

The Economic Club of New York on March 12, 1957, presented its first Gold Medal Award for management to Mr. Cordiner "to recognize excellence in management and to emphasize the contributions made by the executive and his company to the strength of our nation and to the prosperity of our people."

Mr. Cordiner served as chairman of the Defense Advisory Committee on Professional and Technical Compensation in the Armed Forces, 1956-1957. He is currently president of the Employers Labor Relations Information Committee, a member of the Business Advisory Council, the Committee for Economic Development, the National Electrical Manufacturers Association, the Economic Club of New York, the American Management Association, and the Electrical Manufacturers Club.

Mr. Cordiner is the author of the book "New Frontiers for Professional Managers," published in 1956 by McGraw-Hill.

CHAPTER 3

AUTOMATION IN THE AUTOMOTIVE INDUSTRY

D. J. DAVIS, *Vice President, Manufacturing Research, Ford Motor Company*

We at Ford feel that a necessary prelude to any discussion of automation in the automotive industry is a brief review of the history of technological developments within Ford Motor Company itself. Many regard the founder of our company, Henry Ford, as the father of mass production techniques in the manufacture of automobiles. Without these techniques, there would never have been an automotive industry as we know it today.

Shortly before World War I, Mr. Ford set up an assembly line to produce fly wheel magnetos. So far as we know, this was the first installation of progressive assembly on a moving conveyor in the industry. It reduced assembly time of the magnetos from twenty minutes to five minutes. This progressive assembly line principle then was applied to many other small parts, with the result that assembly time of those parts was cut in half.

These successes led to the big test of the new method. Assembly of an automobile chassis in a street alongside the factory at Highland Park, Michigan, was tried by pulling the chassis along a 250-foot route past stock piles of parts spaced at regular intervals. This experiment reduced chassis assembly time from fourteen hours to six hours. It was then decided to mount the wheels on the chassis and roll them along a channel track from one station to the next. Introduction of a moving chain conveyor then eliminated the need for pushing the chassis.

These successes resulted in the installation of conveyors throughout the Ford manufacturing plants to deliver parts to the main assembly line. This advanced assembly technique was a very important factor in enabling the Company to produce the Model "T" car in the large volumes and at the low prices which only a few years before would have been thought impossible.

Conveyors at Ford not only increased productivity by freeing operators from manual handling work, but also decreased overhead costs by reducing aisle and floor storage space. The net result of "conveyorizing" at Ford was a reduction, over a five-year period, of 50 percent in the production costs of the Model "T." The reflection of that reduction in the selling price of the car greatly stimulated demand and opened the Auto Age.

Even in the very early days of the Model "T," the automotive industry had come a considerable way from the time when the first cars were built in back-alley shops and when no one took them very seriously. Those that were built in these little shops were so fabulously expensive that very few people could afford to buy them. Although there were only four cars registered in the city in 1895, Chicago banned them from the streets. Nevertheless, the idea of automotive transportation took hold and, in 1899, output reached four thousand. When, in 1904, production went up to twenty-three thousand, the nation's leading financial experts warned the car makers that they were dangerously over-expanded and soon would go bankrupt.

To us that sounds silly now. As a matter of fact, the bankers probably were right, or they would have been but for one thing. The American automobile manufacturers were to realize the advantages of mass production techniques as established at Ford, and as the cost and prices of the cars went down, the industry grew and America was put on wheels.

From its humble beginnings, it is now a matter of cold statistics that the automobile has been responsible for 90 percent of the employment in the gasoline industry; 80 percent in the rubber industry; 70 percent in the plate glass industry; 60 percent in alloy steel—along with substantial percentages in a great many other industries.

Manufacture of the Model "T," of course, was a relatively simple matter, when compared with the production of our modern cars. The Model "T" had less than half the number of parts that go into our current cars—5,000 as against more than 10,000; they came in any color the customer desired, so long as it was black; there were only a few basic body styles, and those did not change radically from year to year.

In recent years the rapid rate of change in our products has dictated major changes in our productive processes. Our manufacturing facilities now must be sufficiently flexible to permit mass production of hundreds of different models, styles and colors, and these change radically from year to year.

We at Ford have expanded our manufacturing facilities tremendously since the war. By the end of 1958, the Company will have invested more than four billion dollars in modernized and new facilities and plants, and in special tooling. At the beginning of this expansion program, a new department, which we called the Automation Department, was organized to coordinate the planning and in-process handling activities of our stamping operations. This department then branched out and reviewed the in-process handling problems of other operations, such as those for engines and engine components.

These early programs at Ford, however, were concerned chiefly with the rearrangement of standard and existing machines and the means of tying them together with mechanical devices in order to eliminate many of the hazardous and costly manual handling operations. A gradual evolution then began, and at present there are many groups in the various divisions of the Company working on their respective automation programs.

MODERN AUTOMATION AT FORD

All of Ford's new plant lay-outs in recent years have been based upon the use of in-line—or transfer—machines, and mechanical handling devices between them, wherever our studies have shown that their use is justified. In planning and executing this program, we feel that we are doing no more than our predecessors did when they utilized new technology to mass produce the Model "T." We do not believe that automation, as we use it, is a revolutionary development in production technique; rather, it is just another evolutionary phase of our advancing production technology.

Automation is the result of nothing more than better planning, improved tooling, and the application of more efficient manufacturing methods which take full advantage of the progress made by the machine tool and equipment industries.

As used at Ford, automation covers a wide variety of material handling and related devices. In the machining of engine cylinder blocks, for example, automation moves the parts being manufactured into and out of load and unload stations automatically and, at the same time, actuates the machine cycle through electrical interlocking. When it can be applied, automation also is used to index the part, position it, turn it or rotate it, depending upon the requirements of the succeeding operations.

In most cases a finished unit of Ford automation equipment for such machining operations is made up of a combination of standard elements, such as conveyors and simple air, hydraulic and electrical control mechanisms to generate the desired movement necessary for the "in-process" handling.

The electrical interlocking to which I have referred is a signal system, usually accomplished by means of limit switches and relays actuated by the movement or position of the part in the process. The actuation of these limit switches indicates to the automation control panel that parts are in position to be moved by the mechanism, and that all transferring equipment within the machines and equipment are in positions that will not interfere with the movement of the part. To accomplish this, it is necessary that an electrical interlock be established to indicate that the machine or equipment is, as we say, "in the clear"; and also, that it will maintain this condition even though there may be some delay in the completion of the automation cycle. For most machinery and equipment, an interlock is required both at the loading and unloading stations. The loading station interlock must indicate that the space is open and that the transfer bar location will permit the loading of the open station. When the open space has been filled by the automation unit, the interlock must then indicate that the part is in the proper location and that the automatic transfer mechanism is also "in the clear." So long as the machine tool is operating on its automation cycle, it will then permit the next machine cycle to start.

An automated press line is another typical example of the use of automation equipment.

Sheet metal blanks are first positioned at the front of a draw press, where a man loads them into the press loader. The press loader is a mechanical device that loads the blank between the dies and, at the same time, actuates the press cycle. When the press has completed its cycle and the dies open, a lifting mechanism built into the die raises the formed part so that it can be extracted by a horizontal extracting device, or "iron hand," as it is commonly called. The extractor deposits the formed part upon a mechanical device which positions it for trimming in the next press, and at the same time deposits it on an indexing device that conveys the part into the trim die. After trimming, the part is lifted in the die in the same manner as in the draw press and is extracted once again with a horizontal extractor, or "iron hand." These mechanical handling operations are performed at each of as many presses in the line as are dictated by the process. Upon completion of the last press operation, the formed parts are stacked or conveyed to storage or assembly areas.

In our press lines, automation has reduced, but not eliminated, the need for press attendants. One outstanding feature of our new press lines, however, is that attendants are no longer required to have their arms or hands in the hazardous areas of the press operations. Nor do they drain their energy through continuous tugging and hauling on heavy, awkward sheets of metal with raw or sharp edges.

ECONOMICS OF AUTOMATION

The economics of this kind of technological advance are clear. Back in 1908, for example, it took a skilled sheet metal man, working with hand tools, approximately eight hours to shape the upper half of a fuel tank. Today, in our modern stamping operations, it takes approximately twenty seconds. If hand tools were still used to make the upper half of a fuel tank, the labor cost would be approximately twenty-five dollars. Its actual labor cost today is only a few cents. On that same basis, an eighteen hundred dollar car of today would cost approximately eighty thousand dollars.

In today's competitive market we use automation or improved processes wherever they are justified in order to reduce costs or improve our product. If we did not use them, we would soon find ourselves at a competitive disadvantage.

Automation, however, cannot be engineered into every job indiscriminately, since it is not always feasible or profitable. Each application of automation must be carefully analyzed before it can be justified. If daily volume of the part is low, or long-term use of the machine is limited, any possible direct labor savings through automation are reduced, and may be offset by increased maintenance costs and depreciation or obsolescence. For example, we can economically justify the application of automation to the manufacture of engine components for the Ford car engine. The same extensive application of automation, however, cannot be justified on the tractor engine components, due to their lower volume requirements.

As we now see it, there also is little prospect for extensive application of automation in our car assembly operations, where we assemble in twenty different locations and are faced with technical problems and yearly changes in product design. Any automation that is applied here must be capable of ready adaptation to these changing conditions and alterations at a reasonable cost.

In planning for automation, we must be sure that new machines and equipment will produce acceptable parts without excessive down time and maintenance work. In a standard or non-automated production line, operators constantly tend and can adjust each machine. If one machine breaks down, a backlog of materials for the machine can be built up while the repairs are being made, to be worked on later on an accelerated schedule. If one machine in an automated line breaks down, however, all production on that line is halted until the necessary repairs are made. Once the unit is repaired, recovery of lost production must be accomplished on overtime or extra hours.

Although automated equipment and machines may appear to be technically feasible with respect to a particular part, Ford cannot install them unless they can be adapted, modified, or realigned without excessive cost to accommodate the expected changes in the part. Planning for this flexibility requires expenditures of considerable time and money, and when compared with the savings obtainable from automation, we may decide to continue using non-automated equipment, or to use only a limited amount of automated equipment.

If we determine that automated equipment should produce savings in operational costs and that its probable life will permit full depreciation, we must still determine whether its original cost is justified. Automated machinery and equipment, because of its complexity, often costs more (including engineering planning) than non-automated machines and, in any event, is a new investment. Therefore, increased depreciation charges may nullify savings otherwise obtainable and make the risk of installing a new automated production line too great. On the other hand, particularly where new facilities are necessary, automated equipment may cost less than old style machines because of savings in materials from combining several operations in one machine, and indirectly (where plant expansion is involved) because of reductions in floor space, lighting and heating requirements. In some of Ford's new plants, for example, the use of automated equipment required 40 percent less floor space than non-automated machinery and equipment producing the same products in the same quantity.

Thus, while automation is technically possible for many processes, it is feasible for only a portion of them, and requires thorough study before it is applied to any process. Where automation can be economically applied, however, the benefits may be five-fold—increased production, lower accident rate, lower direct labor costs, improved quality in the product, and reduced floor space requirements.

EFFECTS OF AUTOMATION AT FORD

We have been asked, from time to time, what effect automation has had on our labor force. We believe that instead of adversely affecting employment at our Company, automation has created better jobs, while at the same time making them safer and easier.

Our Cleveland Engine Plant, which has been referred to erroneously as an automatic plant by some of the more enthusiastic press and trade magazine writers, actually is a far cry from a fully automatic plant. What we do have at Cleveland, however, and more recently at our Dearborn and Lima Engine Plants, is a marked improvement over our past manufacturing methods. Where once we had two, four, or six separate manually operated machine tools, we now employ a single, multi-purpose machine tool. Where in the past we used chain hoists and conveyors requiring considerable manual handling of heavy, rough pieces, such pieces are now moved automatically from machine-to-machine and are mechanically loaded, positioned and unloaded from the machines. This has resulted in more complete utilization of the machine cycles, which tends to offset the increased investment, and has eliminated many heavy, dangerous, manual handling operations.

Now let us look at what has happened to employment in Ford Motor Company. Installation of automated processes in our stamping, engine, and foundry operations has released a number of employes from unskilled, manual labor jobs. However, automation, while requiring less manual labor, has increased the need for indirect and maintenance personnel.

During the period when installation of automation was progressing, employment also increased. For example, in 1950 we had a monthly average of 115,726 hourly employes on the rolls. These people produced 1,902,488 cars and trucks. In 1957, an average monthly work-force of 143,268 hourly employes produced 2,097,792 cars and trucks.

With respect to the future of automation at Ford, we now can foresee only a limited application in our assembly plants, which currently employ about 33 percent of our total work-force. For instance, the principal difference in our three assembly plants which started production in 1955 and our older assembly plants lies in more adequate floor space, which provides for a smoother and more efficient flow of materials, rather than in any distinct changes in our assembly methods. In addition to the inherent features of automotive assembly plant operations which militate against automation, the competitive demand for continual changes in body structure and trim design currently prevents the extensive introduction of automation in any of our assembly plants.

We have found that, where applicable, automation has supplanted heavy, dangerous and unpleasant work with easier, more pleasant and more interesting work. Moreover, the number of skilled higher paying jobs has increased substantially, in both relative and absolute terms. Finally, and this is of prime importance, these jobs become safer for our employes. As an example, our records for the year 1957 show that on the cylinder block machining operations, which were most significantly affected by automation, the number of accidents decreased 50 percent from 1950. In addition, automation inevitably brings vast new work opportunities to those who are willing to work and learn.

Very frankly, we cannot trace in precise detail the extent and manner in which automation and other measures to improve efficiency have affected our over-all employment figures. Employment volume is affected by a large number of factors, including, among others, product improvements, changes in product mix, and a myriad of make-or-buy decisions on components.

The fact is, however, that Ford Motor Company's nondefense employment has increased, not decreased. During 1957, total hourly man-hours worked were 15.9 percent greater than in

1950—an increase greater than the increase in our unit production.

We do know that without the improved efficiency and cost performance to which automation has made an important contribution, the continuing improvements that we have made in the quality and value of our products would not have been possible within the limits of a competitive price structure. Consequently, we would not now be in the strong competitive position which enables us to maintain our position in the market despite substantial increases during this period in the amount we pay our employes, both directly and in fringe benefits, for an hour's labor.

Looking at automation from the broader viewpoint, all of us recognize that our supremacy as a world power today is due not alone to our natural resources; it is due in major part to our technological progress. This progress, which has been made possible only through a free enterprise system spurred by vigorous competition, has produced for Americans a standard of living envied by the entire world. Automation is just another normal step in our continuous technological progress. Certainly, such progress will create changes. But progress in itself is change —a change always for the better.

We can expect with confidence that automation will gradually bring to our economy the same blessings that all other increases in productivity have brought to us down through the years. In the forward march of technological advances, whole new industries have been and are being created, which provide new job markets for all industrial workers. Automation is but one of the children of a vast family of technological developments now emerging in plastics, electronics, atomic energy, and so forth. All of these developments open unlimited fields for the diversification of industry and for the introduction of many new products that do not now exist, just as television did not exist commercially ten short years ago.

We at Ford do not share the apprehensions of some that the increased use of automation equipment may throw thousands of people out of work, or otherwise dislocate our economy. Indeed, without automation in the steel, chemical refining, food processing and cigarette industries—to mention only a few that are much more highly automated than we ever hope to be— there simply would not be enough production of their products to fill our needs, and certainly not at prices we could afford to pay.

Since, in our opinion, the growth of automation is an evolutionary and not a revolutionary process, it can cause no more than a gradual shift of employment—a shift comparable to that from backward industries into new and growing industries. This type of shift has characterized our dynamic economy for many years, and it is one to which the American people have long been accustomed. The most significant feature of any shift in employment resulting from automation is that much of the shift is from menial labor to higher skilled, better paid, safer and more interesting jobs.

We believe that automation can and will become an important addition to the strength of our nation and to the free world. The growth of our influence in the world has always been directly related to our advances in industrial productivity. We now have only 6 percent of the world's population, yet we turn out one-third of the world's total production of goods and services, and almost one-half of its durable goods. This is because our national genius has found its clearest expression in industrial and scientific technology. We owe it to ourselves and to our country to take full advantage of the skills we possess and to apply any and all technological advances as they become available to us.

D. J. DAVIS has had a life-long career in mass production industry. Most of his experience has been in the field of manufacturing engineering particularly tool engineering. Mr. Davis is a vice president of Ford Motor Company and head of the firm's manufacturing research office. He joined Ford in 1949 as director of the Manufacturing Engineering Office. He was elected to vice president, manufacturing, in 1955 and named to his present position in 1958. During 21 years with the Cadillac Division since 1922, he held various executive staff positions in the production engineering department. He joined Avco Corporation in 1942 as master mechanic of American Propeller Corporation of Toledo, O. In 1944, he became chief designing engineer of the firm's post-war products and later was made staff master mechanic in charge of product manufacturing for Avco's plants. During 1946, Mr. Davis was the operating head of Avco's Lycoming Division and in 1947, he became chief industrial engineer of the Avco Corporation, his last post before joining Ford.

CHAPTER 4

THE AUTOMOBILE INDUSTRY:

A Case Study in Automation

WILLIAM A. FAUNCE, *Assistant Professor, Department of Sociology and Anthropology, The Labor and Industrial Relations Center, Michigan State University*

In 1955 one of the automobile producers decided that, in order to remain competitive with other companies producing automobiles in the same price range, it was necessary to market a model with an eight cylinder engine. Existing production facilities were inadequate to meet this new need and a new plant was purchased and completely retooled with modern automatic transfer machinery. This plant was one of the most highly automated automobile engine plants in operation at the time. In this chapter attention will be focussed upon some effects of the use of automated machinery in this plant. Specifically, comparisons will be made between the automated plant and older non-automated plants of the company in terms of work force characteristics, personnel policies and practices, industrial relations, job content, informal social structure, and work satisfaction.

Automated machines are used primarily in machining rather than assembly operations in this plant. Rough castings of engine blocks, cylinder heads, crank shafts, and gaskets and other small parts are brought into the plant from supplier plants, automatically machined and the engine assembled. Drilling, milling, boring, reaming, broaching, honing, and tapping operations are almost entirely automated. Materials handling, positioning of materials, and, in most cases, machine operation are automatic. There are also automatic indication of tool wear and automatic inspection or quality control. Pieces move automatically from station to station along the line and, except for cases of machine breakdown, the job of most machine operators consists primarily of watching a panel of lights or gauges to see that the operations are continuing to be correctly performed by the machine. Of the 1,600 workers in manufacturing operations in the plant, 1,100 are working in automated departments and the remaining 500 on assembly or other non-automated operations. Through the use of automated equipment, engines were being produced in 1957 in this plant at the rate of 150 per hour.[1]

Few differences were found between characteristics of the work force in this plant and other plants of the same company. Almost all the workers had in fact been transferred from these older non-automated plants. The procedure used in this transfer involved a poll taken by the union to determine which workers in other company plants wished to transfer to the new plant. Approximately seventy-five per cent of the workers polled indicated a desire to transfer. Seniority lists were then used to determine who would go to the automated plant. Except for a few instances where workers with particular skills were needed in setting up operations in the new plant, seniority was strictly adhered to in the transfer.

Some differences in job classification "mix" in the automated and non-automated plants were reported with there being an increase in the proportion of skilled or apprenticable trades in the new plant. Almost 25 per cent of the work force had classifications of this type. Management in the plant did not feel that new skills were required in semi-skilled classifications on automated lines and no training program was set up for machine operators. Some training was given job setters and machine repairmen through special training programs in hydraulics and electrical circuits. The most important personnel problem posed by automation in this plant, however, was the lack of skilled,

---

1. Data were not made available on which a comparison of man-hour productivity between this plant and the older plants could be made. The opinion was expressed, however, by representatives of both union and management that many more workers would have had to be hired by the company to produce the new engines if conventional machining equipment had been used. There had been no lay-offs directly attributable to the new equipment at the time the study was conducted.

salaried technicians and engineers. The personnel manager in the plant described the kind of person needed most as "an engineer who is willing to get his hands dirty." It was felt that a person of this type was needed to locate the source of difficulty and repair it when machine breakdowns occurred. With complex electrical circuits and hydraulic lines spread over a transfer machine which may be a hundred or more yards long, determining the cause of a machine breakdown becomes particularly difficult. In addition, the cost of the machines and the relatively short period of amortization of this cost[2] makes any work stoppage particularly expensive in this plant. It was estimated that thirty salaried technicians who would not have been used with conventional production processes were working in the new plant and the impression was given that many more skilled technicians were needed.

Rates of turn-over and absenteeism in the new engine plant were reported as comparable to rates in non-automated operations of the company. Also, there were no changes in the number or direction of job transfers. Job transfers were discouraged by management in this plant because familiarity with the particular operations on each machine was regarded as more important since mistakes were more costly. The broad seniority units in the plant which facilitated transfers were regarded as a particularly difficult problem by management for this reason. It is the feeling of management generally in the automobile industry that seniority units should be narrowed.[3]

Automation did not have any appreciable effect upon the wage structure in this plant. In automated departments, a larger proportion of the workers were given job classifications with higher hourly wage rates and higher rates for some machine operator jobs were negotiated but no major wage changes were made. There were also few changes in job classifications or descriptions. Only one new job classification attributable to automation was established. The job classification, "console operator," was set up in the automated plant for workers whose primary responsibility was watching a panel of lights signalling

---

2. $55,000,000 over fifteen years.

3. It is the union's contention, however, that seniority units should be broadened. Because automation does not affect all departments or plants equally, it is argued that seniority units should be broadened to protect the jobs of high seniority workers in the automated departments or plants.

operations along a complete transfer line. Although there appear to have been major changes in some jobs on automated lines, job descriptions of existing job classifications were not changed and are still in use.[4]

A few other general effects of automation relating to cost factors in the plant studied may be simply listed here. Automated machinery has produced a lower scrap rate than in the older plants and was reported to have improved both the quality and uniformity of the product. Plant maintenance has been made easier and less costly because of the more effective methods of metal chip disposal built into the transfer machines. Materials storage has been made more efficient and less costly because the rough castings can be centrally located in automated departments rather than having to be placed in small piles near each machining station as in some older plants. Indirect labor costs have been somewhat increased by automation because the increased rate of production has made it necessary to hire more workers in areas such as shipping. Finally, automation has increased the need for preventive maintenance to avoid costly breakdowns in the integrated transfer machines.

While automation has become a major issue in collective bargaining in the automobile industry, it has apparently had little effect upon day to day union-management relations under the existing contracts. As is the case with almost any change in production technology, some problems in establishing work standards and rates of pay on changed jobs have developed. Problems related to transfers or lay-offs resulting from the change could also be expected and have occurred. These problems, however, occur during the period of change-over to the new technology and are not necessarily characteristic of periods of normal operation using the new techniques. The grievance rate in the automated plants studied did not differ appreciably from non-automated plants of the company at the time of the study. An analysis was also made of grievance records in another plant where some departments had been automated. The grievance rate during periods of normal non-automated operation did not

---

4. Negotiations regarding new job classifications are initiated by unions, rather than management. The union, in this case, decided that, in most instances, higher wage rates could be obtained for workers by getting them assigned to existing classifications with higher hourly rates than by negotiating new classifications.

differ from periods of normal automated operation. During the time the new machines were being introduced, however, the grievance rate was significantly higher than in either period. Although automation may be expected to pose important problems for consideration at the bargaining table, it does not appear to have changed existing patterns of in-plant union-management relations.

It was noted above that automation has changed job content on many jobs. The changes regarded as most important by workers in the automated plant were (1) a reduction in the amount of materials handling required, (2) decreasing control of work pace, (3) an increase in amount of attention required by the job, and (4) a change in the type of skill required. In the old plants the worker characteristically controlled work pace in the sense that he pulled down a lever or in some other way actuated the machining process. He had to handle the parts machined either to feed the parts into the machine or to position the part prior to the machining operation. Because he controlled work pace to some extent, the worker was able to vary the tempo of work within the limits of his production quota and could work ahead of schedule and take a break. In addition to creating "breaks" by varying work pace, the extent to which close attention to the job was required varied during different phases of the operation. Finally the skills involved on the job were identified primarily in terms of setting or loading the machine or in inspecting the part machined.

In contrast, in the new plant the job of the machine operation involved, for the most part, pushing a button or watching a panel of gauges or lights. Work pace was established almost exclusively by the transfer machinery. Closer and more constant attention was required by the job and there were almost no jobs on the production line in which it was possible to work ahead and take a break. Ability on these jobs was equated most often with alertness or ability to attend closely to the job.

In addition to the changes in job content, there were differences between the automated plant and the plants using conventional machining techniques in patterns of social interaction on the job. As a result of increasing distance between work stations, closer attention required by the job, inability to work ahead and take a break, and machine noise, many workers in the automated plant were virtually isolated socially. Interaction occurred less frequently within smaller groups and there was. less identification with a particular work group in the automatic plant. While there was less contact with other workers, the worker in the automated plant reported closer supervision by the general foreman and superintendent as well as by his foreman. The increased supervision was a result of both a decrease in the number of workers per foreman and an increase in amount of time spent by each foreman in direct supervision on the line. This increased supervision was regarded by the company as necessary because of the cost of "down time" or work stoppage in the automated departments.

The decreased opportunity for social interaction and the increased supervision in the automated plant were both sources. of dissatisfaction with work in this plant. Workers in the plant also reported increased feelings of tension on the job resulting primarily from the increased rate of production and the amount of attention required by the job. The loss of control of work pace and a feeling of being divorced from the work process in the sense that it was now the machine which did the work were also listed as sources of dissatisfaction with automated jobs.

A majority of workers in the plant, however, preferred their present automated job to non-automated jobs in the older plants. Also, more workers on highly automated jobs in the new plant preferred automation than did workers on less automated jobs. The most frequent reason given for this preference was the decrease in amount of materials handling involved. Workers on some jobs in the new plant, particularly those most highly automated, reported that their jobs were more interesting and more challenging as a result of the greater complexity of the machinery and listed these as sources of satisfaction with automated jobs. Working conditions varied considerably from department to department in both the old and the new plants and appeared, on the whole, to be not much better in the automated than in the older plants. One other factor usually considered in analyses of work satisfaction, i.e., job security, should be mentioned. While almost all the workers interviewed felt that automation was putting people out of work and some listed this as a reason for dissatisfaction with automation, there was little indication that these workers felt their jobs less secure as a result of the transfer to the automated plant. It seems probable that to the extent that automation has increased the concern of workers regarding job security, it is in the non-automated plants where this concern is most likely to be manifested.[5]

While some departments of the plant we have been concerned with here were very highly automated, machinery of this type has only recently been introduced into the automobile industry and it is generally agreed that we stand only at the threshold of the "age of automation." We will conclude this discussion by considering briefly some of the ways in which operations in this plant are likely to become more automatic in the future and some factors which may affect the rate at which these changes will be made. Assembly operations, improvement in electrical controls, and the development of improved electronic inspection equipment were most frequently mentioned as areas where further automation is expected. There have been few major changes in assembly operations since the development of the moving conveyor. With the increased productivity of the automated machining equipment which feeds the assembly line, the need has developed for more efficient assembly operations.

The use of computers to provide a more rapid flow of information to production departments was also mentioned as an area where further automation was needed. It was felt that if, for example, cost data could be made available on a day to day or hour to hour basis, adjustments could be made at any time the production operations were not being performed most economically.

One of the important factors reported as a hindrance to the further development of automation was the shortage of skilled technicians capable of repairing automated equipment. Another was the frequency of design changes in the automobile industry. The constant demand for more powerful and efficient automobile engines has meant annual changes in engine design. These changes can be made with relatively minor adjustments in the automated *machining* operations but would be a more difficult problem in building automated machines for engine

---

5. More detailed analyses of findings of this study regarding social structural changes resulting from automation and attitudes of automobile workers toward these changes can be found in: William A. Faunce, "Automation in the Automobile Industry: Some Consequences for In-Plant Social Structure," (tentatively scheduled to appear in the June issue of the *American Sociological Review*) and "Automation and the Automobile Worker," (prepared for publication).

assembly. The current lack of a rapid flow of information regarding cost to production departments was regarded as a third factor hindering the development of further automation. It was felt, however, that because competitive pressure has resulted in a constant search for techniques of cost reduction in the automotive industry, this plant would undoubtedly become even more highly automated than it is at present.

This acceleration of the rate of technological change has, and will increasingly, make it imperative that knowledge of the consequences of these changes be obtained. Such knowledge is essential if the individual, organizational and economic adjustments required by the changes are to be facilitated. The apparently increasing interest in research dealing with the effects of technological change, which has been spurred by public concern with the consequences of industrial automation, is certainly along deferred step in the right direction.

Selected Bibliography

E. Ayres, "An Automatic Chemical Plant," *Scientific American,* CLXXXVII (September, 1952), 82-96. Discusses effects of feedback control devices in a modern petroleum refinery where automatic control has been very highly developed.

Bureau of Labor Statistics, "A Case Study of A Company Manufacturing Electronic Equipment," *Studies of Automatic Technology, No. 1* (Washington, D. C.: U. S. Government Printing Office, Department of Labor, 1955). A detailed description of the introduction of automatic production methods in an electronic equipment manufacturing company.

Bureau of Labor Statistics, "The Introduction of an Electronic Computer in a Large Insurance Company," *Studies of Automatic Technology, No. 2* (Washington, D. C.: U. S. Government Printing Office, Department of Labor, 1955). A detailed case history covering effects before, during, and after installation of a large electronic computer in an insurance company.

Bureau of Labor Statistics, "A Case Study of a Large Mechanized Bakery," *Studies of Automatic Technology, Report No. 109* (Washington, D. C.: U. S. Government Printing Office, Department of Labor, 1956). A discussion of effects of technological change in a large bakery where automatic transfer and control devices were introduced.

Bureau of Labor Statistics, "A Case Study of A Modernized

Petroleum Refinery," *Studies of Automatic Technology, Report No. 120* (Washington, D. C.: U. S. Government Printing Office, Department of Labor, 1957). A history of an eight year period of change in production techniques including effects of the changes.

E. F. Cooley, "Computer Methods and Application: A Case Study," *The Impact of Computers on Office Management, Office Management Series No. 136* (New York; American Management Association, 1954), 41-46. A description of effects of the introduction of electronic computers in a life insurance company including plans for future use.

W. A. Faunce, "Automation in the Automobile Industry: Some Consequences for In-plant and Union-Management Relationships" (Microfilmed Ph. D. dissertation, Wayne State University, 1957). A detailed analysis of effects of the introduction of automatic transfer machines upon in-plant social structure, job satisfaction, and union-management relations.

F. J. Jasinski, "Technological Delimitation of Reciprocal Relationships: A Study of Interaction Patterns in Industry," *Human Organization,* XV (1956), 24-28. A report of a study of the effects of production technology upon social relations in an automobile assembly plant.

J. W. Kuebler, "The Dixie Cup Company," *Keeping Pace with Automation, Practical Guides for the Company Executive, Special Report No. 7* (New York: American Management Association, 1955). A discussion of reasons for introduction of, nature of, and effects of automated machines in the Dixie Cup Company.

F. C. Mann and L. R. Hoffman, "Individual and Organizational Correlates of Automation," *Journal of Social Issues,* XII (1956), 7-17. A comparison of the effects of different production techniques in an automated and a conventional electric power plant.

A. B. Mooers, "Transfer Devices Extend Automation in Press Shop," *Iron Age* (August 26, 1954) 101-103. A description of the operation and effects of automatic handling devices in the press shop of an automobile plant.

"Push-Button Plant: It's Here," *U. S. News and World Report,* (December 4, 1953), 41-44. A discussion of operation of an automated automobile engine plant including analysis of some effects of use of automated techniques.

W. H. Scott, et al., *Technical Change and Industrial Rela-*

tions (Liverpool: Liverpool University Press, 1956). A report of a detailed study of effects of technological change in a steel plant in England.

J. I. Snyder, "The American Factory and Automation," *The Saturday Review* (January 22, 1955), 16ff. Describes evolution of automation and gives detailed description of the operation of the fully-automated, government owned 155 mm. shell plant in Rockford, Illinois.

C. R. Walker, *Toward the Automatic Factory: A Case Study of Men and Machines* (New Haven: Yale University Press, 1957). A detailed account of an extensive study of the effects of technological change in an automated plant in the steel industry.

WILLIAM A FAUNCE (B.A. in Psychology, Michigan State University, 1950, M.A. in Sociology, Wayne State University, 1951, Ph.D. in Sociology, Wayne State University, 1957) is now Assistant Professor of Sociology and Anthropology and Research Associate, Labor and Industrial Relations Center, Michigan State University. He has recently contributed articles to the *Social Forces, American Sociological Review,* and the *Transactions of the Third World Congress of Sociology* (1956).

CHAPTER 5

AUTOMATION IN THE METAL WORKING INDUSTRIES

M. A. HOLLENGREEN, *President, Landis Tool Company, Waynesboro, Pa.*

The term "automation," as used in the metal-working industries, has been variously defined as;

(1) the increased use of automatic control of handling, machining and inspection of the product, or

(2) more specifically, a machine tool builder regards it as a method of transferring a work piece automatically from one machine to the next, so that as many of the required operations as possible are completed by the time the work piece leaves the end of the line.

The machine tool industry calls them "transfer machines," which is a better name for them.

ECONOMIC FUNCTION OF MACHINE TOOLS

Machine tools are the power-operated machines that cut, forge, extrude, or press metal into the desired shape. They are the very foundation of industrial production.

Machine tools are distinguished by the fact that they are the only inanimate objects that reproduce themselves. All of the consumer needs of today are met either by production on machine tools or on equipment made by machine tools, and machine tools are themselves made on machine tools.

The farmer tills the soil, cultivates and reaps his crop, and sends it to market on equipment made by machine tools. His chemical fertilizers come from plants whose equipment was made on machine tools. The effectiveness of machine tools is therefore a matter of vital concern to our entire economy.

In time of war, they are equally essential, since all of our modern weapons are not only made on machine tools but require a very high degree of accuracy in most of their parts so that only machine tools that can produce within narrow limits of accuracy can be used.

DEVELOPMENT OF PRODUCTIVITY

If we accept the first definition of automation stated above, it is a process that goes back to the very beginnings of the machine tool industry. When engine lathes (that is, lathes operated by power through belts, not by foot-treadle) were first developed, the operator put the work-piece into the machine and removed it by hand after turning it to the required diameter. He made a single outside cut on the work-piece.

Then the turret lathe was developed. Tools could be brought into action from the front or rear cross-slide and six end-working tools mounted on a turret, which rotated on a vertical axis, could be successively brought into play against the end of the work-piece, thus performing eight operations at one setting in the machine.

The production of still greater quantities at one time led to the development of the multiple spindle chucking machine. In this the work-piece is mounted in a chuck on a rotating spindle. This spindle indexes, or moves progressively, from one working position to the next. By the 1920's it was possible to perform 10 or 15 operations on the work-piece without intermediate handling.

Little by little, chucking machines were improved; made heavier, more powerful, with electric and automatic controls, permitting a greater range of feeds and speeds which in turn made possible the use of greatly improved cutting tools. Chucking machines can now be used on a very wide range of work-pieces of various materials, and can work to far more exacting standards of accuracy.

Greatly improved and refined, the lathe, the turret lathe and the multiple spindle chucking machines are universally used in metalworking shops today. Similar developments came in other metal-working arts, in drilling and boring machines of various types, in planers and shapers, in milling machines, in many kinds of grinding, honing, polishing and lapping machines, in machines for generating and finishing gear teeth, in the vast field of forging, bending, extruding, shearing and press work.

In all of these types of machines there were three basic problems. First of all, safety; to reduce danger to the operator to the minimum. Second, greater ease of handling. Originally the operator lifted the work-piece into the machine by hand, later he used a small hoist, electrically or hydraulically operated. More and more he became a source of intelligent control, less and less an itinerant power plant.

The third problem was greater accuracy. After each cut, the operator had to stop his machine, measure the piece with calipers, or a micrometer, then adjust his machine and start up again.

THE BIRTH OF THE TRANSFER MACHINE

It was a natural consequence of this trend that the idea of linking several machines together was developed. We have, then, the second definition of "automation" stated above; the transfer of the work-piece by an automatic conveyor from one machine to the next, so that as many of the required operations as possible are completed by the time the work-piece leaves the end of the line. This production line may be composed of two machines, or of many, it may be as long as a football field. It may be composed of many units of the same kind, thus each station may be for drilling and tapping, or it may include several types of machines of various makes.

AN EXAMPLE OF AUTOMATION

In an early automation line the cylinder block for an automobile engine, of gray iron, was placed on the conveyor. This brought the work-piece into the first position, located it accurately by clamping it against a reference surface which had been machined on the block for that purpose. Then a drilling head, or two drilling heads, drilled certain holes in the work-piece. When the operation was finished, the heads retracted automatically, the piece was unclamped, carried to the next station, and again locked into place, again located by means of the same reference surface. By the time the cylinder block left the line, all of the required holes had been drilled, reamed and tapped.

It will be noted that this type of automation line offers several fundamental advantages. First is the elimination of hard physical work for the operator. Tired men make mistakes and produce scrap. Second is greater safety for the operator; tired men are less alert and more likely to get hurt. In the automation line there is no necessity for the operator to put his hands into the danger zone. Any one of a large number of "stop" buttons brings the entire line to a stop instantly if an adjustment is required. Third: The reduction of non-production time is kept to a minimum. The machine is not distracted from its work; it does not slow down because it is tired. It need not take time out for personal reasons. No time is needed to select the proper feeds and speeds for the next operation. The fourth advantage is greater accuracy. Since the work piece is always clamped against the same reference surface, each operation bears the correct relation to every other operation, thus all of the holes in the automobile cylinder block are the right distance apart, are parallel, or bear the required angular relationship to each other. Fifth, as the result of all of these advantages, we have the increased production of an improved product. Sixth, we have economy of space, an important consideration in many plants.

AUTOMATION IN GRINDING

Another impressive automation line is made up of precision cylindrical grinding machines for the production of automobile valves. Since 16 of these are required for an eight-cylinder engine, and thousands of cars are produced in a day, these are needed in large quantities. If the valve is to seal off the gas pressures of a high compression engine, the valve seat must be machined to a very close tolerance and to a high degree of finish. The stem of the valve must move freely in its guide bushings, yet without play or leakage between the stem and the guide bushing.

The problem is to produce millions of valves for a single model of automobile, all closely identical, so that no subsequent fitting is needed and sufficiently accurate to meet the requirements outlined, yet at a cost which will permit a low price for the car. These exacting requirements are met by an automated line. The rough work-pieces are automatically fed into the line at the first station. They then move up a chute and are loaded into the first precision cylindrical grinding machine. When that operation is complete, they pass automatically to the second grinding machine for the second operation, and so on until the valve is completed. Before that automation line was developed, the pieces had to be loaded, unloaded, and moved to the next machine by hand. This took time—several minutes per piece. On long production runs these apparently unimportant unproductive minutes add up to hours, days and months.

Before this installation was available, a maximum of 275 valves could be ground in an hour. This automation line, utilizing the same number of grinding machines, produces 450 in an hour. In addition, human errors are eliminated, accuracy is improved, scrap has been reduced to a minimum.

THE EFFECT ON THE OPERATOR

Such an automation line is not "automatic" in the sense that no human attention is required. The "push-button factory" is a fantasy. The work pieces must be brought to the beginning of the line. Men assigned to the line must watch the machine constantly; they must be thoroughly familiar with a complex aggregate of machinery and control, and they must be highly trained and efficient.

From many points of view they have better jobs than the men who operated valve-grinding machines before this automation line was installed. Their new job is safer because they can keep their arms and hands away from the hazardous area of the machines. Their jobs are more interesting because they require more technical knowledge and far less manual labor than the old ones did.

The effect on the operator is well described in the following excerpt from the "Supervisory News Letter" of August 1955, published by the E. I. DuPont de Nemours Company; "The upgrading of jobs, necessitated by new tools and modern processes, has greatly increased the responsibility and prestige of the individual employee. Emphasis now is more on skill than muscle.

"Modern jobs are obviously less of a physical burden than

were those of even a few years back, but today frequent efforts are made to picture the modern industrial employee as a mere robot. Foes of automation, for example, contend it will rob workmen of their personality, and make their actions as automatic as the machine they operate.

"DuPont experience is directly contrary. Many an employee today performs duties that were generally exercised only by supervision 30 or 40 years ago. As brains replace brawn, a higher degree of judgment, selectivity and knowledge is required than ever before. The dignity of the employee is enhanced by his greater knowledge, for even in mechanization there is no substitute for intelligence.

"As individual productivity has climbed, and new tools have demanded new skills and given added responsibilities, wages have kept pace. And as plants and processes become more intricate and more productive, skills will be in even greater demand, and will be rewarded accordingly."

This picture is certainly not one of men being thrown out of work by machines.

FEEDBACK

Combined with the automation line may be improved methods of control of accuracy such as feedback. It may also be applied to single machine tools of a wide range of types. The simplest type of feedback, familiar to everyone, is the thermostat used in heating the home. When it signals to the heater that the temperature of the home has fallen below the low limit that has been set, the heater starts up. When the room has reached the higher limit of temperature that has been set, the thermostat signals "enough" and the heater stops.

Automation lines frequently embody the feedback principle. Applied to precision grinding machines, the feedback principle involves the accurate measurement of the work while in process. When it has been ground to the desired diameter, the feed is disengaged. If a piece comes out of the machine under the acceptable minimum diameter the machine is automatically adjusted to grind the next piece to a larger diameter. Conversely, if the piece is too small, the control automatically corrects the setting of the machine and produces the next piece to a larger diameter.

From this it is but a step to the automatic inspection of the parts produced as a further check on accuracy. Oversize pieces automatically pass through a chute to the "re-work" container. Correct pieces are approved and accepted and undersized pieces are passed into the "reject" container.

CONTROL BY RECORDED INFORMATION

Similarly, there is an extensive field for the operation of a machine by recorded information or the digital computer. By means of a perforated tape of a photographic film, a reading device transmits to the machine the proper feeds and speeds, moves the work table or the spindle into the proper position, as the case may require. It does the work of the human operator in preparing the machine for the next operation to be performed, "remembering" a long series of required operations and repeating them without error and without hesitation.

Since the tape can be easily prepared from the blue print of the piece to be produced, and since this type of control can be effectively used in the production of relatively small lots, it has wider possibilities of application than the automation line alone. It will eventually be applied to many general purpose machine tools. The operator is still required to supervise the operation, to replace tools when dull, but he can produce more and better work.

Automation is not an unrelated phenomenon. It is merely the last logical step in the age-old battle against production costs; the constant drive to make a better product. It does not stand alone. It is one of the many resources available to the plant manager. Better cutting tools, better abrasives, more effective coolants, new forms of electric and electronic control, better lubrication of the moving parts of the machine tool, automatic chip discharge—these and many others offer opportunities to produce more work per man-hour at more exacting standards of accuracy and surface finish.

THE LIMITATIONS OF AUTOMATION

It will be obvious that automation is not a panacea. It is not and never will be the solution to all production problems. It has several obvious limitations. In the first place it is costly. An investment so large is for many companies a formidable problem. The investment must be recovered during the profitable life of the equipment. Actually, a company does not show a profit until the savings resulting from the use of a new machine have returned the cost of that equipment and of its installation. But the policy of the Department of Internal Revenue of the United States Treasury which requires the recovery of such invested capital over a period of 15 to 25 years, depending on the type of machine tool, increases the risks involved. We may be able to predict with some certainty that the product will be marketable with but minor changes for 4 or 5 years, but it would be foolhardy to make such a prediction for 15 to 25 years.[1]

Furthermore, in a period of continuing inflation the original cost, even when it has been fully recovered, will not buy the more efficient machine tool that will be available at the end of a "useful life" of even 15 years, to say nothing of 25 years. This means that to keep efficient, the company must build an equipment modernization reserve out of profits after taxes. Long periods of "expected useful life" make this problem more difficult. Since automation requires skilled maintenance men—mechanical, hydraulic, electric, electronic—it presents the company with the problem of finding and training these skills.

Automation is relatively inelastic. Designed to produce so many pieces per day of eight hours, it can be operated for a shorter time when sales decline, but the substantial overhead costs must be distributed over a smaller quantity of pieces. This means a higher cost per piece produced. If sales increase, it brings the penalty of paying time-and-a-half to all of the skilled men required, again increasing the cost per piece. The successful use of automation depends on a stable, predictable demand.

Since idle time is costly, it is necessary to establish a reserve of sharp tools so that a dull or broken tool may be quickly replaced. On an automation line producing airplane engine cylinder heads in the Dodge-Chicago plant during World War II there were 99 cutting tools, all involved in the continuous operation of the line. However, a program of replacing each cutting tool at arbitrary intervals, determined by experience, minimizes down-time from this cause.

Further, unit 2 cannot do its work if unit 1 is not operating.

---

1. For a thorough discussion of this fundamental difficulty see Joel Barlow, "Let's Make Sense in Our Depreciation Policy," *Iron Age* (June 5, 1957), 1-8.

Nor can 3 or 4 or any of the succeeding units. The failure of the machine at any point in the line puts all succeeding units out of operation for the same length of time. This difficulty has been largely overcome by setting up a reserve of parts at each station which can be put into work if a preceding unit is in trouble.

While minor changes in the design of the part being produced can usually be made by re-tooling one or more of the units or even by replacing them by a different size or type of machine, such changes are costly and the successful use of automation depends on stability of design over a substantial period of time.

Notwithstanding these limitations there are still thousands of places where automation is needed in the metal-cutting field. It should be carefully considered wherever there is production in quantity. It is not only large companies which produce in large quantities. The decision must be based on the circumstances in each case.

DOES AUTOMATION THROW MEN OUT OF WORK?

Whenever improved methods of production are discussed, someone raises the issue of unemployment. Since the purpose of automation is to reduce the man-hours required to produce a certain article, it at first seems obvious to many people that it must of necessity "throw people out of work."

What actually happens is that by improving the quality of the product and by reducing the price to the consumer, we create a wider demand for that product, the end result tied with improved production methods being increased employment.

In an address made on February 11, 1955, Benjamin Fairless, Chairman of the Board of the United States Steel Corporation, pointed out that from 1939 to the end of 1953 the population of the United States had increased 22% while the total number of jobs had increased 35%. He emphasized that in the field of manufacturing, where automation and other methods of increasing production have been most widely applied, employment went up 73% during those years, more than three times as much as the increase in population. "The record clearly shows," he added, "that this rapid increase in employment has occurred chiefly because of mechanization, not in spite of it."

When a decided improvement is made in the production line, some men may be shifted from one machine to another, from one department to another, some must be re-trained and learn a new skill or learn to run a new and strange machine. While it can not be dogmatically stated that the introduction of new equipment never leads to lay-offs, this consequence is extremely rare. Usually the equipment is purchased because competition makes it necessary to increase production without increasing costs, without hiring more men. A better product at lower prices increases demand, and the inevitable end-result is further expansion of output and increased employment. It is no mere coincidence that the strongest companies are those that have the best equipment.[2]

Nor can we overlook the fact that what happens on the production line affects men outside of the factory. Manufacturing itself provides about one job in four. But more efficient manufacturing affects the other three jobs. Increased production increases employment in retail establishments, repair and maintenance services, home construction, factory construction, highway and road building, and elsewhere. Even if improved manufacturing had not resulted in the drastic increase in manufacturing employment noted by Mr. Fairless, it would have contributed to widespread employment in every sector of our economy.

A course of unemployment that is real and serious, and that has not been given the attention it deserves, is the unemployment that results when a company fails because it has not kept its plant efficient or has been trying to meet competition with aging equipment. When a company fails everybody in the company loses his job. The stockholders lose their investment. The community loses taxes and competition is lessened. This is happening every day and it represents an economic loss of serious proportions.

AUTOMATION IN THE FUTURE

The problem that faces the nation is not a growing backlog of men laid off because of more efficient manufacturing equipment. It is the difficulty, if not the impossibility, of maintaining our present standard of living with what will shortly be an inadequate work force.

Our population is expected to increase 55 million by 1975, but the labor force—our productive workers—are expected to

---

2. "Automation; Bogey or Bonanza," *Steel* (Oct. 15, 1956), 109-116.

increase only 15 million. The latter is easily estimated, since the productive workers available in 1975 have been born; we know how many there will be. We will have more people over 65 years of age to be supported by these production workers. How will 15 million more workers provide enough for an increased population of 55 million?[3]

There is only one possible answer. The nation's output per man-hour must be increased. The productivity of our machine tools must be increased. Every method that can be devised must be applied, whether it be work simplification, better material flow, control by recorded information, automation, or the modernization of our metal-working plants by replacing the hoary, old machine tools that still encumber them.

What is the alternative? A lower standard of living? Longer hours of work?

Suppose we now tried to manufacture a modern automobile with the machine tools of 1908. There is some doubt whether those old machines had the power and accuracy to build a modern car, but let us assume that they did. The point is that they were much less productive than present-day machines. If they could manufacture a modern car at all, it would take many times the man-hours that are now required. And the cost of the car would be greatly increased. It has been estimated that it would cost $65,000 to manufacture a modern car with the machines of 1908.

Looking ahead, we need not go so far as to predict that an automobile will cost $65,000 in 1975 if it has to be made with the machines of today. But with a larger population and a smaller work force than now, such an automobile would certainly cost much more. The trend toward larger and more powerful cars, the steady increase in wages and the trend toward a shorter work week will ensure that the $2500 car of today will cost at least $10,000 in 1975 if it must be made with the machines of today.

How many automobiles can be sold at $10,000 apiece? How many men would be employed by the automobile industry? Unless we continue to make our machine tools more productive by the use of automation and every other available help, we shall be faced with a situation in which the ordinary man

---

3. Carroll W. Bryce, "What Automation Means to America," *Factory Management & Maintenance* (September 1955).

cannot buy automobiles and other products we now regard as necessities.

THE NEED FOR MODERNIZATION

Unfortunately, in spite of the widespread attention that has been given to automation, in spite of the drastic improvements in machine tools in recent years, the nation's metalworking industries are not replacing their obsolete machine tools at a satisfactory rate. Too many plant managers are still unwilling to replace a machine that has not given them some trouble, overlooking the fact that a machine tool is simply a means of earning a profit and should be replaced whenever that will increase the company's profits. The nation can boast many new factories, beautifully equipped, but American metal-working plants as a whole are suffering from creeping obsolescence.

If we disregard the machine tools owned by the Department of Defense and the new machine tools purchased by that department, and consider only those in use and being purchased by private industry, we find that only about 1% are replaced each year according to National Machine Tool Builders Association. If we assume that the new machine, on the average, is twice as productive as the one being replaced, we are modernizing only 2% of our productive capacity every year. A turnover of once in 50 years, in the face of the tremendous improvements in our machine tools, not in 50 years, nor in 25, but in even the last 10 years, is a sorry record. It has been estimated that it is costing the United States a billion dollars every year for the doubtful pleasure of retaining machine tools that are not giving any trouble.

Automation has been developed in response to a fundamental economic need. If we insist on a constantly improved standard of living, if we propose to defend our republic against the encroachments of a powerful totalitarian state, we must aggressively improve our means of production, for only by increasing our productivity can we safeguard our future.

Selected Bibliography

Automation and Job Trends. (Council for Economic Development, 120 South La Salle Street, Chicago 3 Illinois, 1955.) The case for improved production methods, stated by a group of executives from companies producing capital goods, under the aegis of the Machinery and Allied Products Institute.

Automation; the Magazine for Automatic Operations (Penton Publishing Co., Cleveland, Ohio). A monthly review of new developments, with especial reference to the field of metal-working.

Automation and Technological Change; U.S. Economic Report, Joint Committee on Automation and Technological Change (84th Congress, first session. October 14-28, 1955. Government Printing Office, Washington 25, D.C.). Testimony by labor leaders, economists and industrialists on the social and economic impact of automation on the United States.

E. M. Hugh Jones, Editor. *Automation in Theory and Practice.* (Oxford, England, Basil Blackwell, 1956). A collection of lectures given at Oxford University in 1955, discussing the effects of automation on trade unions, and the social and administrative problems involved.

John Diebold, *Automation; the Advent of the Automatic Factory* (New York, O. Van Nostrand Co., 1952). A scholarly discussion of the background and development of automatic operation; its effects on materials, production processes, employment and labor resistance, stated in general terms.

Paul Einzig, *The Economic Consequences of Automation.* (London: Secker & Warburg, 1956). Employment effects and other aspects of automation, its effect on wages, profits and employment.

Eugene Munter Grabbe, Ed., *Automation in Business and Industry.* (New York: John Wiley, 1957). A thorough analysis of the fundamentals of automation; feed-back controls, analog and digital computers, automation in processing data, flight, the production of petroleum, chemicals, electronic equipment and so on. For digital control of machine tools see pages 494 to 514. For plant automation see pages 547-575.

MILBURN A. HOLLENGREEN was born in Renovo, Pennsylvania and served his time as a machinist in the railroad shops there. On his graduation from Cornell University in 1926 with the degree of Mechanical Engineering, he became Assistant Chief Engineer of the Landis Machine Company in Waynesboro, Pennsylvania. In 1936 he joined the Landis Tool Company as Assistant General Manager. He became president and General Manager in 1948. On the acquisition of the Gardner Machine Company of Beloit, Wisconsin in 1951, he became President and General Manager of that subsidiary. He is the holder of 23 patents. He has been active in the affairs of the machine tool industry, serving as Chairman of the Government Relations Committee of the National Machine Tool Builders Association in 1948 to 1953, handling defense problems of critical importance as liaison between the industry and many Government departments during the Korean War. As the head of that organization in 1954 he was responsible for the success of the Machine Tool Show of 1955, the largest exhibition of its kind ever held in this country.

CHAPTER 6

AUTOMATION IN THE ELECTRONICS INDUSTRY

DON G. MITCHELL, *Chairman of Board and former President,*
Sylvania Electric Products Corp.

Before I discuss some of the effects of *mechanization,* or *automation,* on the electronics industry in general, and my company specifically, I should like to comment briefly on *the current relationship between man and the machine, and the future trends.*

Automation has been deplored as a trend which will disrupt our entire economy; on the other hand, some rather fantastic claims have been made which would have you believe that the machine will one day relegate man to a life of eternal leisure— *monotonous* leisure, I might add—interrupted only by occasionally feeding paper tape to some hungry electronic computer.

I know that the first extreme point of view is wrong; the evidence against it is overwhelming. All I can say about the second point of view is that I don't think that it will ever come to pass, but if it does I hope I am not around to see it. But I have no real fears on that score; the human being has done a pretty good job of taking care of himself in the past, and I cannot see where the future will be any different.

PROBLEMS ARE OPPORTUNITIES

In studying any trend which seems to pose some immediate or potential problems, it seems to me that we could, of course, have what might be termed a "precautionary" point of view— in other words, what is this trend doing to the economy, and should it be moderated, or perhaps stopped altogether? Far more important, however, is the more positive approach—what opportunities does this trend present, and how can we direct it and stimulate it to be even more productive than it has been in the past?

In the pressure of modern-day living, there is all too little opportunity to project a situation into the future and to imagine the existence of circumstances which do not now exist. The constant and completely understandable tendency is to approach some situation in the light of past experience and current facts, but that is only taking a "status quo" approach to the problem. Circumstances change rapidly, as we all know, and many a situation which appears at first to constitute a problem, or even a threat, is actually an opportunity to do something more effectively and to a greater common good than has ever been done before.

So it is with mechanization or automation, which are really one and the same thing. Automation is only a more recent term for mechanization which has really been going on since the Industrial Revolution began.

Certainly the machine presents some short-term dislocations which cannot be ignored by anyone, least of all the persons who are dislocated. It doesn't do much good to try to convince an individual worker who does get displaced from an individual job that over a 25 years' span there is no such thing as technological unemployment. He doesn't care whether there is or is not. All he is worried about is that he lost a job. Without question every technological improvement has brought broader employment and higher living standards. Sometimes in the process some things may happen which either shouldn't happen or could be substantially minimized. Short-term dislocations present a severe test of management. In this respect I should like to point out that a basic policy at Sylvania has been for many years to make every effort to find a new job for anyone displaced by a machine. We have been extremely successful in implementing this policy; in fact, I do not recall any instance that might be termed a serious dislocation of any sort.

In explaining that, I might add that a large percentage of our employees are women because of the high degree of manual dexterity required in our many assembly operations. In the main, women do not intend to stay with you until they reach retirement age. The majority of them hope to get married some day, and, therefore, we perhaps do not have some of the problems encountered in industries that employ a hundred percent male labor.

As it does in so many cases—in fact, *most cases*—the answer to any problem of dislocation seems to lie about mid-point between the extremes, between the strictly emotional point of view, on the one hand, and the overly-practical one, on the other. If we recognize that short-term problems do sometimes exist and that they can be resolved, but do not let the existence of short-term situations becloud the fact that the broader and broader use of the machine is overwhelmingly for the common good, then we will have acted in the traditions of a democratic society.

THE IMPACT OF MECHANIZATION ON INDUSTRY AND SYLVANIA

In approaching mechanization or automation from the standpoint of its influence on the electronics industry, a number of general statements can be made:

1. Without large-scale use of automatic and semi-automatic equipment, the electronics industry, as we know it today, would not exist.

2. Without extensive mechanization, the total working force available to the electronics industry today could not even remotely produce the vast volume and variety of goods needed and demanded by the public, commerce, and industry, and the Armed Services. The machine not only has brought increasingly higher production, steadily decreasing cost to the consumer, and constantly increasing product quality, but *has actually met what would otherwise be a labor shortage.*

3. The increased demand for, and availability of, the products of the electronics industry has brought a great expansion of the basic materials industries—metals, glass, chemicals, plastics.

4. Thousands upon thousands of small businesses have been formed over the past few years, especially the postwar years, to meet the needs of the electronics manufacturers. Even companies which produce most of its own components and materials, place millions of dollars worth of business with small concerns all over the nation.

5. Hundreds of communities have gained new economic strength, either through the expansion of an existing facility, or the advent of a new plant.

6. An enormous new business has sprung up, completely outside the electronics manufacturing business. This is the electronics distribution and service industry, whose distributors,

jobbers, dealers, servicemen and others do an estimated volume of three billion dollars annually—a business which did not exist a few years ago, and which has multiplied manyfold since the war.

These are the ramifications of mechanization. It is not a case of putting a machine to work in one plant, or two plants. It is a case of creating an entire set of industries, hundreds of thousands of jobs that did not exist, millions of dollars of personal income, of buying power, new lifeblood for the entire economy.

THE GROWTH OF THE ELECTRONICS INDUSTRY— "THE WORLD'S MOST PROMISING TECHNOLOGICAL REVOLUTION"

If there ever were an industry which owes its unprecedented growth, its enormous productivity, and its great potential to mechanization, it is the electronics industry. To be sure, this amazing growth over the span of a few short years would not have been possible without rapid strides in the science and engineering of electronics, but advances in the state of the art, no matter how impressive they may be, are in effect meaningless unless they can be translated into *the satisfaction of human needs*. That was and is the contribution of the machine; it translates the ideas of man into end products within the reach of a steadily increasing number of people.

Not many years ago, as recent as the 1930's and the years immediately before the war, the electronics industry, as such, did not exist. It was essentially the radio business, receivers, tubes, and other components, transmitting equipment, and broadcasting. But there was no microwave communication, no electronic navigational equipment for aircraft and ships, no electronic controls. The "billion dollar baby" of the industry, television, was represented by some very small-scale development work in a few laboratories. Then came World War II, and the needs of the fighting forces, especially in the air, brought the word "electronics" into the common vocabulary.

The requirements of the Armed Services were not only met but frequently anticipated far in advance by the scientists and engineers in electronics laboratories. Especially in the early stages of development, much of this revolutionary equipment was assembled by hand. But the hearts of the equipment—the countless types of vacuum tubes—were another story. They were the products of highly intricate machines; without the machines they would not have been available.

As the war progressed the now famous proximity fuse was developed. Again its heart was the vacuum tube, an electron tube capable of being shot out of a gun as the vital part of a projectile, the tube which causes the shell to burst at just the right distance from the target. Then came many other types of tubes, tubes no bigger than the stub of a pencil, yet able to withstand great shock and vibration, as the hearts of communications, navigation, gunnery, and countless other applications.

The rate of mechanization throughout the electronics industry in electron tube manufacturing has been so rapid in the past ten years that new and improved machines would eliminate any necessity for a labor force of 200,000 or 300,000, or even more. In other words, *mechanization would prevent a disastrous labor shortage.* In fact, I dare say that the Armed Services' far greater needs for certain types of electronic devices would be met by fewer people than were needed ten years ago—that is fewer people for the specific job. There are so many total jobs that the total number of people required would, of course, still be very much greater, but for the specific job perhaps fewer people would be needed to make even the very, very many more tubes that would be required.

RECEIVING TUBES ARE "HEART" OF ELECTRONIC EQUIPMENT

All of these tubes belong to the family known as the "receiving tube,"—the tubes which 30 years ago were the heart of your one-tube radio sets, and which today are vital to television sets, computers, which are indispensable tools of this electronic age; microwave telephone and telegraph networks, and the countless other uses of electronics. I dwell on this receiving tube because it is a good instance of this automation and mechanization situation.

The machines which produce receiving tubes are a marvel of ingenuity. That picture is also a marvel of ingenuity because it doesn't give away the trade secrets behind the picture.

With unbelievable accuracy, they take the various metallic and glass components and within a few seconds produce a tube without which some vital job cannot be done. The total output of a single manufacturing plant today is greater than the entire electronics industry before the war. Sylvania is one of the two largest.

LOWER PRICES AND CONSTANTLY IMPROVED QUALITY

Coincident with this steady increase in production have come two consumer benefits which are obviously of paramount importance—*lower prices and constantly improved quality*.

In the early 1930's a typical receiving tube for a radio set cost $6.20. This is the cost if you went in and bought it. If you had someone install it, it might have cost you more. A typical tube today, if there is such a thing, performs a far better job for a longer length of time and costs the consumer about $1.50. That is a reduction of three-fourths, a little more.

Now, all this is in spite of the fact that the average direct labor rate in receiving tube manufacturing, has increased from 34 cents an hour in 1933, to five times that amount today, so the worker gets five times as much money. We get one-fourth as much for the product, and now I will give you a little story about profits.

That $6.20 tube had a profit of over 25 cents in it. The $1.50 tube has a profit of perhaps a nickel. Of course, we are making a lot more of them, but if we didn't pass the lower cost of manufacturing on to the consumer then it must be the referee that is beating us to death because we certainly haven't got it. We are making less than a nickel apiece on that tube today.

That adds up to a lot of money, and don't let me leave you with the impression that receiving tubes, total dollars, are a small-scale business.

The typical receiving tube today not only does a better job for less money than its counterpart of several years ago, but in many instances it does a job that could not be done at all a few years ago. High as production rates may currently be, and convenient though it may be to produce tube types which have become commonplace to the industry, new types are constantly being introduced to do something which has never been done before.

MECHANIZATION BROUGHT TELEVISION TO 36 MILLION HOMES

Still another electronic product which would not be possible without automatic and semi-automatic equipment is the television picture tube, a device which could be found in only a few thousand homes seven or eight years ago, but which today brings entertainment, education, and information to 36 million homes. The current status of color television is a striking example of the importance of mechanization. One of the major obstacles to the widespread availability of color television is the high cost of a television set. This cost will not be reduced to a point where the set is within the reach of a large segment of the public until the cost of producing the various components can be greatly reduced. New automatic machines are needed, for example, to put the thousands of color dots on the face of the tube, and these machines must apply these dots repetitively, at consistently high quality standards, and at high volume. I am happy to say that such machines are now under development.

The most expensive component is the color TV-tube. When I tell you that you have to lay accurately on the face of a color television picture tube hundreds of thousands of separate and individual dots, of three different colors of phosphors, and have each group of three in the exact relationship to each other that the next group of three is, so that when an electronic beam hits the blue dot it also hits the blue dot in the group next door, you have some idea of the complexity and the cost of building such a thing.

Until this is done mechanically, automatically, the cost of color TV sets is not likely to come down to where the color TV sets are available to a large number of the public. I am happy to say such sets are under development.

Here, then, you have a relatively small device, the receiving tube, and the picture tube, heavy and bulky enough to be difficult for a man to carry, both of them mass-produced items, both of them comparatively new from the standpoint of the jobs they are required to do, and both of them the results of mechanization.

THE FLUORESCENT LAMP

Back in 1938 Sylvania introduced a new product, the fluorescent lamp. There was no fluorescent lamp industry at that time; the fluorescent lamp was, in effect, a laboratory device. From this beginning 17 years ago an entirely new industry has grown, not an industry which supplanted the incandescent lamp industry, but which grew along with it.

To show you what is happening in this business, fluorescent

lamp production per operator-hour in 1940 was 3.4, a little less than three and a half lamps per operator hour. Today the equivalent figure is about fifty. I would rather not give you the exact figures because our competition would like very much to know how fast we are making them, but I firmly believe that within the next five years the production per operator-hour will have to more than double again if we are to keep pace with greater public demand for our product. An enormous expansion in lighting will take place during the next several years, and mechanization will be the only way to meet the demand.

Mechanization has not only greatly broadened our business in such original lines as incandescent lamps and receiving tubes, but has permitted us to enter entirely new lines of business— every one of which either equals or surpasses our total sales of 15 years ago.

Such things as automatic assembly of radio sets is rapidly coming. Several of us have automatic machines today which we feel will ultimately produce, with a fraction of our present working force, enough small table model radios to take care of at least 50 percent of what this country uses in a year.

Offhand, it might seem that a tremendous number of people would be thrown out of work.

Well, there are some compensating factors that you must take into consideration. An automatic assembly machine of that kind spews out an awful lot of radio sets, but every one is like the one before, and not everybody wants every radio exactly the same color, shape, size and style. They want different ones. Therefore, you can't afford to set up an automatic machine, to spew out that many radios of the same kind at the same time, so it becomes more expensive to make them this way than it does to make them the way that we are making them today, so we don't make complete units. We make perhaps a little part, which is common to very many television or radio sets, called an assembly, and we can make that thing in great quantities, and supply it to many small assemblers of television and radio sets who could not make their own automatic assemblies. That is the way the industry is developing.

Do not get a picture of a great funnel at one end of a machine into which you pour raw materials and out of which at the other end comes a completely assembled device. We won't live to see that today, and I doubt that our children do.

Automation comes in bits and pieces. First the automating of a single process, and then gradually a tying together of several processes to get a group of sub-assembly complete.

ELECTRONICS — A $9½ BILLION INDUSTRY

Other companies in the electronic industry have equally impressive stories to tell. Some of them are large organizations with considerable diversity of product, whereas others specialize in perhaps one or two components or in some very intricate device, such as the magnetron tube. In total, however, all of us represent an industry of total annual volume of some $9½ billion, employing more than 700,000 persons. A decade and a half-ago, annual volume was about $500 million, and employment of about one-tenth of what we have today.

Keep in mind that those are not solely production workers. Those are the distribution workers, salesmen, broadcasters, etc., everybody who is employed in this electronics industry, whether he is a production worker or not, including the local service man as well.

Now, obviously, the growth of production workers alone in the plants wouldn't be as rapid as that. That $9½ billion in annual sales and revenues includes government purchases, most of it for the Armed Services. It includes TV and radio transmitting equipment, TV and radio receivers, components, industrial and commercial equipment and devices, distribution, service, and the vast broadcasting industry. Each of these areas has a strong growth potential, but without question the field with the greatest fertility is industrial and commercial electronics—the application of the vacuum tube and all of its attendant devices to mass production, to business communications, to transportation. So broad is the horizon, in fact, that the most conservative estimates are that the electronics industry will attain an annual volume of $15 billion, or perhaps more, by 1960, and more than $22 billion by 1964 to 1965, with employment exceeding one million people. Think of it. An industry which has doubled in the post-war years, will double again within the next decade.

Let me hasten to add that these predictions are not wishful thinking or crystal ball gazing. They are based on realistic projections of current trends, population growth, formation of new families, industrial expansion, and other predictable factors. If anything, these predictions are on the conservative side, and the next year may well find us revising our estimates upwards.

One extremely important consideration in making such appraisals of the years ahead is a continuation of the trend in mechanization and automation.

INCREASED MECHANIZATION RATE IS INEVITABLE

It is my frank opinion that the momentum of technological advances in the art of mass production is such that continued mechanization on even a broader scale is inevitable in the years ahead. Whether you have in mind 100 percent mechanization of a single given process, or partial mechanization, finding better ways of doings things is a human trait which cannot be erased. The human mind has a habit of not standing still, and not accepting the way things are done today as the only answer.

As I indicated earlier, increased mechanization is the only answer to satisfying the human needs which increase day after day. Let us assume that by some fiat technological advances were stopped dead in their tracks, the needs of this nation are expanding so steadily that within a very short period of time we would be confronted with a peacetime phenomenon we have never encountered before over any period of time—*a serious labor shortage.*

ONLY INCREASED MECHANIZATION RATE
WILL MEET HUMAN NEEDS

I not only do not even remotely fear that mechanization or automation will cause long-term unemployment, but I am concerned about the strong possibility of a labor shortage in the years ahead, unless the rate of mechanization is increased. The past gives us a good clue to the future. As reported by the Joint Committee on the Economic Report, and the U. S. Department of Labor, at year end 1947 some 44 million persons were employed in industry, in 1954, 50 million were employed. If production techniques had not progressed between 1947 and 1954, 58 million persons would have been required to produce the goods and services actually demanded in 1954. The American population simply could not have furnished that working force. Without increased mechanization, without labor-saving devices, and overall greater production efficiency, the public's needs simply would not have been met.

Now, let us look at the future. I recently read a survey in which a gross national product of $850 billion was predicted for 1975. Now, unless you buy the fact that it is possible to have that gross national product of $850 billion by 1975, then the rest of my argument is specious, but I believe that a gross national product of that magnitude is possible, and I believe that we will have it.

Our gross national product will be this year about $387 billion. A working force in 1975 of 82 million people against 64 million people today is predicted. If the present rate of automation continues, every available worker will have to be putting in 40 hours per week, in order to keep raising our standard of living at the rate it is being increased now. The entire nation's long-term goal of a shorter work week would be impossible. A 32-hour work week, for example, would require an estimated 105 million in the working force, and that large a force simply will not exist.

There is no question but that the rate of mechanization will have to be increased if we are to realize our ambitions of both a steadily rising standard of living and a shorter work week.

I know of no instance of any difficulties with our various unions through our introduction of new machines. The unions know as well as we do that increased mechanization at Sylvania has given us the competitive strength to create broader markets for our goods—and that means more jobs and more security.

SYLVANIA'S DATA PROCESSING SYSTEM

I will show you now what we are doing to carry mechanization one step further to make mechanization really possible. We are going to automate the figure and clerical part of the business via our new Data Processing System, the headquarters of which is under construction near Syracuse, N. Y. A nation-wide 12,000-mile private electronic communications system, engineered jointly by Sylvania and Western Union, will link 51 cities with the Data Processing Center at Syracuse. These specialized communications facilities will tie together all of our plants, laboratories, sales offices, warehouses, divisional headquarters, and executive offices. These various installations will feed financial and production information over the leased network to the Center, where it will be instantly summarized for all levels of management.

The heart of the Center will be a "UNIVAC" machine. The

giant electronic brain will convert a wide variety of information into summarized data on which can be based decisions by management at the corporate, divisional, and plant level. It will gather, record, compute, and classify information concerning production volume, sales, billing, and many other activities. And as we gain experience we fully expect to broaden its activities to such areas as market research, engineering analysis, and other fields.

Western Union, Sperry-Rand, and Sylvania look upon this new project as a revolutionary step in industrial communications. It is, as far as we know, the only existing concept of an entire company tied together communications-wise from a data processing standpoint, and with that function housed in its own facility. Of equal significance is the fact that this project in administrative automation will be of tremendous value to an operating organization whose very foundation is mechanization and automation. Here, too, will be a centralized facility making even more effective a full decentralized operating organization.

I should like to emphasize that ingenious though it may be, UNIVAC *cannot think.* It can only give out what somebody has put into it in the first place. It does the process of manipulating that information very rapidly, but it cannot think; *only human beings think; only human beings make things.*

In summary, if we are going to continue to increase the standard of living of this country at the rate that it has been increased in the past few years, then several things are necessary:

(1) I believe that mechanization must continue to make new jobs faster than it takes people off old ones. So far that has happened, and it is the creation of the new job that is likely to cause the labor shortage that I am worried about in the next 20 years, rather than the army of unemployed which some people fear as of that time.

(2) I feel that unless the rate of automation is greatly increased, we will not have a large enough working force in the future to continue the rate at which we have been raising our standard of living and to reduce working hours.

(3) Far from deploring automation, we should encourage it to increase at a rate we have never before attained.

DON G. MITCHELL, Chairman of the Board of Directors and former President of Sylvania Electric Products Inc., with headquarters in New York City, became one of the country's leading sales executives long before he was 40. He became a recognized authority on corporate organization and policies, especially in the broad area of human relations, and decentralization of operating authority and responsibility.

Mr. Mitchell joined the Sylvania organization as Vice President in charge of sales in 1942, which was the opening of the greatest period of sales expansion in the company's history. After three years in the sales vice presidency, he became Executive Vice President in January 1946, and in May of the same year was elected President. He was elected Chairman of the Board in 1953, and president in 1955, continuing as Chairman.

Mitchell was graduated from the Montclair (N. J.) High School and attended the University of Cincinnati for two years, taking a cooperative course in mechanical engineering, and the University of Florida.

After a short period as an instructor of mathematics in Montclair High School, he joined the McGraw-Hill Publishing Company as an advertising salesman and later became manager of the company's industrial site service bureau. He then moved on to Niagara Hudson Power Company, as Director of Industrial Development for the utility. In 1933 he became Manager of the Marketing Division for American Can Company, where he introduced the paper milk container to the market. Two years later while still with American Can he introduced the metal beer can. He went to Marshall Field and Company in 1937 as General Sales Manager of the Manufacturing Division. In 1939 he became Vice President in charge of sales of the Pepsi-Cola Company and was responsible for that company's rapid sales growth over the next few years. He left Pepsi-Cola in 1942 to join Sylvania. A leader in industry association work, Mitchell is Chairman of the Board of Directors of the American Management Association and of its international affiliate, the International Management Association. He has been a director of the National Association of Manufacturers, National Electrical Manufacturers Association, and National Sales Executives. He is a trustee of the Committee for Economic Development, as well as a member of the Board of Trustees, National Industrial Conference Board, and the Board of Trustees, American Enterprise Association. In December, 1953, Mr. Mitchell was named a member of the Hoover Commission task force, appointed to study the Government's Civil Service program and other personnel responsibilities. In December, 1954, he was named a member of the Committee for the White House Conference on Education, held in November, 1955.

He holds honorary degrees from Northeastern University, Parson's College, Rensselaer Polytechnic Institute, Stevens Institute of Technology and Middlebury College.

CHAPTER 7

AUTOMATED MANUFACTURE OF MACHINES OF COMMUNICATION: *A Case Study*

J. M. HUND, *Associate Professor of Business Administration,
Emory University, Atlanta, Ga.*

During the past six years the electronics industry, long noted for its contributions to the automation of other industries, has embarked on its own program of automated manufacture. In this rather diverse industry are found the familiar products of telephone, radio and television and the less familiar items of electronic hardware such as controls, scientific instruments, and computers. The communications industry, including the first three products, can for purposes of analysis be separated from electronics, and it is to this product group that attention will be directed below. The change which has taken place in manufacturing methods can hardly come within the more restrictive definition of automation, since programming combined with the feedback principle involving automatic correction are only in the developmental stages. Greatly increased mechanization in the production of components and assemblies rather than fabrication of completed items is what the industry is currently witnessing.[1] However, even this degree of automation has had its

---

1. Attributed to Don G. Mitchell, Chairman of the Board of Directors and former President of Sylvania Electric Products, Inc.

impact on the individual worker in the industry, and it is the nature of this impact which will be discussed below. From this one may be able to make some predictions about what further steps of automation may mean to this group of industrial workers. Until recently telephone equipment has been electrical rather than electronic. Even now with new techniques of manufacture and with the new equipment in use, which have involved displacement of workers in plants or at switchboards, the stability inherent in the industry's structure has tended to minimize dislocations. With relatively few firms supplying products to the telephone companies which operate as regulated monopolies, and considering the tremendous growth in service which has taken place since the end of World War II, it is easy to see why greatly increased productivity here has not created serious problems. Accordingly, it is the manufacture of radio and television sets and their components on which this discussion is based.[2]

ELECTRONICS: A DYNAMIC INDUSTRY

Until World War II what is now the heterogeneous industry we call electronics was principally radio and the experiments which led to radar. Even before 1921 radio was considered primarily in the framework of commercial communications (wireless) and marine uses rather than in terms of mass communication and entertainment that we know it today. With the spread of broadcasting, the burgeoning demand for radios attracted scores of firms to their manufacture. Competition soon became keen, as only a small capital investment was required to get into production of the standard circuitry. Market instability resulted from overproduction combined with obsolescence due to a rapidly advancing technology. Failures among the weaker firms brought further dumping and aggravated an already unstable situation.

The advent of television reactivated the same chain of events.

2. In the course of research in the electronics industry the author had occasion to visit several of the firms importantly engaged in the manufacture of radio and television and their allied products and components. Rather than depend solely on the data collected in any individual plant or firm, use will be made of the wider range of observations with the intention of giving the reader a broader view of the topic.

The persistence of unsettled conditions in the industry has meant thin profit margins for many, and labor has found it difficult to keep pace with the wage increases of manufacturing as a whole.[3] The unions with contracts in the industry have failed to organize some of the larger units, permitting employers with union contracts logically to ask for terms which would leave them competitive with the unorganized sectors of the industry. A further reason for relatively low wages in the industry has been the nature of employment requirements. Both in components and in radio and television the work is generally light, but often tedious. A high percentage of female labor is the rule, running from 50% to 75% of the total, depending on the product. This situation has added to the difficulties in union organizing efforts, for women workers often view themselves as only temporary members of the labor force. This may be because they plan to marry and become housewives, because they are expectant mothers, because their children are grown and they wish to add to cash resources pending their husbands' retirement, or because they wish to work for awhile to hasten the retirement of a family financial liability such as a mortgaged house. For whatever reason, they often are indifferent to the publicized values of union membership such as seniority, the promise of strike benefits, or pensions. The result is high labor turnover, itself a barrier to tight union organization.

Since sewing and allied household arts are already known to women, the tedious nature of electronics manufacture and assembly is of but little concern. They perform tasks with screw driver, pliers and soldering iron at which men would be either clumsy or intolerably bored.

This is not to say men are not employed in plants turning out these products, but they are customarily not assigned to the routine assembly operations, but rather to work in packing and shipping, a tool and die work or as technicians. A high proportion of female unskilled or semi-skilled employees, then, helps to account for the lower average hourly earnings noted above. Since increased mechanization has been principally in assembly operations in radio and television plants, it is not the

3. Average hourly earnings in radio and television in 1947 were $1.13 against $1.29 for all durable goods. In 1953 average hourly earnings in electronic tubes were $1.54 and in radio and television $1.62 against $1.88 in all durable goods. *Monthly Labor Review*, October 1953, p. 1050.

men then, but the women who have felt the changes which have taken place.[4]

The recent technological innovations in set manufacturing have revolved around the printed or etched circuit board. With the circuits already determined, the assembly job is reduced to attaching the components on the board at the proper places in the circuitry.[5] It has been estimated that 30% of the labor force in electronics is in assembly operations, or an even more relevant measure, that of the television sets made in the period 1947-50 nearly 60% of direct labor was expended on wiring, lacing and assembling, jobs held principally by women.[6] The component-attaching machine developed by General Mills called Autofab will put together in one minute the number of electronic components that previously took a worker a full day to assemble. Not only are the components attached at the proper places on the etched circuit board, but all of the connections are soldered at once. Since as many as 150 connections were involved, the automatic solder bath operated by one person does the work twelve formerly did in producing 1000 units per day.

In reporting to its members on the effects of automation, the International Union of Electrical Workers used figures on employment and output from the period December 1953 to the same month in 1956.[7] In this three year period, allowing for changes in product mix, 48% more radios and 25% more television sets were produced with one per cent fewer production workers putting in virtually the same number of manhours.[8] Much of this large jump in productivity has been

---

4. Since radio tube manufacture evolved from electric lamp construction, a high degree of mechanization has always existed in this branch of the industry, though recently this has been carried even further. Picture tube manufacture is also highly mechanized, but a larger proportion of male labor is employed.

5. A detailed description of various devices of automation in the electronics industry, and their development, has been written as Ch. 13 in *Automation in Business and Industry* (New York: Wiley, 1957).

6. *Monthly Labor Review*, October 1953, p. 1049, and January 1956, p. 16.

7. In electronics from 1947 to 1952 employment grew by 40% while output expanded by 175%. This was not due to automation, but rather to the flowering of television which reached its peak year in 1950. *Monthly Labor Review*, October 1953, p. 1052.

8. If September figures are used the comparable picture is 7.8% more radios and 10% more television sets with 11% fewer employees putting in 1.1 million less manhours.

in assembly operations manned almost exclusively by women, though improvements in materials handling, layout, and other phases of the manufacturing process have also been made. Just what has been the impact of the changes in assembly methods effected in the past six years or so?

INTRODUCING THE NEW TECHNOLOGY

At the outset one must appreciate the relatively fortunate position of the electronics industry, and more specifically radio and television, compared to other industries where automation has been introduced. This stems from the point made just above concerning the numerical preponderance of women and their contribution to the labor turnover rate. If it can be assumed that most of the women who leave a radio or television plant are also exiting from the labor force, then it should be possible to introduce automation at a pace involving a minimum of displacement both for the plant and the community. This pace can be adjusted to the combined effect of natural turnover (often as great as 10% per month, but probably averaging about 5%), increased personnel needs in other departments as production is expanded, and the transference possibilities created by the new technology. The figures cited above would seem to indicate that this has been fairly well accomplished.

In radio and television manufacturing automation, or more properly, increased mechanization, has been introduced in steps, applying it to a single product first or to one of the many production lines. Labor-management committees in some instances are formed to provide for joint consideration of the problems involved in bringing in the new machinery. The committee meetings are devoted to a study of the machinery itself with comments being solicited from the labor representatives. Such matters as pay rates, new job classifications, and relocation of personnel are discussed and concrete solutions sought. Supervisors are educated on the workings of the new equipment and informed on the progress of its installation so as to be able to answer workers' questions. Other firms have informed only union officers and foremen of the impending changes in an attempt to underplay their importance. In such situations the new equipment has been introduced more slowly.

The use of printed or etched circuits affected the wirers and solderers first, and these people were transferred where possible to final assembly, because even with the new equipment 20-25% of the work is still done by hand. Tubes must be inserted and the chassis put into the proper cabinet. Some new machine-tending jobs were created which required no more basic skill than wiring or soldering, and with two weeks' training women could be put on these tasks. Other new machine operations requiring greater skill were generally filled by men. Since these jobs required more skill and therefore involved greater training costs, the firms sought men whose turnover rate would be expected to be less than for women. The pay rates for these jobs on the new equipment were set from 5-15% above the straight-time rates for unskilled assembly tasks. Even machine-tending jobs involved additional responsibility, as the whole line would be shut down if the supply of components at any point failed.

The use of printed circuits has had two side effects on employment. First, the preparation of the laminated plastic boards has created new jobs which women can handle, whereas the old "inverted cake pan" was turned out by men in metal-working shops. In addition, the use of this lighter material combined with the miniaturization of components (and importantly the concurrent shift to semi-conductor devices) resulted in lighter final products. This has made possible the shift of women to packing departments where until recently male labor was the rule. Because reassignment and retraining combined with normal turnover has softened the impact of displacement, no serious opposition to the introduction of the highly mechanized assembly equipment has been offered by organized labor, though it has firmly announced that it intends to secure for its members a fair portion of the proceeds to the firms arising out of increased productivity. One may well ask what some of the implications for the future of jobs and earnings in the industry may be.

IMPLICATIONS OF THE NEW TECHNOLOGY

Though to date automation in radio and television manufacturing has amounted to increased mechanization, further refinements are on the way which will more nearly qualify as automation in its more restricted sense. Competition in the market for the products will determine when these refinements can be introduced into the commercial production of radio and television sets. No attempt will be made here to describe these new developments; the objective is to evaluate their probable impact in terms of employment and the occupational structure, on job content and the wage structure. To do this requires some knowledge about the economic effects of automation on firm and industry.

It should be pointed out that in radio and television manufacturing factory labor costs range between seven and fifteen per cent of total costs. Approximately 80% of total cost for the non-integrated companies is wrapped up in purchased components. A good deal of capital is sunk into component-making equipment used to turn out millions of units, some of which sell for a penny apiece. Firms doing this work have been slow to render their equipment obsolete by major redesign programs, and major advances toward automation in assembly await component redesign, perhaps along the line of the modules developed in Project Tinkertoy carried on by the National Bureau of Standards.

If only a fraction of a maximum of 15% of total cost can be saved by a machine costing upwards of $100,000, the incentive to automate will not be great unless a market capable of absorbing a high sustained volume of output can be predicted. The erratic and highly competitive nature of the radio and television market does not make such predictions easy, and this undoubtedly accounts for the fact that some firms have moved slowly into automation and have been able to show as favorable a profit picture as those which have embraced it. The inability of present equipment to produce economically a variety of short-run items means that this equipment in commercial use must be devoted to long runs of identical assemblies. "The major economic fact is that, under automation, depreciation rather than labor becomes the major cost. And when labor is relatively cheap, it becomes uneconomical to keep an enormously expensive machine idle."[9]

Organized labor has accused management of pursuing a production policy calling for periods of swift inventory build-up, permitting maximum use of the new machinery, followed by periods of layoffs during which inventories would be absorbed by the market. Since depreciation charges are incurred whether the equipment is running or not, this can hardly be viewed, from the point of view of the firm, as a desirable long-range policy, but one necessary in view of short-term market instability. Since the machinery has an optimum rate of operation, the

9. *Management Review,* January 1957, p. 82.

pursuit of profit maximization would seem to dictate cutting costs where possible. This means that labor, the factor with variable cost, suffers. The major implication to be drawn here is that until the "shakedown" is completed, or until competition has decided what firms are here to stay in this line of manufacturing, there will be no wholesale move to greatly increased mechanization. Because of the barrier of component redesign referred to above, and the often unfavorable cost comparisons between redesigned and "old style" components, automation in the production of radio and television sets may be expected to come fairly slowly.

If the assumption just stated is correct, then a transition to automation involving a minimum of dislocation and hardship can be predicted. Judging from recent progress in this direction, a general upgrading of the occupational structure will be the result, coupled with greater productivity and rising wage rates and average hourly earnings.[10] One cannot blink at the fact that in this revamping of wage and occupational structures, changes in job content will require retraining to provide current personnel whose skills have been rendered obsolete with the means of continued employment. It is also true that firms in a competitive market will try to obtain the maximum return for their expenditures on training. This means a preference for two groups —men and "settled" women. One may predict, then, that the average age of women employees will increase and that the percentage of female labor will fall. An additional reason for the drop in percentage is the increase of technical and maintenance work necessary to keep the new machinery in functioning order, jobs ordinarily held by men.

Not only will the wage structure be subject to upward revision as less unskilled labor is required and more responsibility put onto the individual worker, but the basis for measuring the work and determining payment will also undergo a change. As process methods replace job methods, incentive systems lose their relevance. Under the present state of technology, incentive is given those who process the circuit boards and those who perform final assembly operations. Others are paid on the basis of group incentive, or the number of pieces turned

10. F. K. Shallenberger, "Economics of Plant Automation," *Automation in Business and Industry*, p. 573, states that to maintain the present rate of rise in our standard of living, a 50% rise in productivity over the next ten years will be required, or about twice the average rate of $2\frac{1}{2}\%$ we have witnessed over past years.

out by several workers whose operations are linked, but those on the automatic inserting or assembling machinery are paid on a straight time basis. The team, and not the individual, assumes a new importance, perhaps enhancing the need for social engineering. Incentive, or piece-work systems, then, must be revised to fit the new processes or discarded. As the present machinery evolves into truly automated equipment, permitting programming (giving varied instructions to the machine) and automatic self-adjustment or correction, the individual will regain his importance as a unit of labor, and probably at the same time gain in terms of job satisfaction. This leads into the question of the attitudes held both by management and labor toward the technological changes and their impact which have been discussed above.

ATTITUDES OF MANAGEMENT AND THE WORKER

It is probably safe to say that workers in plants with well-developed communications have not felt as concerned about automation as their elected representatives. Very often the unknown is feared more than the known, and where a period of introduction to the new machinery has preceded its actual operation in the production line, acceptance by the worker has been good. Of course the individual worker performing a job no longer required is properly concerned, but the possibilities of retraining or reassignment have helped to dull the impact in many instances. Adjustment to new work varies with the individual. Some are a good deal less adaptable than others, and hardships in terms of job satisfaction have been experienced.

In early 1957 representatives of organized labor approached the trade association of the industry, the Radio, Electronics, and Television Manufacturers Association (now renamed the Electronic Industries Association) asking that joint industry-labor conferences be arranged to discuss problems arising in the industry as a result of automation. Though at first expressing willingness to carry on such discussions, the association later informed the union men that these were issues for collective bargaining in individual firms. The union, understandably concerned with the prospect of revising its internal structure to fit the new technology, and feeling its area of job control shrinking, was eager to have a voice in determining the overall course of automation in the industry.

The attitudes expressed by management in the firms visited by the author were qualifiedly favorable to automation as of benefit to both the firm and its employees. Executives complained of their inability to take full advantage of increased mechanization during periods of market instability. They deplored the impact of production cutbacks and the concurrent introduction of automation on short-run employment levels. However, assuming stable production, they could see in the long run nothing but benefit accruing to the worker. A part of this benefit is increased job satisfaction. In the sort run, however, some dissatisfactions may be experienced. In conventional assembly methods the women doing wiring, soldering and assembling have control over their work, even though pierced fingers and burns may at times result from it. Inspection presumably assures product uniformity. The new machinery sets the pace of work, and control is lost to the machine which is now "tended." In one operation visited by the author, even final assembly was paced by the rate established by the machine which dip-soldered the circuit boards. To the extent that the chance for some physical discomforts is removed, job satisfaction can be expected to increase. Contrarily, to the extent that control over the job is lost by the individual worker, and to the extent that she has regarded this as important to her, job satisfaction will be found to decrease. It is difficult to say which way the balance will tip except as individual cases are investigated.

If one takes a long-run view of automation in this industry, the outlook is brighter, for, as increased mechanization evolves into automation, worker control returns. The machine takes on the tasks involved in producing the whole product mix. The flexibility introduced by programming and feedback and the ability to turn out different products with the same machinery complex will lend to automation jobs of the future in this industry, an attraction they certainly do not now possess. Even the mental tension, now thought by some to replace muscular fatigue as the current machinery is introduced, will largely disappear.

One further point before summing up. The attitudes of organized labor, sometimes hostile, sometimes cooperative, but always questioning, have been largely governed by the impact of automation on employment. Comparisons with non-automated plants, or even with conventional production lines in the same plants, have dramatized the changes being wrought on employment.[11] The I.U.E. has announced its intention of seeking the guaranteed annual wage. If successful, labor becomes a less variable cost. Since automation provides greater product uniformity with fewer inspection operations, the incentive to replace more labor becomes even greater. However, as this is done, labor's control of its destiny increases. A shut-down is now less tolerable to management, and as labor becomes an even smaller part of total costs, it becomes less worthwhile for management to have collective bargaining end in a strike. The petroleum industry has learned this lesson. One might say, then, that the very automation viewed with hostility, or at least skepticism, by organized labor may be the means to a degree of control at the bargaining table not realized while conventional methods of manufacture were being used.

In this report on the progress of automation in a segment of the electronics industry, emphasis has been placed on the transitional character of the machinery presently in use. From its beginnings the industry has been one of continual change and innovation, with the result that no manufacturer of final products has wished to make capital investments he could not hope to amortize before a new product or technology came along. The same holds true today; machines more closely approximating the flexibility of human labor may be rendered feasible for commercial use. Military electronics has not been treated specifically, so reference has been all along to commercial applications. In military work governmental agencies are willing to pay higher prices to attain increased reliability and flexibility, and it is from these developments, financed by public money, that truly automated processes of manufacture will stem. The incentive for government to sponsor projects designed to modernize the manufacture of all kinds of electronic equipment is evident in the estimate of military requirements in the event of mobilization. It has been stated that even a mild mobilization might run the annual volume of electronics manufacturing up to eight times its current level of something over $6 billion annually.[12]

---

11. One plant manager reported that the increased mechanization had already eliminated 22 assembly operations, and that other savings were anticipated.

12. E. L. Van Deusen, "Electronics Goes Modern," *Fortune*, June 1955, p. 148.

Unstable markets have characterized the radio and television industry in the past, and they continue to plague it today. However, if automated machinery can be developed of the kind foreseen above, the industry will necessarily fall into the hands of relatively few manufacturers—those capable of making the needed capital investments. This will eliminate the dumping operations which have contributed to market instability. As industry-wide production levels are smoothed out over the year, the level of employment will also stabilize.[13]

In the meantime the composition of the labor force will have radically changed with ever-increasing emphasis on the skilled worker and the technician. This transformation in the occupational structure will bring rises in wage rates and average hourly earnings and changes in the bases for making payments. In the short run productivity will race ahead of employment. Job content will be modified, first in the direction of machine-tending or machine repair and maintenance, and then toward reasserted control over the machine. In this first phase worker satisfaction may in many cases decrease, and some displacement and reassignment will undoubtedly cause economic and mental hardships. However, because of the proportion of female labor and the turnover situation outlined above, this industry is more flexible than others. The disenchantment some firms have suffered while trying to introduce a new technology into an unstable market situation will certainly not be calculated to accelerate its introduction. Observation would indicate that when introduction is effected, it can be considerably smoothed by careful preparatory measures taken by both labor and management.

The promise of increased living standards made possible through the wonders of electronics should not be denied. It is for those immediately involved in the social context, management and labor, to call upon whatever agencies may be appropriately concerned for their contributions. The continuous process industries, for example, have gone a long way along the road to automation. Manufacturing industries such as electronics may profit by their experience. They must if we are to live electronically.

---

13. One must admit that the current situation in automobiles, dominated by three large producers, argues against this outcome.

Selected Bibliography

L. K. Lee, "Automatic Production of Electronic Equipment," E. M. Grabbe, ed., *Automation in Business and Industry* (New York: John Wiley and Sons, 1957).

Proceedings of Symposium on Automatic Production of Electronic Equipment (sponsored by the Stanford Research Institute and the United States Air Force, April 1954).

Articles:

A. V. Astin, "Computing Machines and Automation," *Computers and Automation* (April 1956).

G. B. Baldwin and G. P. Shultz, "Automation: A New Dimension to Old Problems," *Monthly Labor Review* (February 1955).

E. W. Engstrom, "Automation," *Commercial and Financial Chronicle* (December 22, 1955).

L. Lee and F. Hom, "Automatic Production of Electronic Components," *Radio and TV News* (December 1953).

Martin Sheridan, "Automation in Television for a Better Industry," Commercial and Financial Chronicle (March 15, 1956).

E. L. Van Deusen, "Electronics Goes Modern," *Fortune* (January 1955).

E. H. Wavering, "Big Step Toward Automatic Assembly," *Factory Management and Maintenance* (October 1952).

E. Weinberg, "Review of Automatic Technology," *Monthly Labor Review* (June 1955).

———, "An Inquiry into the Effects of Automation," *Monthly Labor Review* (January 1956).

"No Sugar Coating From Philco," *Business Week* (April 14, 1956).

"How Automation Hits a Plant," *Factory Management and Maintenance* (November, 1955).

"The First Automatic Radio Factory," *Fortune* (August, 1948).

"The Automatic Factory," *Ibid.* (October, 1953); "The First Automation Strike," *Ibid.* (December, 1955).

"Adjustments to Automation in Two Firms," *Monthly Labor Review* (January, 1956).

J. M. HUND was born in Detroit (1902), attended local school, then Stanford University, and graduated with the A.B.

degree from Amherst College in 1943. Called into the Naval Reserve on active duty in 1944, he was stationed at the Midshipman School at Notre Dame and went through the course in Naval Supply at the Harvard Graduate School of Business Administration.

Worked for three and a half years with Reo Motors, Inc. in Lansing, Michigan in several administrative posts and in the sales department after 1946.

In 1950 he re-entered the Graduate School of Princeton University in the Department of Economics. In 1952 he received the degree of Master of Arts and also became associated with the Organizational Behavior Section at the University. This project was sponsored by the Ford Foundation. In 1954 he completed requirements for the Ph.D. His thesis was entitled "Managerial Decentralization: A Case Study."

From 1954-57 he was associated with the Department of Economics at Clark University in Worcester, Mass. In 1956 he became Director of the Division of Business Administration.

Last year he worked as a research associate with the New York Metropolitan Region Study, a project done through the Harvard Graduate School of Public Administration for the Regional Plan Association, doing a study of the electronics industry in the New York area.

In January 1958 he became an Associate Professor of Business Administration at Emory University, Atlanta, Georgia.

CHAPTER 8

POST OFFICE AUTOMATION PROGRESS

L. ROHE WALTER, *Special Assistant to Postmaster*
General, Washington, D. C.

In 1953 we recognized the need for modernizing the mail
handling methods, equipment and facilities and have launched
upon a vigorous program in that area. We have established a research and engineering organization and have engaged the services of several research and development organizations, both in
government and industry. Our program is divided into the following major areas:

(1) culling, facing and canceling letters,
(2) letter sorting; and
(3) parcel post sorting.

Up to that time all heavy bulk mail, with the exception of a
few large flat belt conveyors and mail chutes in the larger post
offices, was moved manually. Consequently, to move more mail it
was necessary to call in more men.

One of the Postmaster General's first reorganization actions
was immediate attention to surveying post office operations to
determine the potential application of common industrial type
equipment to bulk mail handling operations. Immediate steps
were taken in collaboration with the Bureaus of Operations,
Transportation and Facilities, to provide post offices with bun-
dled mail tying machines, various types of fork lift trucks, portable conveyors, and other machines and devices for facilitating
the handling of bulk mail.

Supplementing the extensive installations of this portable
bulk mail handling equipment, engineering surveys and studies

are continuing to determine further requirements of industrial trailers, electric trucks and tractors, lift trucks, platform skids, and ramps and dockboards. These studies also reveal needed improvements and modifications in the existing portable bulk mail handling equipment.

At the same time, attention was directed toward planning a long range mechanization program encompassing research and the development of machines designed specifically to do post office work.

Each function in the chain of mail operations from collection to delivery was analyzed and improvement projects were established.

Priorities were determined primarily on the principle of "first things first" to effect the greatest cost reductions in the quickest time within fund limitations.

The major problem still is the predominance of manual operations. Improved methods and better scheduling of the work forces are worthwhile programs, but maximum operating economies can be achieved only by mechanization to the point where mail received at the docks is transported through the culling, canceling and postmarking and sorting operations, and to dispatch by machines, with manpower devoted almost exclusively to machine operation.

There is a critical dearth of post office workroom space, and labor costs, because of working mail manually, are excessive. The situation has been constantly aggravated by the rapid increase in mail volume and the great peak load hitting the post offices in the early evening hours.

The mechanization program is a race with catastrophe—the catastrophe of too little equipment too late to handle the increasing mail volume. Even now, much mail is being worked in unsheltered areas and on the streets and the employment of mail handlers is critical in some cities.

Approximately 35%-40% of post office workroom floor space is devoted to working parcel post, using only about 5% of the labor force.

In contrast, about 75% of the post office mail handling labor force is used to work letter size mail.

Consequently, it was obvious that we should attack the parcel post space problem to increase productivity per square foot of floor space, and the letter mail sorting problem to increase manhour productivity.

PARCEL POST HANDLING

Since parcel post processing requires about 35% of the post office workroom space and because space is critical, we have taken a careful look into the potentials of mechanized parcel post handling.

Parcel post sorting has been accomplished by mail clerks lifting packages from a slide or table, reading the address and sorting into 12 to 18 hampers, depending on the distribution system. In some of our larger cities we have used a multiple belt sorting unit. The configuration of this system is such that it is limited to 17 separations which is inadequate for primary parcel post sorting.

Consequently, new systems had to be designed which were flexible to extend the number of sorts up to 100 depending on the volume and breakdown.

The existing system requires much hard, tedious hand labor, including leaning into hampers, lifting heavy parcels and pushing hampers which has resulted in numerous physical injuries.

The conventional flat belt separation system mentioned as being limited to approximately 17 primary sorts costs approximately $15,000 to $20,000 per separation.

Our first approach was the development and procurement of the "Greller" experimental system which involves electro-mechanical controls, belts, keyboards, memory system, diverters, slides and working tables. This system is making 21 sorts at the Baltimore Post Office, and is capable of making as many as space will permit or as required.

This sorting system was developed by Nelson Greller Associates, Washington, D. C., and is a keyboard activated unit using electronic controls. Packages are received and stored on a slide where mail clerks read the address, affix a code (1 thru 11) or (1 thru 10) and place the parcels on moving belts.

These parcels pass a station where the operator reads the code and depresses the keyboard comparable to the code on the parcel.

The information is stored in the memory system and a photoelectric cell measures the length of the parcel.

This information, plus the code information, is transmitted to the diverter which lowers a set of paddles at the exact time equal to the length of the parcel and sweeps the parcel into one of the 21 primary chutes made possible with the new system.

The chutes carry parcels to a waist-high table, where mail clerks sort to sacks for final dispatch.

A second concept is now under development which eliminates the use of paddles and has the same flexibility of making from 15 to 35 separations.

Design of these units includes the use of high ceilings for storage and transporting mechanism with the primary sort directly under the unit. This method conserves floor space which is critical.

Our operation to date verifies the fact that in order to increase production and reduce space limitations we must develop mechanical parcel facing and feeding and increase the number of primary sorts. Furthermore, the system must be capable of accepting any size from 1/4" thickness, weighing 8 ounces up to the prescribed maximum size and weight limits.

More recently we contracted with the Jervis B. Webb Company of Detroit, Michigan, to develop a machine with a mechanical feeder and utilizing a keyboard, which should divert parcels into separators at a rate of 1,200 per hour per man. The pallets will carry their own mechanical memory system which provides the necessary flexibility, and space would be the only limiting factor to provide the number of primary sorts from 15 to 50 or even 100 in the largest post offices where it may be economical to make that many separations.

It is planned to install this unit in the Washington, D. C. post office.

In addition to our loose parcel post handling system we are processing with the installation of a power and free trolley system in the Chicago post office.

This system is designed to readily unload, store and transport incoming and outgoing sack mail. The power and free trolleys will be suspended from the ceiling to utilize the unused overhead space in our post office workrooms, and allow us to store mail in sequence so as to expedite loading of trucks or railroad cars. The entire system will be operated from an electrical console control panel and the mail will be called for and transmitted to the platform upon signalling by personnel.

LETTER SIZE MAIL HANDLING

Another major mechanization area is letter mail processing. Since approximately 75% of the workroom labor force is used in this operation, our first attempt to mechanize was directed toward letter mail facing and canceling. The first experimental machine was tested in 1955 and demonstrated the feasibility of canceling letters at a rate of 30,000 per hour. At the present time we are pursuing two concepts of facing and canceling machines which will feed, sense, cancel and stack at a rate of 30,000 letters per hour. Also each contractor is developing and constructing a culling machine which will enable the post office to dump raw mail in a hopper, cull off all parcels and flats in excess of $\frac{1}{4}$ inch thick and over 6″ high. The culler will eventually stack clean mail for transfer to the automatic facing and canceling machine.

In regard to culling (which is a preliminary sorting process that, for example, separates parcels from letters) all operations in this area are now in crowded quarters. Such conditions are further aggravated by the requirement of a larger staff of men during the peak hours to process the mail as rapidly as it is received. Under the present operation mail handlers cull off various types of mail by hand, and then place all the machine-cancelable letters into a trough so that letters are stacked with the stamps down and forward.

A second operator lifts the envelopes from a "stacker" and places them into a canceling machine for postmarking, canceling and subsequent stacking.

We have engaged two contractors who have analyzed these laborious operations. And, using some ideas from equipment previously developed, we are now well on the way to provide culling, facing (the placing of all letters in the proper position for machine cancellation) and canceling equipment.

In the "culling" operations, collection mail will be emptied into a hopper and then elevated, and subsequently spread over five longitudinal "ports" with two apertures, each equipped with moving belts operating at different speeds.

All letters over 6″ tall will be separated from the letter mail and transferred into a flat mail bin, and the parcels collected in a separate station for movement to parcel sorting center.

In one concept for automatic canceling machine the letters will be stripped off individually and transported past a scanning station which will locate the stamp regardless of position and signal the cancelling unit for postmarking and canceling.

Canceled letters will be stacked with stamps in one of four positions so that letters can be lifted from the stack and placed into the reading machine or keyboard coding station.

We are proceeding with the second concept of canceling where the scanning mechanism searches for stamps along the lower edge of the letter.

Letters with stamps on the lower edge are acknowledged and sent to the canceling station.

All letters with stamps on the upper edge are turned over and sent to a second station for canceling and postmarking.

Canceled letters are stacked in trays with all stamps in one of two positions. Stamps leading or trailing for subsequent processing.

With reference to the facing and canceling of mail—Under the present system letters must be sorted through several steps, all of which requires a great deal of manpower and time.

As an interim measure the Department has procured two proven foreign made keyboard letter mail sorting machines, one known as a Transorma and the other as the Bell. Each of these is capable of sorting into 300 separations. The operator must learn a scheme to determine which of the keys or combination of keys to depress to direct the mail to the proper sort. These were procured to determine the economics of key sort equipment and to test new concepts such as coding and feeding, prior to the time that it would be possible to develop and construct an entirely new concept such as is being undertaken by the National Bureau of Standards.

The keyboard sorters, one of which is in operation in the Silver Spring, Maryland, post office and the other, the Bell machine, installed in the Washington, D.C., post office, are the first machines of this kind ever installed in the American Postal Service, and they represent one of the most significant innovations in American postal history.

The mail is lifted from a stack by vacuum cups and automatically moved to the operator, who reads the address, determines the proper numerical code on the keyboard for that particular address, and depresses the right keys.

When the keyboard is depressed, the proper intelligence is assigned to the conveying system, which carries the letters or receives a signal from the memory to drop the letter in the sorting pocket.

In the code-controlled letter sorting system under development, a binary dot no dot code will be printed on the envelope.

Letters will be stacked in trays and automatically fed into the system which will have 1,000 separations.

An electronic scanner will read the code and transmit the

intelligence to the carrier. The binary letter carrier sensing wheels traverse a system of rectangular code vars and slots associated with each pocket. A coded destination will drop the mating set of memory wheels and allow the letter carrier to be tripped. Our program includes a keyboard operated system where the concept is similar to the Transorma and Bell.

In this case, the machine will have 300 pockets, and canceled letters will be presented to the keyboard operator, who will read the address and convert the information to a numerical code.

The operator depresses the key corresponding to the numerical code, and this action establishes the destination on the carrier so that the letters will be released at the proper destination.

This work is being pursued by one contractor and one government laboratory.

Statistical data has been assembled on the volume of mail, size and destination. With the advances in the state of the art in electronics, considerable emphasis is being placed on coding.

There are two schools of thought, one of single coding at the point of origin and the other of coding only to destination and adding the code for the carrier route at the point of receipt. Emphasis is being placed on developing a simple code and incorporating maximum intelligence in the equipment so that the operator, with little training, can read an address, punch a minimum number of keys which will automatically print the code on the back of the envelope. The code will be machine read and the information matched with the intelligence in the memory system and when the information matches it will send out an impulse to the gate which sorts the mail to the proper destination. Such equipment is now under development.

This unit is designed with a view toward making separations of 100, 200, and 300 and up to 1,000 with the same coding, machine reading and sorting control equipment. Tests are being made to develop criteria for an optimum keyboard design and the ability to use simple codes.

Another area of development which promises to show extensive improvement is the electrical reading of typewritten addresses. During the past three years we have developed one model which is capable of recognizing 18 individual city addresses from all others.

A year ago the contractor demonstrated a jury rig model which was capable of reading the addresses on 300 to 400 letters and successfully sort to 18 distinct pockets or combination of pockets. The letters were addressed with well typed font and the demonstration was made without error. In the interim we are improving on the ability to read poorly typed addresses, abbreviations, window envelopes and to read script.

Concurrent with this development we are developing a presorter which will divert all non-readable mail and allow readable mail to proceed to the electronic reader. One plan under consideration with the successful completion of this development is to process the mail through the electronic reader which will sort out the top 18 to 20 separations and allow the non-readable addresses and the balance of the mail to proceed to a Transorma or Bell machine or to a new mail sorter which will handle the next 300 or 400 separations, or down to a practical limit and the residue will be manually sorted.

In addition to these items we are working on automatic cancellation of flats and considering mechanization for automatic labeling, tying and pouching of letter mail, and a system to make secondary separations automatically into sacks for dispatch.

SPACE UTILIZATION AND MECHANIZATION STUDIES

Space utilization and fixed mechanization layout are being made at all locations where new construction is planned. Also, these studies are being made prior to modifications or additions to existing post offices. The mechanization layout studies are made to determine the best mail handling and processing systems for each facility.

Improvement of fixed mail handling equipment in post offices is a continuing program. Most of the larger postal installations are still equipped with inadequate conveyor and chute systems. In many such situations, improvements in both equipment and buildings are necessary to provide more efficient and economical operation.

Benefits which result from such mechanization include not only better space utilization but improvement in service, improved employee morale and less congestion in work areas.

DETROIT MAIL-FLO SYSTEM

The Detroit intra-office automatic letter mail transport system is designed to move mail expeditiously on the workroom floor with the least amount of time and effort.

The basic concept of this mechanical system is to relieve mail handlers of burdensome tasks heretofore requiring manual effort. This system also clears the workroom floor of all equipment which not only requires manual effort to push, lift, carry or store but also required critical floor space.

The present installation provides a flow of trayed mail to the primary and secondary sorting areas by conveyor. Automatic delivery of trayed mail by means of side conveyor ledges to supply distribution clerks with mail for sorting in both the Primary and Secondary sorting areas, and delivery of mail to points of dispatch from the Secondary sorting area.

It also provides a flow of trayed mail to a storage point in the Detroit city primary sorting area from intake points by conveyors and delivers trayed mail by means of side conveyor ledges to supply distribution clerks with mail for sorting in the city primary sorting area.

The system is adjustable and adaptable to meet changing scheme requirements, and will speed the handling of mail and effect major economies in labor and equipment by bringing about better supervision and a more orderly operation.

It will maintain the dignity of the employee by utilizing his skills more fully and improve employee morale through improved work situations.

Additional installations of the system are being made at several of the other large post offices.

IMPROVED CITY DELIVERY SERVICES

The unprecedented suburban growth of recent years requires continuous expansion of the city delivery service and major changes in mail delivery methods. In making these changes, we have analyzed carriers' routes and developed functional vehicles to meet the needs of suburban carrier delivery. Our objectives are to achieve maximum efficiency from both the service and cost standpoints, to increase safety and to reduce physical labor.

City delivery route analyses were made and approximately 450 functional vehicles were installed in five cities in different parts of the country to help resolve continuing problems such as the best type of vehicle to assign to selected routes; vehicle maintenance; and mechanized route supervision.

Now, city delivery route analyses have been made in many other cities and 8,000 functional vehicles have already been installed in service in approximately 2,000 cities in all states and provinces.

LOBBY POSTAGE SELF-SERVICE

Self service post office lobby equipment is being developed to provide service to the patron after hours, on holidays and Sundays.

In 1948, we installed 1,500 stamp vending machines operating on fixed amounts and dispensing five each of 1c, 2c, or 3c stamps.

In 1956, seven Stampmasters, capable of accumulating credit up to 25c and incorporating the selectivity to permit the buyer to choose one or more of 2c, 3c, or 6c stamps and receive the change, were installed in post offices. These few machines demonstrated that the patron is interested in using stamp vending equipment capable of making change and providing them with the exact number of stamps needed. This equipment is considered suitable for larger post offices.

In order to automate our screenline activities we have developed the following program:

(a) Procure 40 additional coin changing type stamp dispensers for exploratory tests and determine requirements on a nationwide basis.

(b) Acquire a series of inexpensive stamp, stamp book, and postal card vending machines for use in all sizes of post offices where the traffic warrants and the patrons need additional service after hours and over the weekend.

(c) There are available plastic containers capable of holding coils of stamps in units of 100 which also incorporate the ability to dispense stamps from the container.

(d) In addition to the stamp vending equipment, we are studying the need for a semi-automatic parcel post weighing device with attachment to eject the proper postage. The operator will select the zone, depress a key, and the weight registered in the machine with the action of the key will compute the postage required and eject a printed stamp of the correct amount.

Stamp vending machines are planned to vend 3c and 6c stamps, or whatever the regular and airmail postage rates may be. With the deposit of a dime the machine will dispense the stamps and the change. Postal card vending machines are planned to vend single cards with each transaction. The units are scheduled to receive a nickel and release the change. Units will be either electrically driven or hand operated.

Stamp book vending machines are planned to vend two and four pane stamp books and two pane airmail stamp books. Units are scheduled to operate by accumulating change for the item selected and upon pressing the button, the stamps and change will be dispensed. These units will be either electrically driven or hand operated.

Studies are also being made to determine the feasibility of installing stamp vending equipment at entrances to post offices.

By 1968, it is expected the nation's top fifty post offices will have the benefit of most of these new machines. To the letter-dropping public, automation will mean faster service and perhaps lower post office deficits to make up. To postal workers it will mean a much easier way of doing the average day's work.

The appointment of L. ROHE WALTER of New York as Special Assistant to the Postmaster General in Charge of Public Relations was made, according to Postmaster General Arthur E. Summerfield, "to help the Department meet a pressing need for an information program serving both postal employees and the public." He was formerly director of advertising and public relations for The Flintkote Company, Inc., of New York, national building materials and manufacturers. More recently, he served in a creative capacity with two leading advertising agencies —Erwin, Wasey and Company and The Kudner Agency, Inc., and the public relations firm of Hill & Knowlton, Inc. Throughout World War II he was a Commander, U. S. N. R., on the Army-Navy Munitions Board. He is past president of the Direct Mail Advertising Association, Inc.; a former board member of the Association of National Advertisers and the Public Relations Society of America, Inc.; editor-in-chief of the McGraw-Hill Library of Business Management; and previously a faculty member of the Evening School of Business, Columbia University. Born in Dayton, Ohio, Mr. Walter is a graduate of Miami University, Oxford, Ohio, and a member of four fraternities; Phi Delta Theta (national social), Phi Beta Kappa (honorary scholastic), Tau Kappa Alpha (honorary oratorical) and Sigma Delta Chi (honorary journalistic).

CHAPTER 9

RAILROAD AUTOMATION

T. C. SHEDD, *Eastern Editor,*
Modern Railroads Magazine, New York City

The U. S. railroads are one of the Nation's oldest mass-production industries.

True, their product is an intangible one—transportation. And their "production lines," instead of being confined to factory buildings, are spread out across the country in the form of 220,000 miles of railway track.

Even so, some of the problems of railroad operation are similar in principle to those met in manufacturing plants. And, as they drive to improve their operating efficiency, railroads, like other industries, are turning more and more to the techniques of automation.

In recent years, newer forms of transport—notably highway trucks and automobiles—have taken over a substantial part of the freight and passenger traffic once handled by the railroads. Despite this diversion, however, the railroads remain the the pre-eminent form of low-cost, long-haul mass transportation.

This is true because of the inherent "free-rolling" self-guiding characteristics of flanged steel wheels rolling on steel rails. They permit a single locomotive (of relatively modest horsepower) to haul a train of 100 or more cars (each loaded with perhaps 40 tons of freight) over long distances at high speeds. This is done with amazing economy of fuel consumption.

A recent study[1] showed that in 1952 all other forms of transport combined used *eleven times* as much fuel as the railroads—

---

1. *A Ten-Year Projection of Railroad Growth Potential* (Chicago; Railway Progress Institute, 1955), 47, 49.

yet they produced fewer total gross ton-miles of transportation! In 1955, the railroads produced five times as much transportation per employee as the intercity truck lines. And this in spite of the fact that railroads build and maintain their own roadways, which highway carriers do not.

(While highway transport is basically better suited for certain types of short-haul traffic, much of the diversion of freight from rail to truck has roots in political regulation that is applied to the railroads but not in equal measure to other types of transport. This in many cases prevents the railroads from pricing their services to reflect their low costs on long hauls.)

Despite their basic efficiency, however, railroads have a relatively high "labor ratio." They put out roughly 50 percent of their gross revenues in wages and salaries. Thus there is intense pressure to further improve the productivity of railroad workers —to reduce the use of manpower.

MECHANIZATION AND AUTOMATION

Mechanization and automation are the main avenues to increased productivity on the railroads. There is a basic distinction between them—though it often becomes hazy in practice. Mechanization simply means giving workers more efficient tools and machines with which to do their jobs. This usually reduces the number of personnel required in a shop, on track maintenance work, or in an office; but it does not change the basic methods of doing the job.

Automation, as applied to railroads, often requires an entirely new approach to a job—old routines are thrown out. In this concept, a group of inter-related machines perform a number of operations or jobs automatically, using the principle of feedback control.[2] In many cases, the efficiency of work that has already been mechanized can be still further improved by automation.

But whether we distinguish between mechanization or true automation, recent advances in technology have had a tremendous impact on the railroads. In 1956, U. S. Class I railroads (which earn 99 percent of the industry's revenues) produced 9.3 percent more freight transportation, (measured in ton-miles) than they did in 1946. Yet they used 16 percent fewer employees.

---

2. "Now Comes Automation," *Modern Railroads* (September 1953), 35.

The standard measure of railroad operating efficiency is gross ton-miles per freight train hour. The industry figure for 1956 reached 57,012—a 53 percent gain since 1946!

Even many railroad people do not realize the full sweep of recent technological changes in their industry. These changes have permeated line-haul train operations; yard and terminal operations; track and equipment maintenance; and office work. But their implications for the future are even more significant. Railroads, it is clear, have only barely begun to explore the possibilities of automation.

The diesel-electric locomotive has done more than any other single device to improve the efficiency of railroad operations since World War II. The direct effects of the diesel are beyond the scope of this discussion; however it is worth noting that the railroads handled their 1956 traffic with 34 percent fewer locomotives than they used in 1946, when steam locomotives were still dominant.

TRAIN OPERATIONS

The most promising developments in train operations today are in the field of signaling and traffic control. The automatic block signal, first used in 1872, is perhaps one of the earliest examples of automation. From it have stemmed today's advanced forms of train protection and traffic control. The automatic speed control system used on the Long Island Railroad (and on some other railroads as well) is an example.[3] Here, electronic apparatus on the locomotive picks up coded impulses from the running rails. The frequency of these impulses varies, depending on track conditions ahead of the train. The locomotive equipment interprets these pulses and automatically applies the brakes, when required, to keep the train running at a safe speed.

In this case, the speed-control equipment is only an added safety precaution, designed to take control away from the locomotive engineer should he fail to obey a wayside signal indication. But clearly, it would be but a small step (technically) to adapt this type of apparatus to starting and accelerating trains automatically, as well as slowing and stopping them. Then trains

---

3. "Long Island Rail Road Installs Automatic Speed Control," *Modern Railroads* (October 1951), 59.

could operate entirely without human attention, much like the automatic elevators in office buildings.

Such fully-automatic train operation is more likely to come first on rapid transit lines, where the operations are routine, traffic is dense, the right-of-way is inaccessible to trespassers and there are no highway grade crossings.

All trains on a typical line-haul railroad operate on the basis of a "working timetable." Often, however, regular trains are delayed, or extra trains, not in the timetable, must be operated. Thus a train dispatcher is needed to oversee operations and keep traffic moving when the programmed timetable breaks down. Working by telephone through operators at wayside stations, he issues orders that direct trains to advance, or take siding to let other trains pass.

The modern technique of "Centralized Traffic Control" has revolutionized the art of dispatching trains. With "CTC," the dispatcher uses levers on a control machine in his office to change the aspects of the wayside signals and directly operate the track switches on a district that may be hundreds of miles long.

In effect, this means that the dispatcher communicates with the trains through his control machine instead of through the wayside operators. Engineers operate their trains entirely by the indications of the wayside signals, with no need for written orders or even for timetables.

Thus CTC bypasses the train order operators. (Not all of them are displaced, however; at many smaller stations the order operator also has other duties, such as ticket selling.) So efficient is CTC that it permits a single track railroad to handle almost as much traffic as a double track line without the CTC. Already, trains are dispatched under CTC on about 30,000 miles of U. S. railroad line. Almost no conventional block signaling is now being installed.

Indeed the increase of track capacity is one of the biggest virtues of CTC. By using CTC, the New York Central was able to reduce its four-track line between Buffalo and Cleveland to two tracks. Two dispatchers now directly control all operations on this busy 163-mile route, handling some 80 trains a day. It is also evident that with fewer tracks, considerably less track maintenance and fewer track workers are required.

By combining the techniques of automatic speed control and Centralized Traffic Control, it would be possible to obtain full *centralized operations control*. The dispatcher would not only direct the routing of trains, but would also start and stop them as required.

Remote control of train operations was demonstrated on the New Haven Railroad in late 1955. An electric train ran from New Rochelle, N. Y. to Rye, a distance of about eight miles, and return, with a full load of passengers and no one at the throttle. The train was started and stopped through radio control by an operator on the station platform at Larchmont, N. Y.

It would even be possible, through automation, to eliminate the dispatcher himself. The entire operations of a London transit line are "programmed" on tape rolls in the form of a series of punch holes. As the tape moves through a "program machine," feelers scan the holes and close electrical contacts. In turn, these actuate the correct signal, track switch and interlocking moves to route each of 900 daily trains to its correct destination, in timetable sequence. Should a train run late, the program machine "remembers" to dispatch the train, on the proper route, when it shows up and the line is clear.

It is perfectly feasible also to have trains set up their own routes as they go. On a Chicago rapid transit line, trains destined for a branch route are equipped with inert inductor coils. As they approach the branch junction, these coils pass over track-side inductors. The resulting magnetic disturbance is used to operate relays and throw a track switch, thus routing the train to the branch.

Again, there are practical difficulties in applying these techniques of automatic operation to line haul railways. As a result the railroad companies have shown little serious interest in them so far.

YARD OPERATIONS

Automation has made its most obvious impact on the great yards where freight cars are classified. These yards take the cars from incoming freight trains and sort or classify them to form new trains, each with cars for a common destination or moving in the same general direction.

Classification yards are an essential feature of rail transportation; yet their operation is very costly and is often a source of delay to highly competitive traffic.

Most large yards classify cars by gravity. Incoming trains enter a "receiving yard" where the road locomotives and cabooses are removed and the cars inspected. Then a yard engine pushes the cars up a small artificial hill or "hump." At the top of this hump the cars are uncoupled, singly or in groups, and roll down the other side into the classification racks—each of which takes the cars going to a specific destination.

In older hump yards a brakeman rode each car down the hump to control its speed so it would couple into cars already in the classification tracks at a safe speed. And switchmen lined the track switches to route each car into the proper classification track.

Today, "push button yards" have made much of this classifying process automatic. They have tremendously increased the speed of yard operations. Because of that, a single new pushbutton yard often takes over the work formerly done at several other yards elsewhere on a railroad.

U. S. railroads have already built more than two dozen push button yards. Each new yard incorporates advances in automatic classification. And each new yard remains the "most modern" only until the next one is completed.

As of early 1958, New York Central's Robert R. Young yard, at Elkhart, Ind., was the newest and probably the most advanced yard in service.[4]

This yard has 72 classification tracks—each for cars to a different destination. The classification of a train can be programmed in advance in a routing memory unit. This unit is plugged into a switching machine that automatically lines the track switches for each car as it rolls down the hump into the classification yard.

Like other modern hump yards, the Robert R. Young yard uses retarders, or track brakes, to control the speed of cars moving down the hump. Electronic analog computers analyze all the factors bearing on the rolling speed of each car as it starts down the hump. These factors include the car's weight, its "rollability," the distance it must roll, the track pattern over which it must travel—even the effect of weather conditions such as wind and temperature. The computers then determine how much force the retarders must apply to each car so that it will roll into the proper track and couple gently into the car

4. "Central Opens Young Yard," *Railway Signaling & Communications* (March 1958), 28. "CB&Q and NYC Open Major Yards," *Modern Railroads* (April 1958), 75.

ahead of it. This force is automatically applied through the retarder mechanism.

Thus, from the time a car starts down the hump until it reaches its assigned spot in the classification yard it is handled automatically. Actually, two persons are still retained: one, the "hump conductor" monitors the automatic switching machine, and can make last-minute changes in the switching program if he wishes. The other, the retarder operator, monitors the speed control system and can control the retarders manually when required.

The locomotive that shoves the cars up to the crest of the hump has cab signals which tell the engineer when to shove, stop or back up. These signals are normally controlled by the yardmaster, the hump conductor or the retarder operator. Here again, fully-automatic operation of the locomotive is technically feasible.

Even much of the paper work revolved in yard operation is now done automatically. They consists of incoming trains are received at the yard via teleprinter tapes. These tapes are then used to punch cards, to prepare the switching lists and for accounting purposes.

The new Robert R. Young yard enabled the New York Central to eliminate *eleven* other yards (mostly small ones). At the same time the new yard cut by about one-half the time required for freight to pass through the Elkhart terminal.

What this means in terms of labor force can be surmised, though exact figures are not available. Despite the fact that the new yard handles a much larger traffic volume than the old Elkhart yard, it requires probably 30 percent fewer people to operate. This does not take into account the force reductions at outlying yards that were closed or curtailed.

MAINTENANCE

Railroads spend roughly one-third of their income to maintain their locomotives, cars, tracks and buildings. Of this amount, about 60 percent goes for wages. Through concentrating the work, and through mechanization, the railroads have greatly cut the number of maintenance employees. But because of the sharp rise in wage rates the total maintenance wage bill has increased somewhat since the end of World War II.

The diesel has greatly changed the character of locomotive

maintenance. With steam locomotives, railroads required a roundhouse and rather complete servicing facilities every 100 miles or so. Large railroads also required several "back shops" for heavy locomotive repairs.

Diesels may require nothing more than a fueling station every three or four hundred miles and a handful of "running maintenance" (light repair) shops scattered over the railroad. In most cases the railroad needs only one shop for heavy diesel repairs.

Thus by its very nature the diesel has reduced the number of shop jobs on the railroads. Many of these heavy diesel repair shops also make good use of "process line" or "assembly-line" techniques for overhauling the individual components of the diesel engines and electrical equipment. Typical is the setup for overhauling diesel engine cylinder assemblies at the shop of the Seaboard Air Line Railroad in Jacksonville, Fla.[5] The various component parts move through a series of work stations via conveyor systems. These conveyors then bring together the finished parts at stations where they are assembled. The men at each work station use tools or machines especially designed for each operation.

This is the classical production-line method; but it is not automation. Some automation is, however, found in the processline shops where new freight cars are built. Some of them, for instance, use automatic welding machines to fabricate car sides and other parts.

The machining and assembly of wheels and axles are shop jobs that also lend themselves to assembly-line methods. At least one railroad has a wheel shop in the planning stage that will enter the realm of true automation. In this shop, almost all the operations will be automatic—including the machining of axles, the boring of wheels, the checking of both wheels and axles for flaws, and the pressing of the wheels on axles.

Mechanization has brought profound changes in the techniques of railway track maintenance. The day of large "pick and shovel" section gangs, who did almost all the track maintenance work, has passed. One railroad came out of World War II with about 500 section gangs on its 4,000-mile system. Today it has only 132—and each of these has but five men.

The duties of these section gangs now consist largely of

---

5. *Modern Railroads* (May 1958).

inspecting the track and performing a few light maintenance chores. Most railroads now use machine-equipped, specialized gangs to do the major track repair tasks. These tasks include laying rail, replacing ties, raising and aligning the track, and applying, distributing and tamping ballast. These gangs use the process-line method—except that in this case the machines move past the work, unlike manufacturing plants where just the reverse occurs.

The same railroad mentioned above does its heavy track work (except rail laying) with even "fully mechanized" timbering and surfacing gangs. Each of these gangs has 36 men and 12 machines. The work it does formerly required 150 men, who used only a few of the simpler power tools.

A typical mechanized track repair gang uses a machine to remove spikes from the old ties; machines to remove the old ties and insert the new ones; a machine to drive the spikes into the new ties; another machine to raise and level the track; a machine to tamp ballast under the raised ties; a machine to align the track, and a machine to distribute ballast along the shoulders of the roadbed.

Most present-day track machines perform a single operation; each requires one or more men and there is no interconnection between machines. Already, though, some multipurpose machines have appeared. Examples are the combined jacking and leveling machines and the combined jacking and tamping machines. Clearly, the trend is to machines that can do several jobs with but one operator. Perhaps the track gang of the future will use a parade of machines, each doing a specific job, all linked under the control of one operator. Automation may bring to track maintenance further economies comparable to those already achieved by mechanization.

PAPERWORK

In railroad office work, automation has not only arrived but has made substantial progress.

Probably few industries have proportionately as much paperwork as do the railroads. Like other businesses, railroads must perform a large volume of accounting in connection with the sale of their services, the purchase of supplies and materials, and the paying of their employees.

Beyond that, the operation of trains generates huge quantities of paperwork. Still other records and accounts results from the interchange of traffic and of cars among railroads. The close governmental regulation of the railroads produces an avalanche of required reports and statistics. It's not surprising that there are more clerical employees on the railroad than there are track workers.

Railroads were among the pioneer users of the first crude "business machines"; more recently the industry has been quick to adopt the new punch-card, teleprinter and electronic computer equipment. The result has been a "revolution in paper work."

Typical of the approach of progressive railroads to office automation is that of the Canadian Pacific Railway. The CP has adopted the concept of "integrated data processing."[6] Original data are recorded at or as near as possible to the source on punch cards or perforated paper tape. Once in mechanical form, the data are processed entirely by machines.

This concept cuts across departmental lines. The machines prepare, from the same basic data, all the documents that are needed for different purposes in several different departments of the railroad.

To implement this "IDP" system, the CP has set up a transcontinental communications network, linking all the key points on the railroad. The major processing point is the "computer center" at Montreal. Its machines include the largest commercially available electronic computer. Also in the network are eight "district data centers," which accumulate data from local sources and partially process them before transmitting them to the Montreal computer center.

Data sources include some 1500 freight stations and offices, 66 shops, 78 materials store houses and 75 freight yard offices. At the smaller freight stations, data are recorded manually, but are key-punched at the first opportunity thereafter.

The CP aims for a "one-shot" process. For instance, when a car of freight is loaded at a station or yard its waybill becomes the source document. Data from it are transcribed to cards or tape. Fed into the appropriate machines, these cards or tapes then produce the consists and other lists needed for train operations; the reports required for car rental accounting,

---

6. "Paperwork Revolution on the CP," *Modern Railroads* (February 1958), 63.

for revenue accounting, for the sales department, for managerial control and governmental regulation.

The effects of IDP are far-reaching. One project on the CP involved the operations of maintenance shops. From one input of punched cards with basic data, the railroad obtains 15 different output statements. They run all the way from information required to control shop operations to payroll registers and individual paychecks.

On a railroad as large as the CP, full automation of paperwork must be a long-term program. Thus the railroad expects normal attrition to more than compensate for the reduction in personnel. The remaining jobs will require higher-grade personnel and will be higher-paid. While machine operation virtually eliminates errors in transmitting data, it does put a premium on accurate typing and key punching. To improve source accuracy, the CP has set up employee training centers at three points on the railroad.

OVER-ALL EFFECTS OF AUTOMATION

What are the over-all effects of mechanization and automation on a railroad's operations, its finances and its employees? To date, few railroads have aggressively sought to mechanize and automate as fully as possible *in all areas*. One that has is the Southern Railway, a leading carrier operating in the Southeast. The published figures[7] for this railroad are revealing.

The Southern Railway operates about 6200 miles of line. It has been fully-dieselized since 1953. It has been a leader in freight yard automation. Through 1956 it had built three new push-button hump yards; a fourth was to be completed in 1958. Centralized Traffic Control is being extended to help boost road operating efficiency.

The Southern has also pioneered in the mechanization of maintenance work. Car and locomotive work is concentrated in a few modern "process line" shops, with automatic machines used where possible. Even the small shops for light freight car repairs have been revamped on the process-line pattern. The Southern has developed many of its own machines, many of

---

7. *Moody's Transportation Manual, 1957,* (Moody's Investors Service, New York).

them multi-purpose, for maintaining track, roadway and bridges. In office work, this railroad has also pushed ahead rapidly to adopt automation techniques. It too uses one of the largest-size electronic computers, and is programming more and more of its accounting and statistical work on this machine. (This computer saved the system $850,000 in clerical costs during its first year of operation.)

In 1956, the Southern Railway produced 15.3 billion ton-miles of freight transportation, or 6 percent more than in 1946. But its index of operating efficiency (gross ton-miles per freight train hour) increased from about 25,000 to 54,000—a jump of 116 percent. Net railway operating income increased from $19.5 million in 1946 to $40.5 million in 1956.

In 1946, the Southern Railway had just about 40,000 employees. In 1956, the number was about 22,000—a drop of 45 percent. Nationwide, as we have already noted, the number of railroad employees dropped about 16 percent between 1946 and 1956.

This difference probably stems in part from the Southern's greater-than-average aggressiveness in the application of machines and automation. In any case, it appears that the whole industry is due to experience much more intensified technological progress in the years immediately ahead.

THE CHALLENGE TO MANAGEMENT AND LABOR

This presents a real challenge to railroad management—and to railroad labor. Whatever the shape of the technological changes to come, they will without doubt result in a further sharp cut in the total of railroad jobs. The human problems stemming from that fact will have to be solved.

So far, neither management nor the railroad labor unions have faced these problems in an over-all sense. In specific cases, local agreements have been reached to permit the operation of some new facility such as a push-button yard or machine data-processing system. But in more than one instance, strikes or labor unrest have marked the opening of a new yard. Sometimes railroads have not realized the potential benefits of a new facility simply because of inability to work out the "labor angle."

Some of the more progressive unions apparently realize the need to adapt themselves to technological progress, rather than blindly opposing it.[8] But long-held practices and traditions die hard.

Railroads have a long history of trade unionism. The complex structure of working agreements, built up over the years, has in many cases become topheavy and inflexible. Too, the railroads deal with nearly two dozen unions or "brotherhoods." Most of these have jealously-guarded monopolies on jobs in their particular spheres. This often prevents displaced employees of one craft from finding jobs within another craft.

The dawning era of rail automation may well mean the elimination of some of these unions, since the type of work their members do will be wiped out. The merger of some of the smaller brotherhoods into a few larger organizations may result. This would be desirable; but it will call for statesmanship on the part of brotherhood officers and members.

Rail management, too, must exercise much restraint, patience and wisdom in the period just ahead. Further automation—and at a greatly increased rate—is essential if the railroads are even to survive—let alone prosper—as a healthy private enterprise. To realize the promise of automation, management must have more flexibility to adapt its manpower to changing needs.

In the "Washington agreement" of 1936, the railroads and the brotherhoods adopted the principle of protecting, for a period of years, employees adversely affected through railroad consolidations.[9] The same principle is gradually being extended to employees displaced by technological change. This may partially solve the problem of railroad people displaced by automation.

Some observers believe that really extensive automation will actually increase railroad jobs in the long run. They feel it will help the railroads capture more traffic, and thus operate more profitably. There are always more jobs in a growing industry than in a contracting one.

Today the U. S. railroads are hard-pressed by competition, though they remain by far the most efficient form of long-haul mass transportation. The conditions under which the railroads compete must be equalized; but this can be done only through legislation.

---

8. *Maintenance of Way Employment on U. S. Railroads,* (Detroit, Brotherhood of Maintenance of Way Employees, 1957), 161.
   9. *Ibid.,* 165.

Yet basically, the survival and future prosperity of the railroads depends on what they do to help themselves. Further automation is one avenue of self-help—one that has as yet hardly been trod. The fully-automatic railroad isn't likely to materialize any time soon. But the industry is headed in that direction; it will move that way with increasing speed in the coming years.

Selected Bibliography

D. W. Brosnan, "Push Button Yards," *Official Proceedings of the New York Railroad Club,* 64, No. 1, (November 23, 1953) (New York Railroad Club, 30 Church St., New York 7, N. Y.). Describes the first two automatic classification yards built by the Southern Railway, and gives statistics on their operation and economics.

William Haber, (and others), *Maintenance of Way Employment on U. S. Railroads* (Detroit; Brotherhood of Maintenance of Way Employees, 1957). Report of a study made by four economists on the sources of employment instability in maintenance of way work. An excellent picture of employment trends and the effects of mechanization.

R. A. Rice, Jr., *A Ten-Year Projection of Railroad Growth Potential* (Chicago; Railway Progress Institute, 1955). Summarizes study by Transportation Facts, Inc. of trends in travel and transport. Discusses relative economics of each form of transport, projects traffic volumes to year 1965.

The following issues of *Modern Railroads* (Watson Publications, Inc., 201 N. Wells St., Chicago 6, Ill.) discusses in considerable detail recent technological progress on certain railroads:

May, 1958 (Seaboard Air Line Railroad)
April, 1957 (Missouri Pacific Railroad)
November, 1956 (Pennsylvania Railroad)
August, 1955 (Southern Pacific Co.)
November, 1954 (Chesapeake & Ohio Railway).

THOMAS C. SHEDD, JR., is a 1940 graduate of University of Illinois, BS in Railway Electrical Engineering. For one year he was Associate Editor of *Telephony,* Chicago trade journal. In 1941, he joined the Engineering Department, Association of American Railroads, as detector car operator (rail testing). In 1946 he became general supervisor of detector cars. He joined the staff of *Modern Railroads*, Chicago, as Associate Editor in 1951. In 1952 he became Eastern Editor, New York. Member: American Railway Engineering Association, Railroad Public Relations Association, New York and New England Railroad Clubs, American Institute of Electrical Engineers.

CHAPTER 10

THE TEACHING MACHINES

B. F. SKINNER, *Professor of Psychology*
Harvard University

The advances* which have recently been made in our control of the learning process suggest a thorough revision of classroom practices and, fortunately, they tell us how the revision can be brought about. This is not, of course, the first time that the results of an experimental science have been brought to bear upon the practical problems of education. The modern classroom does not, however, offer much evidence that research in the field of learning has been respected or used. This condition is no doubt partly due to the limitations of earlier research. But it has been encouraged by a too hasty conclusion that the laboratory study of learning is inherently limited because it cannot take into account the realities of the classroom. In the light of our increasing knowledge of the learning process we should, instead, insist upon dealing with those realities and forcing a substantial change in them. Education is perhaps the most important branch of scientific technology. It deeply affects the lives of all of us. We can no longer allow the exigencies of a practical situation to suppress the tremendous improvements which are within reach. The practical situation must be changed.

---

* These advances refer to current experimental studies which are concerned with an analysis of the effects of reinforcement in learning and the designing of techniques by which reinforcement can be manipulated with considerable precision, particularly the research by the author in collaboration with Charles B. Ferster and with Ogden R. Lindsley and the applications made by Floyd Ratliff and Donald S. Blough in the Harvard Psychological Laboratories on perceptual and learning processes of a wide range of vertebrate organisms. This chapter is reprinted in part from B. F. Skinner, "The Science of Learning and the Art of Teaching," *Harvard Educational Review*, Vol. XXIV, No. 2, (Spring, 1954), 86-94.

There are certain questions which have to be answered in turning to the study of any new organism. What behavior is to be set up? What reinforcers are at hand? What responses are available in embarking upon a program of progressive approximation which will lead to the final form of the behavior? How can reinforcements be most efficiently scheduled to maintain the behavior in strength? These questions are all relevant in considering the problem of the child in the lower grades.

In the first place, what reinforcements are available? What does the school have in its possession which will reinforce a child? We may look first to the material to be learned, for it is possible that this will provide considerable automatic reinforcement. Children play for hours with mechanical toys, paints, scissors and paper, noise-makers, puzzles—in short, with almost anything which feeds back significant changes in the environment and is reasonably free of aversive properties. The sheer control of nature is itself reinforcing. This effect is not evident in the modern school because it is masked by the emotional responses generated by aversive control. It is true that automatic reinforcement from the manipulation of the environment is probably only a mild reinforcer and may need to be carefully husbanded, but one of the most striking principles to emerge from recent research is that the *net* amount of reinforcement is of little significance. A very slight reinforcement may be tremendously effective in controlling behavior if it is wisely used.

If the natural reinforcement inherent in the subject matter is not enough, other reinforcers must be employed. Even in school the child is occasionally permitted to do "what he wants to do," and access to reinforcements of many sorts may be made contingent upon the more immediate consequences of the behavior to be established. Those who advocate competition as a useful social motive may wish to use the reinforcements which follow from excelling others, although there is the difficulty that in this case the reinforcement of one child is necessarily aversive to another. Next in order we might place the good will and affection of the teacher, and only when that has failed need we turn to the use of aversive stimulation.

In the second place, how are these reinforcements to be made contingent upon the desired behavior? There are two considerations here—the gradual elaboration of extremely complex patterns of behavior and the maintenance of the behavior in strength at each stage. The whole process of becoming competent in any field must be divided into a very large number of very small steps, and reinforcement must be contingent upon the accomplishment of each step. This solution to the problem of creating a complex repertoire of behavior also solves the problem of maintaining the behavior in strength. We could, of course, resort to the techniques of scheduling already developed in the study of other organisms but in the present state of our knowledge of educational practices, scheduling appears to be most effectively arranged through the design of the material to be learned. By making each successive step as small as possible, the frequency of reinforcement can be raised to a maximum, while the possibly aversive consequences of being wrong are reduced to a minimum. Other ways of designing material would yield other programs of reinforcements. Any supplementary reinforcement would probably have to be scheduled in the more traditional way.

These requirements are not excessive, but they are probably incompatible with the current realities of the classroom. In the experimental study of learning it has been found that the contingencies of reinforcement which are most efficient in controlling the organism cannot be arranged through the personal mediation of the experimenter. An organism is affected by subtle details of contingencies which are beyond the capacity of the human organism to arrange. Mechanical and electrical devices must be used. Mechanical help is also demanded by the sheer number of contingencies which may be used efficiently in a single experimental session. We have recorded many millions of responses from a single organism during thousands of experimental hours. Personal arrangements of the contingencies and personal observation of the results are quite unthinkable. Now, the human organism is, if anything, more sensitive to precise contingencies than the other organisms we have studied. We have every reason to expect, therefore, that the most effective control of human learning will require instrumental aid. The simple fact is that, as a mere reinforcing mechanism, the teacher is out of date. This would be true even if a single teacher devoted all her time to a single child, but her inadequacy is multiplied many-fold when she must serve as a reinforcing device to many children at once. If the teacher is to take advantage of recent advances in the study of learning, she must have the help of mechanical devices.

The technical problem of providing the necessary instrumental aid is not particularly difficult. There are many ways in which the necessary contingencies may be arranged, either mechanically or electrically. An inexpensive device which solves most of the principal problems has already been constructed. It is still in the experimental stage, but a description will suggest the kind of instrument which seems to be required. The device consists of a small box about the size of a small record player. On the top surface is a window through which a question or problem printed on a paper tape may be seen. The child answers the question by moving one or more sliders upon which the digits 0 through 9 are printed. The answer appears in square holes punched in the paper upon which the question is printed. When the answer has been set, the child turns a knob. The operation is as simple as adjusting a television set. If the answer is right, the knob turns freely and can be made to ring a bell or provide some other conditioned reinforcement. If the answer is wrong, the knob will not turn. A counter may be added to tally wrong answers. The knob must then be reversed slightly and a second attempt at a right answer made. (Unlike the flash-card, the device reports a wrong answer without giving the right answer.) When the answer is right, a further turn of the knob engages a clutch which moves the next problem into place in the window. This movement cannot be completed, however, until the sliders have been returned to zero.

The important features of the device are these: Reinforcement for the right answer is immediate. The mere manipulation of the device will probably be reinforcing enough to keep the average pupil at work for a suitable period each day, provided traces of earlier aversive control can be wiped out. A teacher may supervise an entire class at work on such devices at the same time, yet each child may progress at his own rate, completing as many problems as possible within the class period. If forced to be away from school, he may return to pick up where he left off. The gifted child will advance rapidly, but can be kept from getting too far ahead either by being excused from arithmetic for a time or by being given special sets of problems which take him into some of the interesting bypaths of mathematics.

The device makes it possible to present carefully designed material in which one problem can depend upon the answer to the preceding and where, therefore, the most efficient progress to an eventually complex repertoire can be made. Provision has been made for recording the commonest mistakes so that the tapes can be modified as experience dictates. Additional steps can be inserted where pupils tend to have trouble, and ultimately the material will reach a point at which the answers of the average child will almost always be right.

If the material itself proves not to be sufficiently reinforcing, other reinforcers in the possession of the teacher or school may be made contingent upon the operation of the device or upon progress through a series of problems. Supplemental reinforcement would not sacrifice the advantages gained from immediate reinforcement and from the possibility of constructing an optimal series of steps which approach the complex repertoire of mathematical behavior most efficiently.

A similar device in which the sliders carry the letters of the alphabet has been designed to teach spelling. In addition to the advantages which can be gained from precise reinforcement and careful programming, the device will teach reading at the same time. It can also be used to establish the large and important repertoire of verbal relationships encountered in logic and science. In short, it can teach verbal thinking. As to content instruction, the device can be operated as a multiple-choice self-rater.

Some objections to the use of such devices in the classroom can easily be foreseen. The cry will be raised that the child is being treated as a mere animal and that an essentially human intellectual achievement is being analyzed in unduly mechanistic terms. Mathematical behavior is usually regarded, not as a repertoire of responses involving numbers and numerical operations, but as evidences of mathematical ability or the exercise of the power of reason. It is true that the techniques which are emerging from the experimental study of learning are not designed to "develop the mind" or to further some vague "understanding" of mathematical relationships. They are designed, on the contrary, to establish the very behaviors which are taken to be the evidences of such mental states or processes. This is only a special case of the general change which is under way in the interpretation of human affairs. An advancing science continues to offer more and more convincing alternatives to traditional formulations. The behavior in terms of which human thinking must eventually be defined is worth treating in its own right as the substantial goal of education.

Of course the teacher has a more important function than to say right or wrong. The changes proposed would free her for the effective exercise of that function. Marking a set of papers in arithmetic—"Yes, nine and six *are* fifteen; no, nine and seven *are not* eighteen"—is beneath the dignity of any intelligent individual. There is more important work to be done—in which the teacher's relations to the pupil cannot be duplicated by a mechanical device. Instrumental help would merely improve these relations. One might say that the main trouble with education in the lower grades today is that the child is obviously not competent and *knows it* and that the teacher is unable to do anything about it and *knows that too*. If the advances which have recently been made in our control of behavior can give the child a genuine competence in reading, writing, spelling, and arithmetic, then the teacher may begin to function, not in lieu of a cheap machine, but through intellectual, cultural, and emotional contacts of that distinctive sort which testify to her status as a human being.

Another possible objection is that mechanized instruction will mean technological unemployment. We need not worry about this until there are enough teachers to go around and until the hours and energy demanded of the teacher are comparable to those in other fields of employment. Mechanical devices will eliminate the more tiresome labors of the teacher but they will not necessary shorten the time during which she remains in contact with the pupil.

A more practical objection: Can we afford to mechanize our schools? The answer is clearly yes. The device I have just described could be produced as cheaply as a small radio or phonograph. There would need to be far fewer devices than pupils, for they could be used in rotation. But even if we suppose that the instrument eventually found to be most effective would cost several hundred dollars and that large numbers of them would be required, our economy should be able to stand the strain. Once we have accepted the possibility and the necessity of mechanical help in the classroom, the economic problem can easily be surmounted.[1] There is no reason why the school room should be

---

1. *Status of Current Research on Teaching Machines, January 1, 1958:* Under a grant from the Fund for the Advancement of Education, Skinner is developing and testing a high-school and college level teaching machine in which the student writes his answer and compares it with an answer uncovered by the machine. Material is inserted in the machine on twelve-inch discs, each of which contains thirty radial "frames," one of which is exposed at a time. The machine automatically eliminates items which the student has answered correctly, either once or twice according to the nature of the material. The student's answer is covered before the machine's answer is revealed, so that it cannot be changed and the student's decision that his answer is right is recorded in permanent form. After completing the material the student returns the disc, together with a paper tape bear-

any less mechanized than, for example, the kitchen. A country which annually produces millions of refrigerators, dish-washers, automatic washing-machines, automatic clothes-driers, and automatic garbage disposers can certainly afford the equipment necessary to educate its citizens to high standards of competence in the most effective way.

There is a simple job to be done. The task can be stated in concrete terms. The necessary techniques are known. The equipment needed can easily be provided. Nothing stands in the way but cultural inertia. But what is more characteristic of America than an unwillingness to accept the traditional as inevitable? We are on the threshold of an exciting and revolutionary period, in which the scientific study of man will be put to work in man's best interests. Education must play its part. It must accept the fact that a sweeping revision of educational practices is possible and

---

ing all his answers, to a file. Programs are being constructed for parts of Introductory Physics, parts of beginning French, and a substantial portion of a General Education course on Human Behavior. The latter was tested in the Spring Term of 1958 in Skinner's course at Harvard and Radcliffe. (Demonstration material in geography, literary appreciation, and high-school trigonometry has also been prepared.)

Such a machine has the advantage of (1) informing the student immediately of correctness of his response, (2) permitting him to work at his own rate and to advance as rapidly as possible toward completion of the work of the term, (3) permitting the design of material in a carefully arranged program of small steps in which the student progresses rapidly and is practically inevitably successful.

A different type of machine is also being developed for use in the lower grades. The student composes an answer by moving sliders; the machine senses the positions of the sliders and compares them with a code on the material which has been inserted into the machine. Unlike the high-school and college level machines, such a device tells the student that he is wrong without telling him what is right. However, it demands a certain rigidity in the form of response which is not advisable at the higher levels.

A third "preverbal" machine is being developed to make the kindergarten pupil sensitive to color, shape, three-dimensional forms, and so on. The machine can be used to teach reading. The pupil's vocal response is not evaluated (the machine does not teach him to talk) but it does tell him whether or not he has selected a textual pattern appropriate to a picture or other textual material. This is the only one of the three machines which uses multiple choice.

Aside from the practical problem of developing adequate sets of material for instruction in various fields—a task which must in the last analysis fall to specialist in those fields—Skinner is concentrating on the psychological problems involved in the analysis of knowledge and in the construction of programs which will be successful in imparting competent behavior to students at all age levels. Actual classroom tests are planned.

inevitable. When it has done this, we may look forward with confidence to a school system which is aware of the nature of its tasks, secure in its methods, and generously supported by the informed and effective citizens whom education itself will create.

[*SKINNER'S TEACHING MACHINES: A supplementary note by Ward Edwards of the Operator Skills Branch, Operator Laboratory, of the Air Force Personnel and Training Research Center, originally intended to inform Air Force personnel working on similar problems of Skinner's work.*]

For the last couple of years B. F. Skinner has been intensively concerned with the design and use of machines intended to teach arithmetic, spelling, and similar subjects to elementary school children. This report describes some of the principles he uses in building and programming such machines.

Machines. Skinner has designed four teaching machines so far. No. 1 is a very simple multiple-choice machine. The problem appears in a window; the student presses one of five buttons. If he is right, the machine moves on to the next problem; otherwise he must continue pushing until he hits the correct button. This, the simplest of Skinner's machines, is used with pre-reading problems for very young children. I doubt if Skinner is much interested in this machine, since it violates one of his fundamental principles of teaching machine construction, the one which says that the student should construct rather than merely recognize the answer.

Machine No. 2 is intended for teaching arithmetic. The problems, written on tape, appear in a window. The student has four sliders; operation of each causes a digit from 0 through 9 to appear in a window; thus he can produce all numbers from 0 to 9999. He sets the appropriate answer to the problem into the sliders. When he feels his answer is right, after any corrections he may care to make, he turns a crank. This first locks the sliders into place, so that the student may no longer change his answer. Then he sets all sliders to zero. If the answer was correct, it rings a bell and advances the question tape into the next position; otherwise it leaves the student facing the same question again and operates an error counter. In this as in the previous machine the machine knows the right answer because it reads holes punched in a part of the tape which doesn't appear in the window.

Machine No. 3 is a multi-purpose machine intended for more advanced students, and for subject-matter areas in which simple numerical (or literal) answers are inappropriate or hard to come by. It has a large disc on which questions are written; these appear one at a time in a window. There is a separate window which exposes a piece of paper tape on which the answer is to be written. When the student has written an answer which satisfies him, he operates a control. The control covers his answer with a transparent cover so that he can no longer change it, and then exposes the next position on the disc, in which the correct answer is written. The student compares the correct answer with his own, and decides whether he was right or wrong. If he was right, he operates a control which makes a mark on his answer tape indicating that he has scored himself right, operates a detent which reduces the number of times this position will recur, and advances the disc to the next question and the tape to the space for writing the next answer. If he was wrong, he operates another control which simply advances disc and tape to their respective next positions. The disc repeats its cycle as often as necessary; any question which the student has, in his own judgment, answered correctly twice is thereafter omitted. Practice continues until all questions have been answered correctly twice. The difficulty with this machine, in Skinner's view, is that it reveals the right answer in the course of telling the student that his answer is wrong. This difficulty is inherent in any self-scoring system.

Machine No. 4 is still only a gleam in Skinner's eye. He hopes to persuade a well-known manufacturer of computers to build it for him. It will have a window in which questions appear. The student punches his answers out on a keyboard like a typewriter keyboard, and they appear in another window. He can change or edit them as much as he wishes after originally typing them. When he is satisfied, he operates a control which locks his answer into the machine, scores his answer as right or wrong, records the result, and, if the student was right, goes on to the next question. Right, for this machine, must mean right symbol by symbol—a fact which severely limits the kind of question which can be used with it. The machine is a glorified version of Machine No. 2.

Principles of Machine Design. Underlying the design of these machines are a number of reasons for using teaching machines and several principles about how they should work.

The advantages of using machines for Socratic-type teaching, which of course could also be done by a teacher or a book, are that they provide immediate feedback (information about whether the student is right or wrong), that they permit the student to work at his own rate and thus are adaptable to wide ranges of intelligence, that they are self-motivating, and that they offer convenient methods for providing reinforcement. Skinner reports that all students on whom he has so far tried the machines find operating them so much fun that no further reinforcement is needed, and he is not now doing any work on the question of how to use reinforcements in connection with such machines. From his extensive work on reinforcement schedules, however, he feels that the most advantageous reinforcement schedule for such a purpose would be a variable—ratio schedule (i.e. the student is given a nickel or some other intrinsic reward every nth trial, with n varying around some mean value). The typical slot-machine pattern, in which there are fairly frequent small rewards and a very infrequent big reward seems especially well suited to maintaining a high rate of machine operation.

The most vigorous restriction Skinner feels should be imposed on the design of teaching machines is that the student should compose his answer, rather than recognizing it. This means that multiple-choice questions are inappropriate—a fact which greatly complicates the problem of designing appropriate machines. The reason for this is that reproduction rather than recognition is what you wish to train for, and he does not believe that recognition training transfers to reproduction very well. A self-scoring machine like Machine No. 3 solves the problem of accommodating compose-the-answer questions on complicated subjects. Unfortunately, it violates yet another principle, namely that the machine shouldn't reveal the right answer in the course of telling the student that his answer is wrong.

Principles of Programming. Of course the most important question about teaching machines is: how should they be programmed? Skinner feels that he doesn't have any very satisfactory answers to this question; this is where research is needed. But some principles do seem clear.

Teaching machines, Skinner feels, should teach by a Socratic method of questioning, with a minimum of exposition and a maximum of active participation by the student. The step between each question or item of information and the next should be very small, so that the student is seldom wrong. This is important both to maintain motivation and to make sure that the student gets a complete understanding of the material being taught. The material should be presented in a logical progression so far as possible, and in such a way that the student understands what he is learning, rather than learning by rote. (These are Skinner's words, not mine. If you ask him what "understanding" means, he will probably say that it means having a very rich associative context for each concept taught before the next one comes along. The implications of this principle for programming teaching machines are obvious, even though the concept of "understanding" is one of the least well understood concepts in psychology.)

If you ask Skinner how to produce a program, he would say: Decide what behavior you wish to produce in the student, analyze it into responses, and then teach those responses one by one. It is from this fundamental principle that Skinner deduces the principles which say that the material should be presented in a logical step-by-step progression and that successive steps should be very small.

In the course of deciding on a succession of steps, Skinner uses another principle. Elements should not be taught before operations; rather operations should be taught as early as possible. In arithmetic, for example, Skinner starts with the numbers 0 and 1. Then he teaches 2, and immediately teaches that $1 + 1$ is 2, and that $2 - 1$ is 1. Then he teaches 3 and 4, and immediately starts teaching multiplication and division. Then he teaches 5, and starts teaching the concept of prime numbers. In short, he teaches quite complicated mathematical concepts about numbers before the student knows what all the numbers are. This is a technique for producing the "rich associative context" I have mentioned above.

In the course of trying to ensure that the student is very seldom wrong, Skinner encourages lavish initial use of crutches (my term, to which Skinner has some objections). For example, he may set out the answer in a distinguishing color in the early stages of learning something, or he may write the question as a rhymed couplet with the right answer being the rhyming word in the second line. He feels that such artificial aids, which of course must eventually drop out, are essential in the early stages of learning about something. The principle underlying this is that any road sign, however artificial, is helpful in a trackless wilderness, but becomes unnecessary when the wilderness becomes familiar.

I believe that the general notion of using teaching machines is the most important development in applied psychology since aptitude tests—indeed it is potentially the most important development since intelligence tests. Most of the principles Skinner presents seem very reasonable to me. In particular the notion of composing the answer rather than recognizing it seems to me important enough to justify the inconvenience in machine design which it imposes. The only major principle Skinner uses which seems somewhat doubtful to me is the use of crutches. It would be easy to design research to find out whether crutches help or hinder.

Another fact about Skinner's work on teaching machines should be recorded, though it is hard to make a general principle out of it. Skinner uses a very great deal of ingenuity, a kind of ingenuity well worth copying, in making up programs. A sample program, intended to teach children all about the word "manufacture," follows. Each box represents one frame on the teaching machine (machine No. 4 would be necessary).

1. MANUFACTURE
means make or build
Chair factories manufacture chairs
Copy the word here:

....

2. Part of the word is like part of the word
FACTORY
Both parts come from an old word meaning *make* or *build*
M A N U .... U R E

3. Part of the word is like part of the word
MANUAL
Both parts come from an old word for *hand*. Most things used to be made by hand.
.... F A C T U R E

4. One letter goes in both spaces:
M .... N U F .... C T U R E

5. One letter goes in both spaces:
M A N .... F A C T .... R E

6. The word ends like
PICTURE, LECTURE, FRACTURE
M A N U F A ....

7. Chair factories

....

Many of Skinner's other programs are similarly ingenious. A tentative standard for good programs is that they should interest experts in their field by their imaginativeness without causing the students they are intended for to make too many wrong answers.

Since Skinner has been mostly concerned with training grade

school students, some modification of his principles to apply to older students and more difficult material is probably called for. But some of these principles still make sense, notably those of small steps, extensive use of crutches in early stages (if this actually is helpful), low probability of error, and operations mingled with elements. Skinner feels that even in electronic training composing rather than recognizing the answer is crucial; to me this seems to be an important and straightforward experimental problem.

The greatest need in the further development of teaching machine techniques is, of course, the development of principles of programming. Skinner is surely right in saying that there is no substitute in programming for digging into the actual subject matter to be taught; "You'll never teach electronics until you have analyzed what electronic behavior is." But there must be many more as yet unknown general principles of optimal programming; research to discover them is badly needed.

Another kind of needed work is development of a compose-the-answer machine which can interpret answers, rather than checking them symbol by symbol. The problem is essentially the same as that of developing a translating machine to translate one language into another, and is probably at least as difficult as the translating machine problem.

B. F. SKINNER, author of *Walden Two* and *Science and Human Behavior,* is Professor of Psychology at Harvard University. He has been president of the Eastern Psychological Association and is actively associated with the Harvard Psychological Laboratories.

CHAPTER 11

AUTOMATION AND THE ACCOUNTANT

CHARLES L. KEENOY, *Vice-President of Engineering and Development, National Cash Register Co., Dayton, Ohio*

What will increased office automation mean to the accountant? I should like to point out at the start that I am not in a position now—nor do I expect to be in the near future—to present to you, wrapped up in a neat little package, all of the ramifications of office automation. Nor do I believe that any single person can hope to do so.

Actually, the advent of electronics in record-keeping and data-processing has placed the office equipment industry in the most challenging yet potentially rewarding, role in its history. And I think the same thing can be said for accountants. We are like the would-be interplanetary explorer who previously had to confine his efforts to gazing into telescopes and dreaming, but who has suddenly been presented with a means for visiting any planet he desires.

Electronics is that new mode of transportation for you, and for us. Because of its tremendous speed in carrying out our errands, the electron can take us just about anywhere, statistically speaking. But we are still in the process of deciding just where we want to go and whether each given trip is worth the cost and effort involved.

We do know of course that we are in the early stages of an accounting and business management revolution which, in some respects, will rival the industrial revolution in its effect on the lives of everyone. This revolution in office procedures and data-processing, promises to do for man's mind what the industrial revolution did for his body. By harnessing the electron, we will be able to multiply a single clerical employee's efforts many times.

Economically speaking, we are making the electron our servant in the nick of time. In a few more years American business would have been figuratively snowed under by its ever mounting volume of paperwork. At the start of the 20th century, for example, only one man in every 40 employees was a paper worker. With pen and ink as his chief tools, this early-day clerical was able to keep up with all of the demands placed on accounting by business.

Forty years later, however, just prior to Word War II, our record-keeping task had mushroomed into awesome proportions. Despite the advent of electro-mechanical accounting machines, the development of punched-card tabulating systems, and the invention of other posting and analytical office machinery, by 1940 it took one employee in every 10—instead of one in every 40—to keep up with the necessary paperwork.

But still the tide of record-keeping continued to rise. On the crest of this tide, the office equipment industry achieved new sales records. Business spent hundreds of millions of dollars a year for labor-saving office equipment. Despite this investment, despite the concentrated developmental efforts of the entire office equipment industry, the volume of paperwork has multiplied faster than the tools for handling it. Last year, according to Census Bureau estimates, one out of every six employees— instead of one in every 10—was a paper worker. There are actually more clerical employees in our country today than agricultural workers. Clearly, the cost of record-keeping has become a major economic and management problem. It is a dilemma which faces businessmen everywhere.

Fortunately, our technology has a pair of trump cards with which to meet this challenge.

The first of these is the concept of Integrated Data Processing, which in our alphabetical era quickly became IDP. Integrated Data Processing is built around a common machine language which can link together, in a continuous automatic chain, different machines and operations. Its goal is to minimize manual effort in data processing. The punched card was the earliest means of using the IDP concept on an extensive scale, and will remain an important tool in this area.

Our second trump card is the tiny and often misunderstood electron. Prior to a decade ago, its major service had been in bringing entertainment to us—through radio and later television —instead of helping to keep our businesses economically solvent.

You also have seen the spectacular rise of the electronic computer—the fair-haired child of our laboratories and drawing boards. Volumes have been written and hundreds of speeches made about the giant brain and how it will help the large corporation or government agency solve many of its problems. For this reason alone, I should like to concentrate on some areas of business automation with which you may be less familiar. I refer to the impact that electronic computers and their associated equipment will have on the small manufacturer, the small retailer, the small wholesaler, and indeed the average housewife. After all, the number of small businesses in the country exceeds the large ones by many times. With over 170,000,000 people in the United States, anything that affects the lives of these people directly, will have a profound influence on our way of life and can't help but influence our present accounting and auditing practices.

The first of these might be described as the extension of the punched card concept—and therefore the IDP concept—to additional types of media. For example, if data can be punched into a card and thereafter used and re-used whenever needed, then it is conceivable that similar means of preserving data can be developed. From this realization came the punched price tag for the garment industry, which is essentially a miniature punched card designed for a specific purpose.

It also seemed probable that if data could be preserved in various types of cards through different configurations of holes, then by the same token information could be punched into a strip of paper tape, to be deciphered and extracted later when needed. From this concept has emerged the punched paper tape recorder. This device—a plain-looking box about a foot square—seems likely within the next decade to become one of the most important machines in tomorrow's automatic office.

What are some of the potentialities of these extensions of the Integrated Data Processing concept?

First, consider retailing. There are approximately 1,500,000 small, single-owner retail and service establishments in the United States. The list includes just about any kind of enterprise you could think of—service stations, bakeries, apparel shops, shoe, hardware, camera, drug, small food stores, and so on.

In practically all of these enterprises, the person responsible for managing the business is primarily a merchandiser. At least we can be fairly sure that he is not often an accountant. And yet non-specialist that he is, this man, because of the complexities of doing business today, must have at his fingertips virtually the same type of information as a multi-million-dollar corporation with large accounting departments and expensive analytical machinery. The result is tedious, time-consuming and costly pencil-and-paper analysis. The small businessman, if he is to keep afloat in a sea of government reports and other bookkeeping chores, works long and hard after store hours; he may even enlist the fumbling help of members of his family. When paper work becomes too burdensome it is usually the first thing to be sacrificed. This can be fatal since present tax rates, forgotten expenses, purchases, charges, etc. can be the death blow to a struggling young business. While competition grows more intense, while the costs of doing business rises, the retailer fights the battle of paper work with his inept and inadequate hand methods. And all of this is done at a severe cost in time—time better spent in merchandising, sales promotion and other aspects of retailing.

Consider the possibilities offered to this retailer by extension of the Integrated Data Processing concept to which I referred earlier. Let us assume that the price tags on his merchandise contain—in addition to *printed* information on size, style, color, manufacturer and so forth—this same data in the form of punched holes. Let us also assume that each salesperson in his store has his or her individual token or card, again with essential information punched in code; and to go a step further, that each credit customer also carries a token or card with punched data including the customer's account number, the kind of account, and so forth.

If, at the time of a transaction, all of this information—about merchandise, salesperson and customer—can be collected and preserved automatically, then clearly we. have the beginning of truly automatic accounting in retailing.

However, we have considered so far only the fixed or non-variable information involved in a retail sale. What about all the factors that change from transaction to transaction, factors which cannot be covered by such inflexible media as pre-punched price tags and salesperson or customer tokens? What about prices which are subject to change in the form of markups or mark-downs, charges, discounts, taxes, gift wrapping, other charges, and so forth?

The answer will be a sales registering and recording device, a new and versatile version of the traditional cash register whose horizons have broadened as the concept of Integrated Data Processing has grown. For by means of such a cash register, we can record all variable information about any transaction in a matter of seconds. This includes items other than sales, such as the recording of charges, expenses, purchases, etc.

Thus, out of the need for more automatic data-processing in retailing will evolve what might be called a "Sales-Tronic" system. In its simplest form, a Sales-Tronic system could be made up of a special register linked electrically with a punched paper tape recorder. The recorder punches a paper tape record of all information entered into the register. The data on the tape can later be transferred automatically to punched cards and then processed on tabulating machines, or it can be fed directly into electronic data-processing systems equipped with a paper tape reader.

In more complete form, a Sales-Tronic system could include —in addition to the register and recorder—a media reader. This is a small sensing device which can "read" the information on pre-punched price tags and salespersons' and customers' tokens. The media reader then causes this information to be punched into paper tape by the recorder, along with the data entered through the register.

The implications of these developments are obvious. For the first time in the history of merchandising, the average retailer can possess the key to freedom from bookkeeping drudgery. Through sales registering equipment linked with automatic recording equipment, his entire day's operations can be preserved on a strip of inexpensive paper tape.

At the end of the day the tape can be delivered or mailed to a nearby processing center. There, either on tabulating equipment or through an electronic computer, it is processed. Back to the retailer comes just about any kind of report he requires. Certainly this would include a statement of his sales, purchases, expenses, accounts receivable and accounts payable figures broken down by department or classification. The reports might also contain cumulative and comparative information. The retailer might want to know, for example, how his sales and expenses compare with those of a year ago. The possibilities as you can see, are virtually unlimited. For the first time at a price he can afford the little fellow would have access to the same type of information that is usually reserved for the large size corporation.

You will also note that the sales recording system of the future is built upon a single basic assumption—the correctness of the original entry. From this fact we can draw a conclusion which is of considerable significance to accounting. It is simply this: Never in the history of accounting have sound audit and rigid control at the point of original entry assumed such importance. A Sales-Tronic system places in the hands of one person—the sales clerk—the only manual handling of entries. All subsequent data perpetuation is automatic.

Clearly then, it is imperative that the basic documents for control and audit be sufficiently detailed to compensate for the lack of checks and balances of earlier, more traditional data-processing systems. Otherwise, we shall have built our house on a poor foundation indeed.

To meet the requirements for auditing and control of data entered into the system the device which makes the record of original entry takes on added importance. First it must produce a printed detailed journal of each entry. It must also have the ability to print the same information on the original document i.e. sales slip, purchase invoices, expense vouchers, etc. By printing the same reference number on the original document and the detailed journal we can audit the accuracy and authenticity of the information entering the system.

The new original entry machine should also provide daily control information for the merchant. This does two important things. It provides him with a daily statement of business so that he knows at all times where he stands in relation by sales, purchases, expenses, etc. Secondly, since his detailed information will be supplied by an outside concern it is important that the owner of the business possess the control figures against which all other figures must balance. I don't believe many merchants would be content to wait several days before they knew if their cash was in balance, or if their sales for the day had reached the breakeven point. Even more important, since the detailed report regarding sales, expenses, purchases, accounts receivable, etc. will be supplied by a service bureau the control figures must be determined and controlled by the merchant and not determined as a by-product of adding up the detail information.

In many respects, the Sales-Tronic system will be similar to conventional sales registering systems of today. The point is, however, that if audit and control at the point of original entry are urgently required today, they will be even more essential tomorrow.

So far, I have mentioned only a single area in which the highly versatile paper tape recorder can integrate the processing of data—that is, the recording of sales in almost every type of retail establishment, from the small shoe outlet around the corner to the giant department store. The paper tape principle is equally applicable to almost every type of original-entry business machine. In each case it preserves, as a by-product of the creation of necessary documents, a complete record of original entries for subsequent audit, and analytical or statistical breakdowns.

I believe you will agree that although less famous than its bigger brother, the electronic computer, the paper tape recorder is destined to play a steadily growing role in the field of record-keeping. Many of these recorders have been installed and are working successfully, with every indication that the trend will spread and grow at a very rapid rate.

Up to now I have talked about punched cards, punched price tags and punched paper tape. You may have wondered why it is necessary to print arabic numerals for us to read and produce the same information in the form of coded holes for machines to read. The answer is fairly simple. When the Arabs invented their numerals they didn't know anything about electronics, and the electronic brains that have been developed to date are not smart enough to read the arabic figures in their present form.

Though little-heralded, a major breakthrough in data processing is just around the corner. I refer to direct character recognition. This of course is the ability of a machine to literally read the same Arabic figures you and I read when these figures are printed in magnetic ink. As a technological advance, direct character recognition clearly goes far beyond the principle of magnetic tape recording with which we are all familiar.

Direct character recognition is achieved through electronic reading heads which scan magnetic-ink figures. The reading heads then send wave patterns to a device which compares the patterns with those stored in a magnetic "memory." Here, the patterns are translated back to figures in machine language.

Again, the implications of such an advance in record-keeping techniques are almost self-evident. Certainly, the banking industry has been quick to turn to this principle as the long-sought key to automatic handling of checks and other documents. It is not surprising that bankers are excited by this potentiality. During the past 15 years, for example, the number of checking accounts in the United States has doubled—to 50 million. More than 9 billion checks are being written each year. And with each check being handled five or six times you can imagine the fond expectations which bankers have attached to the concept of direct character recognition.

Just as in the recording of a retail sale, a commercial bank transaction involves both fixed and variable information. For example, an identifying number can be given both the bank and the depositor. (Incidentally, I believe that the time is not far off when one of the first things that will happen to our children after birth is that they will be fingerprinted and given a number. This number will be synonymous with their name and will be used to identify them insofar as checking accounts, charge accounts, social security accounts, etc. are concerned.) This information will not change, and therefore can be pre-printed— with magnetic ink—on the depositor's check and deposit slips.

However, the amount for which each check is written is a variable factor. It cannot be printed on the depositor's check with magnetic ink until after the check has been written. The ideal time for imprinting this amount is during the check's processing through a bank's central control and proof operation. Following this step, the check is ready for truly automatic sorting and processing. The possibilities here for greater efficiency in paper handling are enormous. Nor is that the end of the story. The advent of Integrated Data Processing including electronic character recognition, will make it feasible for banks to offer hitherto undreamed-of bookkeeping services to their depositors.

Clearly, the miracles of electronics are about to be extended to millions of average Americans who are, and will remain, far removed from the realm of giant electronic "brains."

Truly, management would like to translate into practical benefits the new concepts in record-keeping which technology is producing. Thus, although bookkeeping will tend to become more and more automated in the years ahead, the functions of the accountant will increase in stature. There will be a growing demand for intelligent counseling in the use of our new electronic tools—their advantages and disadvantages. Top management will find a pressing need for constructive recommendations on reports and reporting methods. And if the accountant is found inadequate in knowledge, or capacity to adjust to new techniques, there is a very real danger that his traditional functions will be usurped by the technician.

I do not believe that many accountants will find themselves in that undesirable position.

The teaching profession must be equally alert to the potentialities that have been opened in the field of data-processing during the past few years. There is a widespread realization that the student coming from the high school, college or university must be instructed in accounting machinery as well as in basic accounting principles. Otherwise, business will founder in the quagmire of mounting paper work.

All of us—the business machine manufacturer, the accountant and the teacher—must go down this road together. Our futures are interdependent. We must comprehend and then weigh the desirability of the new reports which for the first time in accounting history are becoming economically feasible for the average business. There has long been need, for example, for more analytical information on sales, inventory and usage, and manufacturing and distribution costs. At the same time, we must resist the temptation to over-report, to produce such a bulk of information that management is left with a severe case of statistical indigestion and resultant disillusionment.

We must also realize that we can no longer guide the accounting process through every single step as it takes place. We shall be dealing increasingly with invisible figures and statistics. We shall have to decide where we are going and why; set up the mechanism for capturing and auditing necessary data at their source; and then defend our integrated system against such inroads as outmoded company traditions, inter-departmental rivalries or frictions, and fear of change caused by misunderstanding of what the system has been designed to accomplish.

CHARLES L. KEENOY is Vice-President of Engineering and Product Development at The National Cash Register Company, Dayton, Ohio. Born in Pittsburgh (1914) he attended both the University of Pittsburgh (B.S.) and Columbia University from 1932-38. From 1942-1945, he served as Lieutenant in the United States Navy. He received the Executive Training course at Equitable Life Assurance Society of the United States in 1939, remaining there until 1942 when he became associated with The

National Cash Register Company. Since joining NCR, Mr. Keenoy has held such positions as Special Representative, Accounting Machine Manager, Manager of Product Development, and in January, 1953 was appointed to his present position as Vice President of Engineering & Product Development.

CHAPTER 12

AUTOMATION IN DATA PROCESSING FOR THE SMALL TO MEDIUM SIZED BUSINESS: *A Case Study*

Curtis Jansky, *Product Planner,*
Royal McBee Corp., Portchester, New York

The purpose of this chapter is to show how one medium sized corporation has used a partial approach to automatic data processing in order to achieve certain economic and operational advantages over a purely manual data system in the field of sales analysis. To achieve this purpose, the general nature of and needs for more advanced data processing in business will be analyzed first. In such an appraisal, certain general guide posts can be established for achieving the maximum advantages with the minimum expense using partially automated data processing.

THE PURPOSE OF DATA PROCESSING IN BUSINESS

The first and most familiar objective of data processing is to record and summarize the every-day transactions to establish internal controls. The procedure involves the processing of detailed records of individual transactions to obtain the totals required in each step of data processing in order that the postings or tabulations of each can be "balanced." As an example, the record of a sales transaction is processed to create a bill, to develop an account statement, to adjust inventory and to produce a summarization of all sales by desired classifications. Likewise, the records of production in a factory have a discrete effect on inventory, as well as upon costs and production scheduling.

The second purpose of data processing is to create summarizations of all transactions over a prescribed period of time. A summarization of total sales, costs of sales and corporation expenses are needed to develop a pictured corporate earnings in an operating statement with its supporting schedules such as, for example, a report of total sales by products.

The case discussed in this chapter describes the development of management control information, in this second area of data processing, from detailed records of sales transactions.

A FUNCTIONAL BREAKDOWN OF DATA PROCESSING

The procedures in data processing can be broken down into four general operational categories. These are:—

1. The recording of the transaction.
2. Organization and sorting of the data.
3. Calculation and summarization of the transaction data which develops new classified data.
4. The organization and recording of this developed data into summarized totals in management report form.

TWO MAJOR REQUIREMENTS FOR DATA PROCESSING

Of particular interest are two major requirements which business places on data processing systems. The first of these is that the data must be available for audit at any time. The second is that means must be provided for processing "exceptions" in the procedures and exceptional transactions in an orderly and efficient manner.

While some advances have been made in auditing machine-language media (such as punched cards, magnetic tape, etc.), small to medium sized companies find it more practical to maintain the economic and operational advantages of making auditing possible at any time and any place in the system by human observation and investigation. Therefore, the media used for data processing should be easy for the human being to read.

A fully mechanized or "integrated" data processing system is a procedure for handling data along restricted lines. The problems of programming a fully automatic data processing system are complicated by the fact that such a system must provide for all possible "exceptions" to the "normal course of events." In a partially automated system, the operator has the facility of taking the "exception" out of the "normal course of events" routine at any point and handling it manually. Here again, it is apparent that the data carrying document should be easily read.

HUMAN ABILITIES AS APPLIED TO DATA PROCESSING

The objective of this chapter is to tackle the problem confronting the business which cannot afford the luxury of a fully automated or integrated data processing system. To what degree, then, can the small to medium sized business use automation in data processing? The writer believes these businesses can profitably use mechanized data processing to aid the human operator in those functions which are difficult or time consuming for the human being to perform. Here, the mechanized units should be integrated with the inherent capabilities of the human being.

What are these inherent capabilities of the human being? Practitioners in the field of human factors[1] tell us that man has unique abilities in the field of judgment, reasoning power and flexibility. On the other hand, man's ability in speed, power and durability is sadly lacking when compared with a machine. For our particular purposes, man should not be encumbered with highly precise operations requiring unflagging attention and precision of action.

Designing data processing systems, which adhere to such principles, achieve two ends:—1. The operator is relieved of those duties and activities which she is least capable of performing reliably and rapidly. 2. The activities assigned to the operator generally are of such a nature as to stimulate her intellectual activities.

A CASE OF SALES ANALYSIS BY PARTIALLY AUTOMATED DATA PROCESSING

Middlesize Corporation is a manufacturer located in an industrial community. They manufacture and distribute 5,000 catalog items through 150 or more wholesale salesmen in about fifty distribution points throughout the United States.

Middlesize Corporation sells their products on a relatively close profit margin in a keenly competitive field. The overall

---

1. John D. Vanderberg, "Man and Machine," *Machine Design* (April 17, 1958).

profitability of the Corporation's operations depends upon the marketing of the proper ratio of product mix in sufficient volume to maintain a profitability picture for each product. Three basic needs for sales analysis in Middlesize Corporation are as follows:—1. Corporate management has known from experience that salesmen in Middlesize have a tendency to concentrate on a few classes of products in the catalog. The classes of products which are sometimes ignored by salesmen are low-profit items which are important primarily because they help keep competitors out of the offices of Middlesize Corporation's best customers. For this reason, sales analysis reports are developed to show the breakdown of a hundred classes of products sold through each of the fifty distribution points and by each salesman. In this way, errant salesmen can be identified early and this practice corrected. 2. A breakdown of products sold by major customers affords a picture of whether or not competitors are making important inroads in Middlesize's field of business. This information is particularly helpful in determining where to send specialist salesmen from the home office on what products. Note that the information developed here is different and is handled differently from the information developed in 1. above. 3. Sales for the entire range of items in the catalog fluctuate seasonally and with locality. Since these seasonal fluctuations have a trend from year to year, an important use of sales analysis is to develop raw data for determining economic levels of inventory and for long-range production planning.

Outside of the sales analysis field, these *same* unit documents are used for updating inventory records and for calculating salesman's commission at the end of each month.

MIDDLESIZE CORPORATION'S NEED FOR PARTIALLY AUTOMATED DATA PROCESSING PROCEDURES

Before the installation of new equipment, Middlesize Corporation employed twelve adding machine operators and two supervisors on two shifts (8:00 a.m. to 5:00 p.m. and 5:00 p.m. to 11:30 p.m.) to develop eleven different monthly sales analysis reports. These fourteen girls had to create and process 80,000 unit documents a month. Since six of the reports involved the use of all the unit documents on each report, the total handling of unit documents actually exceeded 480,000 units per month, not including sorting operations. Therefore, each girl had to handle over 120 unit documents per hour for every working hour of the month.

Referring again to the four functional areas in data processing described above, prior to the installation of new equipment, the procedures used in Middlesize Corporation were as follows:—1. Marginally punched cards were pre-edge-notched (batch grooved) for product number (fixed information) and placed in a tub file. These cards were then pulled from the files, one prenotched document for each product item on the invoice. When an accounting control figure is prepared for a batch of invoices, the quantity and extended price (variable information) for each product item were printed on each unit document using a split platen adding machine. This control figure, incidentally, was used for control of sales analysis operation, as well as for control of accounts receivable and control of original orders. The remaining variable information (customer's number sales district and salesman's number and date of transaction) was edge-notched in the unit document on a key-driven notching machine.

2. Sorting of unit documents was performed by standard edge-notched procedures. These procedures are more than sufficient for any level of partially automated data processing.

3. Accumulations on the data on these unit documents to develop sales analysis data were made by adding batches of cards using adding machines. In this operation, the girls had to read each number on each unit document at least six times to produce all eleven sales analysis reports. It was the boredom of handling each card, reading it, and punching the number just read into an adding machine that was the major problem of this data processing system. Operators could not be expected to handle a job of this nature with speed and accuracy. Above all, these adding machine operators were being used like machines, performing a simple routine which required the minimum of judgment and intelligent mental activity.

4. Sales analysis reports were transcribed onto standard forms for use by management.

MIDDLESIZE CORPORATION'S PREPARATION FOR MECHANIZED DATA PROCESSING EQUIPMENT

A single piece of equipment, which can mechanically add data directly from a unit document, caught the attention of the Manager of the Sales Analysis Department. His concept of this equipment was that it was "automatic" and "would save people." On this basis, alone, the equipment was ordered without any serious investigation into the use of this equipment to handle his particular problems in developing sales analysis data. The Manager of the Sales Analysis Department developed procedures to be used on this new equipment without consultation with the manufacturer. As a result, the first three weeks after installation were wasted by attempting a procedure of performing sales analysis which did not adequately account for the capacity and limitations of the equipment.

After this three-week period, the Manager looked for assistance from the manufacturer of the equipment in organizing his procedures to develop the sales analysis reports the corporation wanted. At this point, progress started to be made in using the equipment efficiently, hampered rather seriously, however, by the personal relationships that prevailed in the Department. It should be pointed out here that the Manager's inability to develop an efficient system by himself using the comparatively simple piece of equipment is best laid to the highly emotional quality of personal relationships in the Department, as well as his unfamiliarity with the equipment, rather than to the equipment or to the competence of the Manager in systems and procedures.

PERSONAL RELATIONSHIPS IN THE SALES ANALYSIS DEPARTMENT

The Manager of the Sales Analysis Department has made a habit, for a number of years, of timing the efforts of his day shift and his night shift, and comparing the results for each supervisor to see. Unfortunately, this engendered competition was not conducive to harmonious relations in the department.

The Manager was also most intolerant of errors in the work, resulting in the girls' shying away from responsibility where errors might be charged to them. In view of the volume of work required and the nature of the work, it is uncharitable to expect this performance without errors.

These and other examples of the tension which existed in this office caused the writer to check the validity of this case as being a typical example of a partially automated data processing system. This checking showed that, although this tension is not representative of a majority of offices, it is present in so many offices that this case cannot be called atypical.

THE EQUIPMENT AND ITS USE IN MIDDLESIZE CORPORATION

Each of the two pieces of equipment installed in the Sales Analysis Department can code (punch) numbers into a unit document which later may be "read" directly into the tabulating (adding) section of the machine. For the purpose of this chapter this machine will be called a tabulating punch.

Quantity and extended price (rather than being printed on the unit document as described above) are coded in the unit document in machine language (punched holes) without effort and as a by-product of the development of the accounting control figure. This control figure is now developed in much less time on the tabulating punch than the time previously required on an adding machine.

Unit documents, thus prepared, need never be read by the human being in normal operating procedure. The tabulating punches are used to accumulate groups of unit documents that have been organized and batched by the operator.

For each batch of unit documents, the tabulating punch is used to produce a "total" or sumary card. These cards are then tabulated with the summary cards of succeeding reports to develop "year-to-date" totals.

While this technique considerably reduces the handling of the unit documents, each of the 80,000 unit documents must still be handled at least six times to produce all eleven of the monthly sales analysis reports. Therefore, each machine operator handles more than 850 unit documents for each working hour of the month.

BENEFITS ACCRUED BY MIDDLESIZE CORPORATION USING PARTIALLY AUTOMATED DATA PROCESSING

It has been estimated that about half of the economic benefits to Middlesize Corporation resulting from the installation of two tabulating punches is due to the performance of the machine with its operator. The other half of the benefits is due to improvements in data processing procedures brought about by the installation and use of the tabulating punches.

The Management of Middlesize Corporation has been very pleased with the installation of the tabulating punches because they now are receiving error-free sales analysis reports on schedule. Up to this writing, only two of the reports have been analyzed on a cost basis. One report, which previously required $81.00 of man-hour expenditures, is now developed for $3.10 of labor. A second report, which originally cost $139.00, has been reduced to a cost of $6.00 in labor.

Probably a more meaningful statement of the economic benefits to Middlesize Corporation is that the installation of two tabulating punches has released six girls for other work. Here, then, the expenditure of $2,000 a year for equipment has saved the Sales Analysis Department a salary charge of over $22,000 a year.

It would seem, however, that there is a benefit to Middlesize Corporation, its employees and the community which, in the long run, may be far more important than the dollar savings mentioned above. This is the saving of human energies and talent that accrue from the use of humans in operations which they can easily perform. For the employees in the Sales Analysis Department, frustration has been replaced with the satisfaction of achievement, which extends into their social relationships as well as throughout their work situation.

Selected Bibliography

Ethel Bailey, *Modernize Your Business Methods* (from *The Trader and Canadian Jeweller*, March, 1956).

S. J. Crause, *How a Simple Control System Cut Production Costs* (from *Office Equipment and Methods*, II,1, January, 1956).

E. A. Henry, *Production Line Technique Applied to Foundry Labor Accounting* (from *Foundry*).

S. D. Kaufman, *We Should At Least Try to Collect* (from *The Modern Hospital*).

D. W. Kern, *Use of Keysort in the Registrar's Office* (from *College and University*, XXXI, 1, October, 1955).

L. A. McPherson, *Inventory Reduced by 38%* (from *Purchasing*); *Punch Cards for Inventory Control* (from *Hospital Management*, June, 1957).

T. G. Mohney, C.P.A., *3 Tools of Business for Distribution and Statistical Problems* (from *New York Certified Public Accountant,* XVII, 5, May, 1957, 295-303).

Plant Wide Control of Every Job (from *Production,* XXXIV, 3, September, 1954, 81-83).

T. P. Rutter, *Why Canadian Western Chose Manual Accounting at New Plant* (from *Office Equipment News,* January, 1957).

Harold Schneider, *How Grooved Cards Give Us Simple, Accurate Good Control* (from *Restaurant Management,* March, 1953).

Hiram Sibley, *Charge Ticket Presents the Evidence that Enables Connecticut Hospitals to Recover Full Costs for Service* (from *The Modern Hospital*).

CURTIS M. JANSKY was born February 23, 1923, in Minneapolis, Minnesota. He graduated from the University of Illinois in June, 1949, with a Bachelor of Science Degree in Electrical Engineering. From 1949 until 1956 he was Project Engineer, at Sperry Gyroscope Company, Great Neck, New York. He had responsibility for design of electronic data processing and electronic data communication systems, and technical and human factors for design of radar indicator systems and radar system operator controls. In 1956 he became Assistant Director of Research, Nowland & Company, Greenwich, Connecticut. Directing market research projects in the field of electronic data processing. He now works as a Product Planner of Advanced Product Planning Department, for Royal McBee Corporation, Port Chester, New York. He holds membership in the Institute of Radio Engineers, and is a Vice Chairman of Professional Group of Human Factors in Electronics.

Chapter 13

AUTOMATIC DATA PROCESSING IN THE LARGE COMPANY: *A Case Study*

David G. Osborn, *Market Research Staff, Data Processing Division, International Business Machine Corp.*

Highly advanced automatic control may not yet be common in industrial operations generally. However, automation in the form of "computers" has already made an impact on the non-manufacturing operations of many companies and organizations. Computers have been installed with great fanfare in many medium-sized and large companies. Their accomplishments in performing at phenomenal speeds have been dramatically heralded. Few tools have had such a rapid widespread effect on concepts of management.

CLERICAL AND SCIENTIFIC COMPUTATION BOTTLENECKS

High-speed, electronic, digital computers have found speedy acceptance for a variety of reasons. Paralleling each other during record years have been several trends which have combined to create a sizeable market for computers. (1) Business and other organizations such as government departments have become increasingly aware that clerical costs are high and going still higher. (2) The growing complexity of business has generated a need for more information in less time. (3) The frontiers of scientific endeavors are being pushed out with a quickening tempo so that the need for performing scientific computations at reasonable costs has been greatly accelerated. Some of this computation was already being performed, but slowly and at high cost. Much of it has long been regarded as desirable, but the mechanical means were not available for doing it. A large part of the scientific computation involves relatively new research areas which hardly existed until recently. (4) A shortage of properly qualified personnel in both scientific and business-data fields has been aggravated by such stark realities as the low birth rate of the pre-war depression decade. (5) Coupled with all of the above trends has been the development of equipment which could meet the needs of organizations requiring vastly improved techniques for computation and data processing. Rapid recent progress in computer development is based on the pioneering work of a number of distinguished scientists—Norbert Wiener, Claude Shannon, Howard Aiken, and von Neumann, to mention a few of the more prominent.

In most businesses the volume of clerical or business data operations far exceeds that required for the scientific type of calculations. There is plenty of proof of the acuteness of the clerical problem. "(1) In 1920 statistics showed a ratio of 11 clerical employees for every 100 factory workers: today for the same 100 factory employees 25 are employed in the office. "(2) Production costs have gone steadily downward, while office costs are soaring to new highs."[1]

SOLUTION OFFERED BY ELECTRONIC COMPUTERS

To help in solving many of the cost and productivity problems suggested by the above trends, electronic computers have arrived at a most opportune time. ". . . computers not only can handle mountains of business paper work in hours or days, instead of weeks or months, but, more important they are providing answers to problems business could never before solve in *any* length of time."[2] ". . . the cost of a million multiplications by electronic computer is about $3. (This refers to operating charges and does not take into account planning and other overhead costs.) The same job done by clerks and hand calculators would cost thousands of dollars."[3]

An increase in accuracy is another of the benefits that com-

---

1. Ralph W. Fairbanks, "Electronics in the Modern Office," *Harvard Business Review*, XXX (September-October, 1952), 83.
2. William B. Harris, "The Astonishing Computers," *Fortune*, LV (June, 1957), 136-137.
3. George O. Von Frank, *The Electronic Industry in New York State* (Albany: State of New York, Department of Commerce 1957), 15.

panies derive from the use of computers. The speed with which answers can be supplied means a much smaller penalty to be paid for delays in getting information or for the inability to obtain it at all.

Computers make a vital contribution to business in three main areas. In *management control,* they facilitate the communication of data to executives in usable form at speeds scarcely dreamed of before. In improving the processing of data for management decisions, electronic computers are ". . . contributing greatly to the spawning management tool known as 'operations research.' This is the application of scientific principles and complex mathematical procedures to administrative studies. By analyzing a particular sphere of a firm's operations in terms of an integrated system, by including all diverse interrelated factors that impinge on that system and by use of appropriate mathematical equations, the optimum combination of factors can often be determined. This technique can be applied, say, to the purchase and maintenance of company equipment, to forecasts of demand, to the pricing system or to personnel utilization. Machine research can never fully replace the businessman's ultimate judgment. However, use of this technique will give him more complete and more timely facts on which to base his decisions. In fact, the new system can often incorporate selected control points which will automatically signal when some action should be taken."[4]

In *record keeping,* computers help eliminate a great deal of tedious drudgery. Maintenance of ledgers, journals, and inventory files are typical record keeping applications.

Operational processing of documents and information embraces such functions as preparation of invoices, writing of checks, and handling of insurance premiums.

ACME AND AUTOMATIC PROCESSING

The case study described in this chapter summarizes the actual experience of a company with automation of its data processing. The company will be referred to as the Acme Diversified Products Corporation, although that is not its real name. Acme is a manufacturing firm with its main plant and head offices located in the eastern United States. Its products include both consumer and industrial products, but many of them are specially engineered

---

4. *Ibid.,* 15.

products tailored to the unique requirements of individual customers. Acme has a number of plants located in various sections of the United States and it has sales offices in major cities of the country as well as abroad. During the postwar era it has grown rapidly. Achieving some of its growth by acquiring smaller companies, it now has annual sales of between $100-million and $500-million.

ORIGIN OF ACME'S INTEREST IN A COMPUTER

Acme's interest in computers was spurred by the president's reading of an article in the *Wall Street Journal* about them. At lunch one day he asked some of the other officers of the company, "Why don't we have one?" The question was only a half-serious, needling one, but did play an important part in the company's decision to formally investigate computers and the contribution they might make to improved company performance.

Actually, in a number of sectors of the company there had been some curiosity about possible benefits from a computer. Most of the reasons stated above for the popularity of computers played a role in the growing interest at Acme. The accounting department, for example, was beginning to have difficulty in meeting deadlines with existing equipment because of the scarcity and rapid turnover among the young women who were doing the bulk of the clerical work. The chief engineer had been dissatisfied for a long time with the company's inability to compute the design of certain products with all possible factors taken into account. He knew that to do so would require computation capacity far beyond any existing resources in the company. He also knew that competition was improvising its designs and lowering its costs. He gave some thought to the use of an analog computer before the growing needs of other departments and the president's interest forced the investigation of computer requirements for the company as a whole. As it worked out subsequently, an analog computer would not have been versatile enough for the varied applications that were ultimately put on a computer.

Acme had been using punched card equipment for several years in a number of applications. Its engineering department had used a card programmed calculator at a service bureau from time to time for applications that were essentially scientific—requiring long and involved calculations instead of high-volume processing of routine business data. Acme was thus in a fairly good position to explore the possibilities of much larger capacity electronic computers.

INVESTIGATION OF COMPUTING SYSTEMS

A decision to make a sizeable investment in a computer and all of the related accessory equipment was not taken lightly by Acme. The vice president of organization planning was assigned to form a committee which was to study the feasibility of computers generally, and then decide which computer to recommend. The head of the study group was the systems and procedures manager of the accounting department. Serving with him in the evaluation group were a senior engineer and a department head in manufacturing.

Both the feasibility study and the purchase decision were very involved processes. Like many other companies which have looked into data processing, Acme's team made an intensive series of visits to other computer installations in related industries and to the schools and plants of computer manufacturers. These trips consumed many months. They were aimed primarily at getting a clear idea of what the various types of equipment had to offer and how the available equipment would fit in with Acme's needs. Many more months were spent in detailed studies of the needs of various Acme divisions and departments.

Because the computer field is quite new to many companies, reliance is often placed on outside consultants for advice in numerous phases of selection and use. Acme used three different consulting organizations: the data processing division of its certified public accounting firm; a management consultant specializing in problems of data processing; and a research institute.

More than a year was spent travelling, studying, and conferring. Management was a little surprised at the cost of these endeavors, including the consultant fees. But before the decision was finally made, several of the senior officers of the company joined in the junkets too.

Acme made its first screening of available computers and concluded that the desk-size type with external memory, selling for under $50,000, was too small for its needs. At this stage, the very large computers with purchase prices on the order of $1-million (or rental on the order of $20,000 or $30,000 per month) were also eliminated. A second screening eliminated the models which Acme felt could probably not be delivered on a satisfactory schedule. The final decision was narrowed down to a medium-sized computer which had the performance characteristics and features that seemed to best suit Acme's anticipated needs. Acme needed a computer which was suitable for both scientific and business data problems. It also wanted a computer which would be compatible with various pieces of tabulating equipment being used by the company.

More than a year elapsed between the original decision to explore the potentialities of computers and the subsequent decision to acquire the particular one that was selected. Basically, the decision to lease a computer was made by the team members who were middle and lower management men. Their decision was accepted with only slight resistance by the executive committee of the company which consists of the chairman, president, and executive vice president.

Management had a vague realization that it might be some time before the computer would be clearly justified on an economic basis. The officers knew from the experience of other companies that investment in study and analysis preparatory to getting the computer would be large. Acme's top echelon knew also that mechanical problems might plague the computer at the beginning. But the management liked to think of itself as being progressive and was willing to commit itself to a little experimentation as the price of progress.

Acme's computer was leased rather than purchased. This question had been a vital one, perhaps the most important single element that had to be considered. Acme's reasons for leasing can be summarized as follows: (1) In a field which is so new and so rapidly changing, Acme wanted to avoid the risk of investment in expensive equipment. (2) Under existing tax laws, leasing was more desirable. (3) Acme can get rid of the equipment on a certain amount of notice to the manufacturer and have it replaced with more advanced equipment as it is developed. (4) Acme was accustomed to leasing its punched card equipment and had enjoyed letting the manufacturer worry about maintenance problems. (5) Acme has expanded rapidly and has nearly always been hungry for capital. It believes that money invested elsewhere in the company (*e.g.*, in raw material inventory, management development, etc.) will bring a greater return than the same capital applied to computer purchase.

PREPARATION FOR THE COMPUTER

Having taken the big step of deciding to go into automated data processing in a big way, Acme continued the same careful planning that had gone into the purchase decision. Preparations were outlined in meticulous detail. The head of the feasibility team was appointed to the newly created post of manager of data processing. Each of the two other members of the team had hoped to get the nod, but they were returned to the departments from which they had been loaned. As consolation, however, each was promoted to a position of higher responsibility. They brought to their new positions a valuable perspective which was part of the benefit to the company.

Preparations for installation of the computer embraced a number of major tasks: physical preparation of a site, organization of the data processing department, training of personnel, integrating the performance of the computer into company operations, and "selling" the computer to the rest of the company. Physical site preparation was the least of Acme's problems, because a new headquarters office building was being built. It was necessary to modify the plans only slightly to accommodate the tons of equipment which would be delivered in a little more than a year. Some of the wiring had to be altered, partitions moved, and air conditioning capacity increased.

INTEGRATING THE COMPUTER

Integrating the computer into company operations was a major project. Programs had to be worked out for the applications which would be put on the computer. The programs, which were the ultimate purpose of the systems analysis, involved the creation of the detailed series of commands and instructions to the computer, telling it what to do and when.

Product codings were changed to make them compatible with the computer. This was only one of many changes in the company's operations as a result of the introduction of the computer. Much of the modification of operations resulted from the necessity of standardizing the "inputs" to the computer. The elimination of irregularities was the biggest problem. The programmers had to ferret them out and make them "regular."

Management now feels that one of the main improvements

brought by the computer was a forced re-examination of many methods and procedures that had been followed unquestioningly for years. This re-examination, however, posed a thorny dilemma to the data processing manager, who was confronted with the unenviable chore of getting some of the company practices modified. In this task he had the assistance of the planning vice president who continued his interest in the project. A few of the old timers resisted the changes, but ultimately went along with most of them. To keep peace, the computer team did reluctantly modify some computer routines to accommodate existing practices.

In converting one application from a manual operation to the computer, much trouble was caused by the resistance of a middle-aged accounting supervisor. It was necessary for the programmers to find out from him the exact details of the existing intricate computation system. This supervisor would omit important details of the computation in his explanations to the programmers. When they ran the program for debugging,[5] they would get incorrect answers and have to try to persuade him to reveal more information. He was, of course, deliberately sabotaging their efforts. Before this application was finally worked out, several programmers had quit. Eventually the situation was resolved when the supervisor resigned after he had refused a transfer elsewhere in the company.

CHANGES IN EMPLOYMENT STRUCTURE AND FUNCTIONS

Impact of the computer on such areas as worker readjustment, reassignment, and training has not been revolutionary in Acme. There has been a slight reduction in the number of personnel involved, but this has been more than compensated by a drastic increase in clerical and engineering productivity. Clerical reductions have been nearly balanced by increases in computer operating personnel such as programmers, and some personnel problems have been solved. A smaller number of higher paid, more skilled workers who operate the computer replaces a larger group of lower paid, unskilled, clerical women employees. The selection and training of computer operating personnel has been a major concern of the data processing manager,

---

5. Debugging is the correction and testing of a computer program prior to regular running of the application.

but has concerned the rest of the company only slightly. The question of programmers will be discussed in more detail later.

Several engineers were permanently transferred from the engineering department to the data processing department to work as programmers on the computer. They continue to work in close contact with the engineering department. The design section of the engineering department now is able to achieve more and better results with approximately the same number of people. In other departments the computer has not caused any sensational reductions of personnel. This is partly because some of the business data applications on the computer are new functions which were not performed before—new reports, for example. Increase in the programming staff has somewhat compensated for clerical reductions in a couple of departments, but total salary payments have been reduced; most of the programmers had been transferred from other departments.

As in many other companies, the clerical staff reduction has been no problem at all. Some of the reduction took place in types of work which are done by girls who remain in the labor force only temporarily before marriage and family raising. Transfer to the computer of some of the functions formerly performed by such girls has benefited the company by helping solve the problems of turnover, recruiting, and training of new girls. The company has also continued to grow rapidly enough so that a few clerical people were transferred to other work.

The number of employees has been cut slightly but a great deal more work is accomplished by a total group of employees slightly smaller than was previously doing the engineering and business data work. The punched card tabulating department was simply moved into the data processing department. This was a logical move because punched cards constitute one of the important inputs and outputs of this computer system.

PROGRAMMING AND OPERATING PERSONNEL

Installation of a computer in Acme resulted in an urgent need for persons equipped to operate it, mainly programmers. Generally, Acme wanted to follow a policy of transferring persons already in its employ to the new data processing department. The data processing manager believed that better results would be obtained by teaching programming to Acme employees than by orienting new programmers from outside to Acme's business.

It has not always been possible to apply this policy, however. At times when it was difficult to obtain enough properly qualified insiders, it was necessary to go outside.

At one time the manager thought that mathematicians would make excellent programmers. Subsequent experience has shown that a certain combination of analytical ability and practicality works better than a purely mathematical background. That a mathematics or physical science background is not necessary was proven to Acme by its experience with a Ph.D. mathematician who was an expert in number theory. After unsuccessfully attempting a certain problem for three months, he gave up. A relatively untrained girl worked out the analysis in a week. Any mathematicians used in the future will be teamed with "practical" people.

Acme's experience bears out much of what an executive in another industry has said: ". . . scientists, engineers, and mathematicians are not indispensable, or necessary. And Ivory Tower geniuses need not apply; the people you will want must have the ability to get along with other people; they should be pleasant, cooperative, and friendly. They should have a high degree of analytical ability—the ability to get information, analyze it, and reach sound conclusions based on it . . . They should be intelligent. Theirs will be the task of learning a new and complex art and of relating that art to complex and detailed business procedures in a way never before possible. Further, your people should have wisdom, an interest in your business, and an interest in and capacity for hard work. . . . both males and females will serve; youth is certainly not a requisite."[6]

The engineers function well as programmers in the engineering department applications. But for the business data applications, people with diverse backgrounds were obtained from accounting, marketing, manufacturing, and auditing. The manager tested more people than he intended to use. His own experience as well as that of others proved the wisdom of using aptitude tests for screening out unsuitable candidates.

One of the programmers is a former production worker who taught himself in preparation for working in a computer age. Foreseeing the need for somewhat mathematically trained personnel, he took night school and correspondence courses in both

6. Stevens L. Shea, "Organizing for Electronics," *Advanced Management,* XXII (December, 1957), 6-7.

mathematics and accounting. Two programmers were hired from the outside and trained by Acme. One was a young woman liberal arts graduate and the other a former professional musician. Both have adapted superbly.

Training for the programming and operating staff consisted of sessions at the manufacturer's school for several weeks, attendance at electronics seminars, visits to other companies, and a great deal of study. The "curriculum" included such technical systems work as the designing of flow charts, the construction of forms for both input and output, the intricacies of punched cards, the elements of the particular machine language, and the logical processes of the computer. The balance of the training was done on-the-job in actual programming, analysis of applications, debugging, and helping with the installation of the computer.

More than a year was devoted to training and preparations. Within the year the new building was completed and a site was ready for the computer. Slightly more than two years elapsed between the original decision to explore computer possibilities and its actual installation.

EXPLANATION OF COMPUTER TO OTHER EMPLOYEES

The manager of data processing planned carefully for the proper introduction of the computer to company personnel. Because of the novelty of the computer and of the general unfamiliarity with it, most of the company headquarters personnel at all levels were included in educational programs. In some cases, for employees only slightly concerned with the computer, brief explanations were given to large groups. They had a chance to look at the computer, to hear it described, and to be put at ease about technological unemployment. The senior officers of the company had to be informed of the potentialities and opportunities represented by the computer. They were taken in quite small groups and a great deal of time was taken to answer their questions. These meetings helped enlist their support for the computer group in the future.

Meetings with the heads of departments or their representatives were important for getting them to see ways the computer could help them. A bid was made for them to submit problems for the computer group to solve. This part of the plan was too successful, because the avalanche of problems was too much to attempt for the foreseeable future.

A booklet describing the computer was prepared for distribution to all main office employees. Articles about the computer were written for the internal employee publication, and one was included in the external company publication going regularly to opinion leaders and community organizations.

PROGRESSION THROUGH PHASES OF COMPUTER UTILIZATION

One significant difference between a computer and other simpler data processing equipment a company may have used previously is that management may have to be somewhat constantly aware of the system's utilization. Utilization may start at a low rate and rise with the passage of time. Demand for computer time is likely to grow so that the system tends to expand. At least, that is what happened in Acme's case.

Acme started with a medium-sized computer and a minimum of input and output equipment. An engineering application and two in business data were regarded as the initial justification for the computer on a cost basis. The computer was initially used on a one-shift basis. The justifying applications had been programmed before installation and were put on the computer as quickly as possible. Unscheduled "down time" for the computer was a little higher at the beginning than had originally been anticipated, but the manufacturer's resident maintenance engineer was able to work with Acme's staff to reduce this to almost zero within a few weeks.

As the computer performed successfully on the initial jobs, more and more uses were considered for it. Each of these, of course, had to be programmed and debugged before being run. As time went by, the basic first shift was fully used. This necessitated a decision by the data processing manager to rent an additional shift from the manufacturer. An advantage of a second shift was that the rental was only 50% of the basic shift rental, so that the computer investment could be "paid off" earlier. Economic justification could be more easily proven for subsequent applications.

Performance of the computer has pleased people in the company at various levels, so that the work load has grown steadily.

More accessory input-output and conversion equipment has been added to the system and a third shift has been added. The computing budget has increased gradually, but the efficiency has improved as the manager's skill in administration has grown and the overall productivity of the department has risen.

As more and more applications are being put on the computer, a point is being reached where there will have to be some change in the equipment. Acme has obtained from the operations research group of one of the major engineering schools a formula for predicting computer demand. With this formula it expects to be able to plan future utilization and acquisitions more accurately.

Already, nearly the maximum of input-output and conversion and storage accessory equipment has been added to make the computer as productive as possible. Other steps which can be taken are (1) to replace the present medium-sized computer with a larger-scale one with faster processing and larger memory, (2) to acquire another medium-sized computer to supplement the present one, (3) to screen out the least economic applications from the present computer, (4) to get a computer for one of the branch plants or regional offices relatively close to the main offices and transfer some applications to it, (5) to have some work performed by a computer service bureau nearby, or, (6) to rent time on another company's computer.

Acme's data processing manager looks forward to the time when there will be other installations of similar computers in the company. With at least two computers, the dependence on either one is lessened. If one breaks down, programs and input data can be flown to the other one and run off there. A tremendous potential for this and other computers is seen in the company. In branch plants and regional offices, Acme will be able eventually to use computers to accomplish a vast number of tasks.

OTHER UNIQUE ASPECTS OF COMPUTER

Aside from growth of internal demand for it, impact of the computer on Acme has been felt in some unexpected ways. Most of the changes are seen as comparisons with the conventional office equipment Acme has known before—hand operated calculators, accounting machines, punched card sorters, etc. Essentially the transition has forced Acme to take a new view of data processing equipment as office capital equipment, which is both so large and in such a technologically volatile field that Acme leases it instead of buying, and has made all sorts of adjustments to get used to it.

Before leasing the computer Acme had been accustomed to spending a few hundred dollars for a calculator and two thousand or so for an accounting machine. By contrast, rental for the computer has passed the figure of $10,000 per month. To be spending such sums for office equipment was a little difficult for some of Acme's executives to get used to. The magnitude of the equipment, together with the frequency of new developments in computers, was important in the decision to lease the equipment. For Acme this was an innovation; the only office equipment which it has ever leased before was the punched card equipment which was relatively low in rental cost. Because of its size and cost, some of the other executives kid the data processing manager about the little factory that he runs "in competition with the plant manager." There was some substance to their jokes inasmuch as the computer did have to have a special area of the building adapted for its installation and special conditions maintained for its operation. Acme had never before constructed space so carefully around its office equipment.

Because of the newness and uniqueness of the computer field, some new perspectives were thrust upon managment which it had not anticipated. In this rapidly changing field, the people associated with the computer have to allocate a certain portion of their time to merely keeping informed of developments and progress.

The data processing manager is in frequent contact with other companies using similar computers. He consulted with some of them during the feasibility study, and learned that a frequent exchange of experiences is useful. This exchange continues despite the fact that some of these other computers are in companies which compete strenuously with his.

Acme's manager of data processing does not even wait for the manufacturers' salesmen to come around to see him. He visits their plants to find out what developments are being planned. In this way he feels he stays better informed than some of the salesmen who call on him.

Certain professional organizations have also arisen to meet the needs of computer personnel. Typical of these are the Association for Computing Machinery and the conferences of users of particular brands of computers. A whole new field of publications to interest the computer personnel is now available—*Datamation, Computers and Automation,* and the *A.C.M. Journal*—to mention a few. Active communication with their colleagues is much more necessary for computer operators and administrators than for those concerned with other types of equipment. There is also a host of new subscription and consulting services which, for a fee, will keep the data processing manager informed of developments and provide him with advice on operation and planning.

For Acme's executive group generally there is available an almost unending series of conferences, courses, and seminars on electronics, data processing, and operations research. Many of these are operated or sponsored by such organizations as the American Management Association or the Society for the Advancement of Management, but other professional organizations and consultants also operate them.

Another innovation that sets the computer apart from older accounting and calculating machinery is the presence of a fulltime engineer. He represents the computer manufacturer, taking care of preventive maintenance and any emergencies that occur. If the computer were larger, the manufacturer would probably supply more than one engineer. In the scheduling of computer operations a certain number of hours daily are set aside for routine checking to uncover potential trouble before it can interfere with processing.

BENEFITS FROM COMPUTER

Like a number of other companies with computers, Acme's management believes that the main benefits have not been in reduced payroll costs of clerical work, but in the improved ability to have information for control, decision-making, and planning. In short, the computer has become a tool for improved managing.

Accuracy of demand forecasts has enabled Acme's marketing executives to shape more effective strategy. Sales analysis has been expanded to give sales executives a closer look at more factors in each product line and geographical area. Elimination of errors and the cost of correcting them has improved most of the reports, records, and operational processing which have been put on the computer.

One of the important benefits is a vast saving of time. Results are available today instead of next month; next Monday instead of in six months. With the computer, management now has end-of-the-month reports seven working days sooner than before. Inventory reports which formerly were quite time-consuming and slow are now produced by the computer more accurately, in greater detail, and faster.

The engineering department, with no increase in staff, has been able to turn out in minutes product design specifications that formerly required more than twelve hours. More variables are checked and the products are more efficient and economical. Another dividend is the avoidance of mistakes in design which would be expensive to correct. Thus, thousands of engineering man-hours are saved.

ATTITUDES TOWARD THE COMPUTER

Management, as indicated, above, is extremely pleased with the computer. Other computers will be acquired later and many more applications will be put on them. These will include such tasks as a closer analysis of costs and the application of operations research techniques to future planning. The president is so pleased with the computer that he has had its layout altered for better display to visitors. To make it more impressive, the observation window has been enlarged. If the volume of applications and the centralization of functions warranted several computers at the headquarters offices, the president would prefer to do as some other companies have done—array the computers in two lines down a large corridor with plate glass along both sides to create a massive impact. The computer has been praised in the annual report to stockholders. It has been shown on a local television program.

Company personnel at middle and lower levels have been generally pleased by the installation of the computer. Many of them are as proud of it as top management is. Some of them have obtained special permission to bring their families or friends to observe the computer. For the computer operating personnel, of course, working with the computer has been a source of great prestige. They are quite proud of the opportunity. Feeling on the part of some other office employees that the programmers were "eggheads" or "screwballs" has been completely dispelled.

Approval of other computers to be acquired by the company later will be relatively simple. This first computer has proven to management the clear desirability of having such equipment, and lower level decisions will be more common later on. However, with the decentralization of decision-making in Acme, each plant manager or sales division manager will be able to decide whether he wants to allocate his limited capital funds for a computer. Acme's experience to date indicates that with competent analysis of the potential applications and on the basis of the first computer's success, there will not be much difficulty "selling" these other executives on them.

The chairman of the board has become such a computer enthusiast that he wants to indoctrinate all future managers with the philosophy of data processing. Future computer installations will be staffed in such a way that groups of young accountants, engineers, sales supervisors, and others likely to rise to positions of responsibility will have an opportunity to operate the computer systems for at least a portion of their careers. The experience is expected to familiarize them with the expanded horizons of management assisted by modern data processing systems.

NEW HORIZONS IN AUTOMATIC SYSTEMS CONTROL

Experience with its automatic data processing system has symbolized Acme's entrance into a new era in which evolving technology will assist management on a vast scale. The computer has become a valuable tool to management, solving problems in many phases of corporate organization and functioning. Improvements in management control, record keeping, and operational processing have been accompanied by more efficient designs, greater clerical productivity, and a number of other benefits.

Beyond these impressive results, though, the computer system has begun to give company executives a glimpse of the powerful techniques which can tremendously increase their comprehension of the operation of their sizeable organization. They already have an excellent foundation for visualizing their company as a single system functioning within a broader and infinitely more complex system—the society in which it exists and which to a degree it serves.

Selected Bibliography

American Management Association, *Electronics in Action* (New York: American Management Association, Inc. 1957).

Thirteen papers originally presented as talks to an A.M.A. electronics conference. Emphasis is on actual experiences.

Richard G. Canning, *Electronic Data Processing for Business and Industry* (New York: John Wiley & Sons, Inc. 1956). Relates the capabilities of data processing equipment to the needs they fill for users. Explains programming, the systems study, electronic systems, and other facets.

P. E. Cleator, *The Robot Era* (New York: Thomas Y. Crowell, Inc. 1955). An English view of automation, including computers.

Eugene M Grabbe, Ed., *Automation in Business and Industry* (New York: John Wiley & Sons, Inc. 1957). Prominent engineers and scientists in a lecture series at the University of California outline fundamentals, techniques, and developments in automation systems with a number of contributions dealing with computing.

Pierre de Latil, *Thinking by Machine* (Boston: Houghton Mifflin Company 1957). Translated from the French. Describes the principles that make the most complex automatic machines possible. Shows how the concept of feedback or retroaction is related not only to machine design but to the natural world. Chapter XI deals with "calculating machines."

Cuthbert C. Hurd, "Computing in Management Science," *Management Science,* I (January, 1955), 103-114. Points up the dimensions of extremely high-speed computing and the implications for management of powerful tools for mathematical analysis.

George Kozmetsky and Paul Kircher, *Electronics Computers and Management Control* (New York: McGraw-Hill Book Company, Inc. 1956). Written for the businessman, describes the fundamental characteristics of electronics systems and the basic concepts of scientific methods of analysis.

Howard S. Levin, *Office Work and Automation* (New York: John Wiley & Sons, Inc. 1956). Outlines computer hardware and its use in management.

David G. Osborn, *Geographical Features of the Automation of Industry* (Chicago: The University of Chicago, Department of Geography 1953). Analysis of changes in space and manpower requirements in industrial operations which become automated. Case studies include factory examples as well as data processing, transportation, and energy conversion. Some implications of automation are explored.

David O. Woodbury, *Let Erma Do It* (New York: Harcourt,

Brace and Company 1956) . Discusses automation technology and its social implications, including several chapters on data processing. Extremely readable.

DAVID OSBORN, as assistant director of research for a market research and management consultant firm, has been responsible for studies in the area of data processing systems and business machines. Recently he joined the market research staff of International Business Machine Corporation's Data Processing Division. His doctoral dissertation at the University of Chicago was on automation. He has served as an assistant to the operations vice-president of The Kroger Company. Other experience includes teaching at Indiana University, city planning, and electronics work in the U.S. Navy during World War II.

CHAPTER 14

AUTOMATION AND THE BELL SYSTEM

C. W. PHALEN, *Executive Vice-President,*
American Telephone and Telegraph Co.

"Automation" is a relatively new word. It is variously defined. To me it means general technological progress of the kind that has been taking place in our industry and in others for many years. It is in this broad sense that I would prefer to use the word.

Automation, however, apparently suggests to some people that factories and even some industries of the future will run automatically, with only a few persons needed to control operations.

I wish to emphasize at the very beginning of my remarks that nothing resembling this result is in prospect for the telephone industry. We see changes ahead, but not changes of a revolutionary nature. Although we use large quantities of automatic equipment, the dial telephone being a good example, we are convinced that our business will continue to be operated by very large numbers of employees.

Before entering into a detailed discussion of automation in the Bell System, I should like to stress several basic thoughts.

The first has to do with the reasons for automation in our companies. We have introduced dial equipment and many other scientific and technological improvements throughout the years for three basic reasons: to improve the quality and usefulness of telephone service, to satisfy the demand for service, and to keep the cost of our product at a reasonable level.

The result has been a marked expansion of telephone usage over the years. Telephones and telephone calls have increased many times more rapidly than population, and even more rapidly than Gross National Product. I will give the facts in detail later.

The increased telephone usage resulting from scientific and technological progress has, in turn, helped to expand employment. This will be clear by comparing the present-day situation with that in 1920, the year conversions of telephones to dial operation first began in the Bell System. At the end of 1958 about 94% of Bell System telephones were dial-operated, and customers could dial directly about 40% of all calls outside local areas. Yet employees numbered nearly three times as many as in 1920, when there were no dial telephones, and over twice as many as in 1940, when about 60% of telephones were dial.

In studying automation in the Bell System it should be borne in mind that we render an essential public utility service, subject to regulation as to operations and rates by public authorities. It follows that technical advances in the business and economies in operation are necessarily and properly shared in large measure with the public.

With these preliminary remarks I will go to a review of past technological developments in the Bell System. I will then discuss the effects they have produced, having in mind distribution of gains from such developments among workers, investors and consumers. Finally I will give our views as to possible future developments.

PAST SCIENTIFIC AND TECHNOLOGICAL DEVELOPMENTS

Undoubtedly the most important technological change that has taken place in our business has been the change from manual to dial switching. This is the only development that I will discuss in any detail.

It is important to remember that dial switching equipment is just one component in what is really a single integrated plant, comprising a vast network of communication channels and related equipment.

Telephone service of present-day quality and scope is therefore dependent not only upon modern dial equipment but also upon improved transmitters and receivers, on improved cables, on the loading coil which reduces transmission losses, and on the telephone repeater which amplifies voice currents at intervals along the route. It is dependent upon apparatus which makes possible numerous telephone circuits on two pairs of wires, upon the modern coaxial cable—a small cable which can carry many hundreds of telephone calls simultaneously, on the radio relay with its beaming of long-distance telephone calls and television programs without the use of wires or cables, and on many other devices and techniques which have been developed and improved over the years for the carrying on of our business.

DEVELOPMENTS IN AUTOMATION

Now as to the development of dial switching.

Early in the telephone business the complex problem of connecting customers with each other was met by means of switchboards, where customer lines would terminate and where any two could be linked together. The early switchboards were all manually operated.

As the business grew, it became evident that manual operation would be less and less suited to handle the volumes of traffic being developed and foreseen for the future, to say nothing of permitting improvement in quality of service. After much investigation and experimentation, a dial system designed to meet our service requirements was introduced in 1920, permitting customer dialing of local telephone calls. Dial apparatus has undergone substantial change and improvement since then.

Dial switching has been important to long-distance service as well as to local service. By the end of 1958, an estimated 8,500,000 customers in 900 communities could dial long distance calls directly. By 1965 this will be the general practice for nearly all telephone users in the nation.

Nationwide dialing of station-to-station long-distance calls requires a uniform numbering plan, a large network of dial switching centers, and automatic equipment for registering certain information as to the calls. The automatic equipment, which also keeps a count of certain chargeable local calls, is the so-called "AMA" apparatus—automatic message accounting. It employs punched tapes which register the calling telephone, the called telephone, the time connection was established and the time connection ended. A machine takes the information off the tape and assembles it for each customer.

The long-distance calls which are customer dialed, and serviced by the AMA apparatus, are station-to-station calls not involving operator assistance. Operators will still be needed on other types of long-distance calls, including person-to-person calls. These other types of calls are the ones which require the longer periods of operator time.

EFFECTS ON THE USERS OF SERVICE

Let us first look at what has happened to our customers—the users of the service.

Scientific and technological progress has improved the quality of service and has extended the scope and usefulness of service. A few examples will illustrate the improvements in quality. Dial switching provides a faster and more uniform service, with full load carrying capacity available 24 hours a day. The speed and accuracy of the remaining operator-handled services have been improved through better apparatus and equipment. In the early 1920's, the average time required to complete a long-distance call was about 10 minutes; in 1958 it took a little over a minute, with some calls dialed directly going through in as little as 15 seconds. Today more than 95% of the long-distance calls are handled while the customer remains on the telephone; in 1920, less than 10% were so handled. In 1920, carrying on an average long-distance conversation was like conversing with a person in an open field at a distance of 80 feet; today the equivalent distance has been reduced to six feet. Subscribers' trouble reports—something wrong with the equipment which prevents satisfactory service—have been reduced 70%.

Technological progress has not only improved the quality of service, but has also increased the variety of the services we offer. As of today, there are more than 400 separate services available to our customers. Let me mention just a few by way of example: telephone service for ships and automobiles, service to foreign countries, teletypewriter exchange service, dial switchboards for individual customers, inter-communication systems and wiring plans, automatic answering devices, time and weather service, conference service, picture transmission service, radio and telephone network service, "speakerphones" (hands-free telephones), volume control telephones, and school-to-home services for convalescing children.

Automation, therefore, has not only given the customer better service but a wider and wider variety of services.

COST OF SERVICE HAS REMAINED REASONABLE

Now let us look at what has happened to the cost of service. The Bureau of Labor Statistics started in 1935 to price the cost of local residence telephone service as a part of the so-called market basket of goods and services included in the Consumer Price Index.

While the cost of all goods and services has gone up 95%, the cost of local telephone service rose less than one-third as much, or thirty per cent since that time.

I believe, therefore, that the technological changes which the period since 1920 has brought have affected the customer in these ways: He has received constantly improving service, he has received a wider variety of services, and the cost of service has been kept at reasonable levels.

RESULT HAS BEEN EXPANDED USAGE

The customers have reacted to this in a very natural but most significant way. Their usage of the service has increased tremendously. Progress and reasonable prices have stimulated demand. Bell System telephones in 1958 numbered more than 54,000,000—more than six times as many as in 1920.

More than 78% of American households in 1958 had telephones, as against less than half that percentage in 1920. The number of long-distance conversations has increased many fold since 1920.

The increase of telephones has been much more rapid than the increase in population. The growth in telephones has even exceeded the large increase in Gross National Product.

Technological progress has played an important part in this increase.

EFFECT ON EMPLOYEES

Let us now see how employees have fared.

More usage has created more jobs. Before going into the details I would like to explain what the Bell system does to meet the problems of individual employees affected by automation. For example, here is what we do in connection with the introduction of dial systems:

The Bell Companies have developed, over the years that dial conversions have taken place, a guide for carrying out conversions. At the heart of all plans is the awareness that the company has a social responsibility to eliminate or alleviate adverse effects on its personnel.

In establishing the date for a conversion, the controlling consideration is to make the date sufficiently far in advance, generally about three years, to provide ample time for human as well as technical planning.

The conversion having been scheduled, the first step is to inform the employees. Usually, most of the employees on the payroll at that time can be given assurance of continued employment; this is partly because the force progressively reduces itself by resignations in the normal course and partly because of the opportunities for reassignments. During the pre-conversion period, all losses from the force are replaced, so far as possible, with people hired for temporary employment. These temporary employees are chiefly people who desire work for only limited periods, such as young women expecting to be married soon or former employees who are willing to return to telephone work to help out for a short period of time. Regular employees who state their intention of resigning or taking an early service pension prior to the cutover date are urged to defer this action wherever possible until the dial conversion. These steps are all designed to reduce the number of potential displacements. Advance planning is also done to provide transfer opportunities for employees. These transfers may be to other types of work or to other offices in the same or other communities. Any retraining is done at the expense of the telephone company.

In most cases, as a result of these measures, few, if any, regular employees must be laid off. The employees principally affected are telephone operators. The number of such layoffs is relatively small. And usually those laid off have been offered transfers. To the individuals involved, however, the layoff can be a source of much difficulty. To meet such conditions, the Bell System companies for many years have had severance pay plans under which employees who are laid off receive lump sum payments varying in accordance with their length of service and wage rates.

An observation which I think will interest you was made

some years ago about the Bell System treatment of employees in connection with dial conversions. When she was Secretary of Labor, Frances Perkins wrote: "Of the hundreds of occupations in which women are listed in the Census of Occupations, only about a dozen employ more women than do the telephone companies. The human problem of the displaced worker when the cutover was made from the manual to the dial system telephone exchanges is an almost perfect example of technological change made with a minimum of disaster. It was accomplished through human as well as technical planning."[1]

With this background, I would like to take a broader look at the effect on the employee body of the conversions to dial switching.

The introduction of a dial system reduces the number of operators required for a given number of calls at a given location. It would be a mistake, however, to conclude that dial conversions have reduced the number of employees needed in the telephone business. To do so would be to ignore the important factor of growth in our business and to ignore the further fact that dial systems and other technological changes have been most important in producing growth.

Large numbers of telephone operators are still required with dial operation. Many types of calls cannot be handled automatically, such as information calls, calls needing assistance, some calls from coin telephones, person-to-person calls, and reverse charge calls.

Let us look at the facts. At the beginning of 1920, when there were no dial telephones, 115,000 persons were in the employ of the Bell System as telephone operators. At the end of September, 1958, the telephone operating forces numbered 200,000.

These figures are significant. However, the facts I will now give illustrate far better the over-all situation with respect to employment.

The number of persons employed by Bell System telephone companies has increased almost three times since 1920. The Bell System's proportion of the total civilian employment was about 75% higher in 1958 than in 1920.

Payroll of the Bell System telephone companies has increased more than 10 times—from $261 million in 1920 to $2,970 million

---

1. Francis Perkins, *People at Work* (New York: John Day, 1934), 209.

in 1957. Moreover, earnings of Bell System employees have kept pace with those of workers in industry generally.

I would sum up the effect of technological improvement on our employees this way. It has brought increasing customer usage which, in turn, has created more jobs. And they are better jobs—jobs that pay much higher wages and jobs that compare favorably with those in other industries. Moreover the "real wages" of telephone employees have increased appreciably and their purchasing power is at new high levels. This increased purchasing power benefits not only our people but is good for the economy as a whole.

EFFECTS OF DEVELOPMENTS ON CAPITAL INVESTMENT

I will now discuss briefly how investors in the telephone business have fared.

It has been through the incorporation of improvements into plant that the savings of the telephone investors have been profitably employed and the integrity of the money invested in the business has been maintained.

Scientific and technological developments have also created a continuing need for further investment in the business. In the 12 years following World War II the Bell System added capital at the rate of almost a billion dollars a year.

Share owners of American Telephone and Telegraph Company numbered more than 1,600,000 by 1958.

EFFECT OF TECHNOLOGICAL PROGRESS ON NATION AS A WHOLE

I have commented on the effect of progress in the telephone art on three groups: customers, employees and investors. I now wish to make some observations as to the benefits this progress has brought to the nation as a whole. These general benefits are in addition to the increased purchasing power of our employees already mentioned.

The fast, accurate and dependable telephone service now furnished in this country is vital to the common life of the nation and to the national defense. Further, developments in the telephone art have made important contributions to such projects as the DEW line, the air-warning system across the Arctic. Telephone techniques have also led to better military equipment such as the weapons systems used for the automatic aiming and firing of guns, the control of bombs, and the control of guided missiles such as the NIKE.

The nation as a whole has also benefited from developments in the telephone field which have been of appreciable importance to products of other industries. Examples are motion pictures, hearing aids, radio, television, and automatic devices of many kinds.

There are a number of new developments which hold considerable promise for improved service, for further expansion of the telephone business, and for keeping the cost of telephone service at reasonable levels.

Transistors are an important example of the newer developments. These devices generally perform the same functions as vacuum tubes, but with the major advantages of smaller size, greater durability and greatly reduced power consumption. It is expected that the Bell System will make substantial use of transistors in many ways.

Another new development is the hollow-tube wave guide. Unlike wires and coaxial cables, these tubes possess the unique property of diminishing transmission losses as frequencies rise, thus permitting use of much higher and a consequently wider range of frequencies. Our scientists tell us they believe that one day a single one of these wave guides will carry simultaneously tens of thousands of cross-country telephone conversations as well as hundreds of television channels.

The solar battery is another pioneering Bell System development. It is the first device efficiently to convert energy from the sun directly into electrical power. At this early stage its ultimate potential can only be conjectured, but it has obvious advantages as a power-supply when the requirement is very small and commercial power is unavailable.

Awaiting its first installation is an entirely new dial switching system which, unlike our present dial equipment, will make little or no use of electro-mechanical switches. Instead, transistors and other electronic gear will provide the nerve system and direct the switching of calls.

These are some of the new things under active study or development in the Bell System. I believe that these will illustrate the direction of our efforts toward improved service, wider scope of services offered, and improved efficiency. However, I again want to emphasize that none of the improvements now being made and none of the developments that we foresee is of a nature that will revolutionize the business or substantially change its present character. They will be a continuation of the evolutionary changes that have been going on over the years.

One final word as to future employment prospects.

In the future, as in the past, the nature of some telephone jobs will change. But, as in the past, we believe that scientific and technological improvements will come with ample time for adjustment and retraining of workers. Force adjustments will be handled so as to produce a minimum of hardship, applying procedures which have been used successfully in the Bell System over many years.

I think the facts of the past provide the best clue to the future. Technological changes in our business will continue. They will make telephone service better, will widen its scope and will keep the price reasonable. Coupled with aggressive selling, which is a part of our program, such developments should further increase the usage of the service, and it is increase in usage that creates jobs. There will probably be some scattered adverse effects on personnel and these we will try to minimize. In the future as in the past there will probably also be fluctuations in employment. However, the long-term trend in our business has been one of increasing markets, increasing job opportunities and substantial capital investments. We see nothing in the present picture which should change this trend.

CLIFTON W. PHALEN, born in Washington, D. C., and a graduate of Yale University, has spent his entire business career in the Bell System. He started in 1928 in the Plant Department of the New York Telephone Company in Syracuse, N. Y. In 1948 he was elected Vice President of the American Telephone and Telegraph Company serving first in charge of Public Relations, then in charge of Rates and Revenues, and later in charge of Personnel Relations. He was elected President of the Michigan Bell Telephone Company September 1, 1952 and an Executive Vice President of American Telephone and Telegraph Company effective March 1, 1956.

Mr. Phalen is Director and Member of the Executive Committee of the American Telephone and Telegraph Company; Director The Chesapeake and Potomac Telephone Company, Southern Bell Telephone and Telegraph Company, The Pacific Telephone and Telegraph Company, Long Lines Department of A. T. & T. Co., National Safety Council and The Greater New York Fund.

CHAPTER 15

THE NATURE OF AUTOMATION IN THE
TELEPHONE INDUSTRY

JOSEPH A. BEIRNE, *President,*
Communications Workers of America

Mechanization and telephone workers have worked side by side for the past twenty years. Changing from manually operated to dialing local calls has been a long and familiar procedure. Recent years, however, have witnessed the introduction of even more penetrating mechanization and the advent of what we now call automation. You are all no doubt well acquainted with so-called local dial conversion. It has been in use for many years. Perhaps you are less familiar with some of the more complicated mechanical devices developed by the Bell System. For that reason, this paper will describe some of the Bell System's automation devices. Briefly, they can be summarized as follows:

1. *Operator toll dialing and customer toll dialing.* At present approximately 50 per cent of all long distance calls are dialed by an operator directly to a subscriber. This is known as operator toll dial. Ultimate plans are for the subscriber to dial directly the long distance number he is trying to reach without any intervention or assistance by an operator. This is known as direct subscriber long distance dialing. It exists to a limited degree and is increasing steadily. This means that anyone desiring to place a long distance call will be able to do so without the use of an operator except for person-to-person calls, where the use of a single operator will still be required. It is our understanding that ultimately even person-to-person calls may be completely automated.

2. *Automatic message accounting.* An inseparable part of cus-

tomer toll dialing is mechanical equipment known as automatic message accounting machines. When you dial a toll call, this equipment records your telephone number, the number you are calling, how long you talked, etc., on a punched tape. A busy signal or no answer is likewise recorded. These perforated tapes are then decoded by additional machines which finally assemble, translate, sort, and summarize all billing information. As we understand it, in time, you will place a telephone call and be billed for it without personal intervention by any telephone worker. Telephone calls will soon be "100 per cent pure," "untouched by human hands"—other than your own.

3. *"M-4."* Bell Laboratories recently announced the introduction of a new machine designated starkly as "M-4" which puts together complex wiring circuits mechanically under the direction of punched tape, which an electronic brain reads and translates.

The picture would not be complete without reference to new electronic test sets which automatically test cables during periods of wet weather; T-type terminals that enable one man splicing crews; Murphy test sets that enable cable splicers to designate cable pairs without the use of a helper; microwave stations that eliminate countless miles of poles, cable, wire, crossarms and maintenance on these facilities; new types of central office equipment that virtually eliminate maintenance, automatic power plants wherein auxiliary power supplies are cut in and out without aid of manual assistance; telephones on which customers can adjust the volume of bell-ringing; the Ryan plow which lays cable mile after mile in trenches without assistance.

I have never seen this Ryan plow, but I understand it is something to observe in operation. It has a long metal finger extending down some 40 inches into the earth. It is pulled by two or three caterpillar tractors. It digs the trench, lays the telephone cable, and covers it in a single operation. The plow can lift out of the way any rock less than 40 inches in diameter. On a recent cable job in Missouri we understand a single Ryan plow laid 60 miles of cable in four days with a crew of less than seven men. Can you envision how many workers this job formerly required when trenches were dug separately, cable laid and trenches then filled in?

We have no desire to be alarmists. Quite the contrary; we

point with pride and gratitude to miracles we have witnessed during our lifetime. The Bell System, since the end of World War II, has undergone possibly the largest expansion of any corporation in the history of this country.

The Bell System is composed, among other things, of twenty-one operating telephone companies, the Long Lines Department of the American Telephone and Telegraph Company, the Western Electric Company and Bell Laboratories. This System's primary function is to furnish local and long-distance telephone service. When we talk about the Bell System in the United States we are, in effect, talking about practically the entire telephone industry. It is estimated that the Bell System provides approximately 90 per cent of telephone service in the United States as measured by annual gross revenues. The remaining 10 per cent is shared by some five thousand independent telephone companies.

It is our considered judgment that employment levels and job opportunities in the Bell System are more significant than those in any other single company in the United States. The Bell System with its 578,436 operating employees (as of the end of 1954) is the largest single private employer of labor in the United States. In addition, the System employs another 100,000 people in its Western Electric and Bell Laboratories subsidiaries. The employment of these 100,000 people bears directly on employment levels in the operating portion of the System since they perform the research, manufacturing, distribution and installation of telephone equipment. Except for the United States Government, no employer controls the destinies of as many workers as does this single corporation known as the Bell System. Approximately one out of every sixty-two nonagricultural and nongovernmental civilian workers in the United States is a Bell System worker.[1] Moreover, these workers are found in almost every community throughout the Nation. It must of necessity follow, because of the size and the geographical scope of this large corporation, that employment levels in this single Company will have some effect upon employment levels in the communities in which this Company operates and ultimately upon the Nation as a whole. Employment levels in the Bell System are also sensitive barometers of this country's general economic activity since

---

1. As of December 1954 there were 685,944 Bell System employees and 42,269,000 nonagricultural, nongovernmental civilian workers.

telephone business and employment levels reflect over-all business fluctuations.

The number of telephones alone in the United States has more than doubled in the past decade. We acknowledge the fact that despite intensive mechanization of local telephone calls there are over 150 per cent more people employed in the telephone industry today than there were twenty-five years ago.

There are, however, some very important lessons which we can learn from history and which should serve as the basis for planning the future. During the six-year period between 1929 and 1935 a relatively small decrease in telephone service resulted in a 33 per cent decrease in employment. In the single year 1953-54 substantial increases in telephone service had the net effect of a decrease in employment. In other words, it appears that telephone employment can be maintained during increasing mechanization periods only by tremendous increases in telephone business. We think it unlikely that the kind of increases which have taken place in the last decade can continue into the future. For this reason we think we are faced in the telephone industry with constantly decreasing job opportunities.

CWA has many suggestions regarding what can be done to minimize undesirable effects of decreasing employment levels in this key industry. It has been our program for many years to seek a reduced workweek in the telephone industry not only because of the tedious and tiring nature of some of the jobs but also to spread employment opportunities at times when it was desirable to maintain rather than decrease employment levels. The necessity, for example, for operators to have a shorter workweek has been intensified as equipment they handle requires more numerous simultaneous operations resulting in increased nervous tensions. Telephone operating has always been a job requiring constant attention, unrelieved concentration and the ability to do several diverse operations at the same time. Automation has intensified and accelerated these pressures. It is our hope that as job opportunities decrease the Bell System will become more receptive to CWA's shortened workweek demands.

On an over-all basis it would also contribute toward a more orderly method of meeting day-to-day automation job displacement problems if there were uniform, or more nearly uniform, wage and working condition practices throughout the Bell System. As the situation exists today, a worker may find his job nonexistent in an area where CWA has been successful in negotiating good wages and working conditions. The only job available to him on a transfer basis may be one covered by a trade union contract or no contract at all, and where the resulting wages and working conditions are not as good.

CHANGING NATURE OF JOBS IN THE INDUSTRY

Mechanization has also had a direct and significant effect upon the over-all structure in the telephone industry. For example in 1945, immediately prior to postwar mechanization and expansion, traffic operating employees—that is workers handling calls—comprised 52 per cent of the total labor force in Class A telephone carriers. At that time the industry was approximately 65 per cent local dial and 5 per cent toll dial. In 1953 with local dial around 80 per cent and toll dial 44 per cent, and with the number of telephones and telephone calls almost doubled, the traffic operating force shrank to 41 per cent of the total force. During that time professional, sales, clerical, and maintenance people increased as proportions of the total labor force.

It is extremely significant, moreover, that between 1945 and 1953 the number of operating employees increased only 24.5 per cent compared with increases of 97.2 per cent among professional workers, almost 100 per cent among business and sales forces, 89.3 per cent among clerical workers, and over 109 per cent among maintenance workers. This is not peculiar, we think, to the telephone industry alone, but seems to be characteristic of shifts in the labor force in industry generally as automation advances.

TELEPHONE OPERATORS

Perhaps the hardest-hit single job has been and will continue to be that of telephone operator. In 1921, with approximately 3 per cent of Bell telephones dial, there were 118,470 people in that single job. That year the Bell System handled approximately 1.3 billion calls per month, or around 10,640 calls per operator per month. In 1954, the latest year for which data are available, there were approximately 175,200 telephone operators in Class A telephone companies in the United States.[2] These

---

2. Class A telephone companies are those having annual revenues exceeding $250,000.

companies handled approximately 6.3 billion calls per month, or 35,800 calls per operator per month.

The peak number of operators, 182,500, was reached in 1949 and has been declining almost steadily since. Had there been no change in dial status since 1921, and assuming other factors remained constant, the industry would require approximately 589,749 operators to handle current telephone call volume, or over three times the number of operators actually employed.[3] Perhaps the hardest stories to tell about technological changes are ones centering around individuals rather than statistical totals. We could tell you, for example, that during the past ten years in the single city of Milwaukee the number of operators was reduced from 3,500 to 1,000 in what, to outward appearances, was a relatively orderly manner with very few actual layoffs required and practically no transfers out of the city. But this is not the entire story. Among the thousand operators who remained on the job you will find some of the deepest problems inherent in technological change. A telephone operator who has been doing essentially the same basic operation for thirty years suddenly finds herself confronted at age 50 with the necessity to learn entirely different and more complicated work procedures. From an experienced, confident, efficient local operator, she suddenly becomes an inexperienced toll operator. Within this framework lies the real human drama of automation. With proper retraining this mature worker could become as valuable as she was formerly, and possibly more so. Her knowledge of the industry, company practices and traditions, her loyalty, her willingness are invaluable assets. But her retraining requires more than just a mechanical, routine approach.

We know of cases where some workers have gotten sick on the steps of the new toll center; others developed various illnesses which could be traced to fear of new work operations. We have been told of mature women crying in restrooms, improperly prepared for new methods and fearful of losing their jobs or being pressured into unwanted, early retirement with inadequate pensions. The tragedy of the mature worker whose skill area suddenly disintegrates and is incorrectly retrained is profound.

We don't want to leave the impression that telephone man-

---

3. Estimate obtained by dividing 1954 average monthly telephone calls by number of telephone calls handled per month per operator in 1921.

agement has been unaware of some of these problems. Reports to us indicate that some of them have been understood and genuine attempts have been made to ease the transition. However, in our opinion, not enough has been done. Certainly the same zeal and perfection which the Bell System uses in its programs relating to machinery has not been equaled in its human relations programs. A retraining program designed for workers in their middle years who have been accustomed and, in fact, carefully instructed to do only a particular operation for as long as thirty years, must make allowances for certain psychological and emotional problems. We think the world's richest corporation, the Nation's largest single private employer, can do a better job than it has been doing to date. The Union can be of assistance to the Company in this regard and, in fact, has had to fill the vacuum in many localities when the Company's retraining program was mechanically sound but humanly inadequate.

We could cite you dramatic example after another of towns with 100 operators working one day, and just 30 or fewer left the next. Transferring workers from one town to another may seem at first glance a reasonable solution to a difficult problem. But let us keep in mind that such transfers may mean the uprooting of family ties and, in most cases, the very town to which the telephone operator is transferring will be cut to dial in another year, necessitating still another move.

Making transfers available to displaced workers is, of course, frequently no solution at all to the problem. This is particularly true of women workers. Many women workers play a dual role, that of worker and household manager. Frequently they have still a third responsibility, that of mother. This means they are not mobile workers, because their future is tied to their family and their husbands' jobs. These additional duties generally make this type of worker responsible and dependable and we think she deserves something better than the false offer of a transfer or the unemployment market.

The young, single woman worker poses still another problem in job displacement situations. She may be tempted to accept the Company's transfer offer to nearby or distant towns away from the protection of her home and parents. This, of course, has deep social implications.

The statistics and general statements offered should not obscure the much more important problem of human beings finding their entire lives shaken up by so-called "progress." If telephone business continues to expand only at the modest 1954 rate, that is annual increases of 4.6 per cent in telephones, and 3.8 per cent in telephone calls, we estimate conservatively that by 1965 there will be anywhere from 100,000 to 115,000 fewer people employed by the Bell System. By that date A. T. & T. boasts the System will be 90 per cent local dial. Our estimate is conservative because it is based on a constant rate of job displacement, that is 1.6 per cent per year, the drop between 1953 and 1954. A more realistic figure should assume an accelerated rate. Our experience indicates that job loss seems to snowball rather than proceed at a fixed rate.

We would not be surprised if there were 200,000 fewer Bell System employees by 1965 unless there are tremendous business increase compensating factors.

The Communications Workers of America has always realized the importance of watching the job opportunity situation in companies in which workers represented by CWA are employed. Union committees, working with elected Union officers and specialists, have from time to time reported on employment levels in the telephone industry and these reports have served as the basis for CWA policies. We in the telephone industry have lived with "mechanization" and its successor "automation" for many years. They have been accompanied by fear and job insecurity. They have been our constant companions and loom larger today than ever before.

We should like to state quite clearly and emphatically that we have never resisted mechanization for the sole purpose of maintaining jobs for workers we represent. Our basic attitude has been twofold:

First, that any employer, particularly one as large and as far-flung as the Bell System, has a responsibility to introduce conversion to machinery in such a way and at times that such changes will have the least possible detrimental effects upon the economy of the industry and the country. Second, it has been our position that benefits derived from mechanization should be shared by owners of the industry, employees in the industry, and the public at large.

We feel these two basic attitudes have special merit in a government-regulated, guaranteed "fair profit," monopoly industry.

LACK OF RELIABLE DATA

Consistent with our first basic position we have attempted, from time to time, to discuss with Bell System officials their program of replacing workers by machines. It was our desire to use such information to answer the many fears expressed by CWA members and to assist us in developing intelligent, realistic collective bargaining programs. We wish we could say to you that our fears have been allayed. We wish we could say to you that this giant Bell System is as up-to-date in its labor-management relations as it claims to be in the mechanical equipment field. Unfortunately we cannot do so. All overtures made by the Union to date to discuss, in a mature fashion, the nature and timing of anticipated equipment changes and their possible effects upon employment in the telephone industry have been categorically rebuffed. The Bell System is willing, in fact eager, to discuss with bankers, life insurance groups, and almost anyone else who will listen and not ask questions, their program for so-called progress. The line is drawn, however, squarely in front of CWA.

Certainly, before intelligent decisions can be made in a field as complicated as technological changes, a groundwork of fact and cooperation is necessary. What does it really matter if it takes 30 seconds instead of only 15 to complete most long distance calls if we gain this speed at the price of unemployment and its accompanying domestic and international human misery? Why the headlong rush into mechanization if slower movement gives us time to contemplate what we're doing and where we're headed? Certainly we ought to be talking about the problem.

We are not pessimistic about the future. This country, and indeed this planet, will meet its problems with ever-increasing abundance for its people. We think most elements in our society understand this and will show their appreciation to the past and their responsibility for the future by making human values paramount even in the age of automation.

JOSEPH A. BEIRNE, Washington, president of the Communications Workers of America, has been a leader in the telephone labor movement since 1937. He has been president of CWA since it was formed in June 1947, and was president of its predecessor organization, the National Federation of Telephone Workers, from 1943 on. Beirne started work at the age of 16, continuing his education nights. He graduated from St. John's Parochial Grammar School in Jersey City, then attended Dickinson High School, evenings, from which he graduated.

He attended Hudson College of St. Peter (nights) in Jersey City for three years and then did evening work at New York University. When he left school, Beirne had over three years' college credit.

His first job was as an office boy for the F. W. Woolworth Company. At the age of 18 he went to work for the Western Electric Company at its Kearny, New Jersey, manufacturing plant—as a utility boy, earning 32 cents an hour. Later, he transferred to Western Electric's distributing department, working as an instrument repairman in New York and Brooklyn.

He has been in full-time union work since 1939, and has been one of the prime movers in the evolution of the telephone labor movement. In 1937, he led in formation of the National Association of Telephone Equipment Workers, operating in the distributing department of the Western Electric Company. Beirne was its first president, a job he held until he took over leadership of the national union in 1943. Beirne is a vice-president of AFL-CIO and of its Industrial Union Department. He is chairman of AFL-CIO's community services committee and is a member of the executive board of the political action arm of the federation. He became one of the three CIO members on the Wage Stabilization Board when the board was reconstituted May 8, 1951 until 1952. He is a vice-president of the United Community Funds and Councils of America, also a member of the executive committee. He is a member of the national board of Americans for Democratic Action and of the National Religion and Labor Foundation. He has been named by Labor Secretary James P. Mitchell to membership on the nine-member Labor Advisory Committee for the Department of Labor. Beirne is also a Director of the American Arbitration Association.

Section III

Automation and Responsibility

CHAPTER 16

WHAT'S NEW ABOUT AUTOMATION

EDWIN G. NOURSE, *former Chairman, Council of Economic Advisers, Executive office of the President of the United States*

The development of technology by mankind has moved in stages of accelerating change, with later steps so different in magnitude as to amount to difference in kind. Man was, in his own body, endowed with simple mechanical equipment: teeth and fingers for cutting or tearing, fists for pounding, legs, arms and back muscles for lifting vertically or moving horizontally. Over many ages he developed dexterity in using this bodily equipment and in magnifying it by the fashioning of simple and sometimes not so simple hand tools—knife, saw, hammer and axe.

It was another long, slow step from hand tools to power machine when the ox, the horse, and the water wheel were harnessed to give us simple power machines. It is only 200 years since we leaped forward into the age of steam, and then into electric power late in the 19th century. The internal combustion engine is essentially the gift of the last 50 years, and synthetic chemistry and electronics the product of yesterday. Experts tell us that the passing from the electric age to the age of electronics was a greater cultural leap than that from the age of steam to the electric age.

Automation is a child of the vacuum tube and the transistor, and as a new development in our industrial life, must be considered on two levels: (1) the technological facts, and (2) the economic (and hence social) implications.

In its technological aspects, automation consists in a large and growing array of mechanical and electronic devices that give producers new machines for doing new things or old things better and equipping old machines with new gadgets. The essential feature of this new step in mechanization is the application of electronics to the control of mechanical and chemical processes. The "mechanical brain," or electronic computer and tester, is the central feature of this development, and the practical result is the substitution of mechanical for manual controls at many points in the physical process of production. As has repeatedly been pointed out, these controls also make it possible to do things we could not do at all by manual methods—notably the production of atomic power.

Many "practical" industrialists have stated that automation is just a little more of the mechanization that we have had for some two centuries. And, they add, we always have assimilated such change without any trouble and will do so in this instance. This, although it has an element of truth, is a considerable oversimplification.

To understand the nature of this progressive evolution and what automation has added that is new, we need to recognize four aspects of the productive process—power, dexterity, judgment, and foresight. The initial element, power, may be that of the human body, the draft animal, steam, or what-not up to atomic energy. How this power is applied depends on the manual dexterity of the worker. Such was the case with the handicraft stage of industry which relied heavily on trainings of workers for skill in a trade. Today we talk more of the "technician"—the super-mechanic and the laboratory assistant. He or she often needs higher mathematics or even a graduate degree.

Overlapping this area of manual dexterity is the quality of judgment. The bench worker at a trade must exercise judgment, good or less good, as to how he will cut material, when metal is at right heat, and when a tool must be sharpened. The laboratory technician must judge the proper time of x-ray exposure, the best way of staining tissue, or how to put data on a punch card. Finally, in this exercise of judgment, the time element must be taken into account. The shop worker and laboratory technician, and still more, the engineer and the manager, must think ahead as to what order will follow in processes, what time flow will make materials ready at the right time and place, perhaps what side-effects or consequences remote in time may flow from a current operation. Planfulness and organization are the formal ingredients in what we properly refer to as the industrial "process."

With this background, we can see the meaning of automation in larger terms than old fashioned "mechanization." The power machine lifted the burden of physical toil from the back of the tool wielder. The automated battery of machines lifts the burden of fallible human judgment and unremitting attention from routine workers. At the same time it enlarges the demand for scientists and engineers (including social engineers) .

The computer or "mechanical brain" is the dramatic symbol of automation. By its use we can organize the production of a complicated finished product as a continuous process, with the packaged article "untouched by human hands" from raw material to delivery truck. Of course "continuous process" began back in the days of simple mechanization. Probably the best example was the flour mill, where wheat was poured into hoppers at the top of the mill and moved by gravity till it emerged as sacked flour on the shipping floor. But this process was not continuous in the refined modern sense. It required human intervention every time anything went wrong or whenever a change in the operation had to be made. It was a "dumb brute" process, not a process in which judgment or pre-thought-out control was built in. This further step was achieved only when electronic control mechanisms introduced the principle of "feed-back."

Under this principle, electronic mechanisms make it possible to conduct more elaborate, more economical and more precise continuous productive operations because the outcome of the process controls the process itself, starting, altering, or stopping it so as to make it produce a desired result. A complicated machine tool can be pre-set—that is "told" by means of punchcard or recorded tape—to continue running in a certain way until the article produced (and being continuously tested or inspected by the machine itself) shows that the cutting element has become too dull to turn out perfect work. Then the machine, or battery of machines, stops itself, inserts a fresh, sharp cutting tool, and starts up again. Such a machine can also follow "instructions" to produce a given size or design of product until a certain number has been turned out and then switch itself over to another pattern of operation.

ECONOMIC FEED-BACK

Besides such purely mechanical aspects of the feed-back principle, it has economic applications that are even more impressive. An electronically equipped warehouse or toolroom keeps a continuous record of all articles that move in or out and flashes a warning signal whenever stocks become either too large or too small, the rate of movement too fast or too slow for best results. This makes possible "scientific inventory control" instead of periodic tallies and more-or-less informed guesses. This sort of flow-measurement can be extended to record and classify data about consumer incomes, intentions to spend, taste preferences, and the like, so that production will be adjusted to future magnitudes and changes in demand, long before the fact of the completed sale. Then our economy would have the law of supply and demand "wired for sound"—the authoritative voice of the ultimate consumer.

These wider ramifications of the principle of feed-back in its economic applications move us on to a fourth level of automation, namely "rationalization." It has been defined as follows: Rationalization of a productive system means that the entire process from the raw material to the finished product is carefully analyzed so that every operation can be designed to contribute in the most efficient way to the achievement of clearly enunciated goals of the enterprise. The scientific, rationalist philosophy takes on numerous new implications when it can be implemented by modern electronic machinery. In this aspect also automation is not something altogether new. It is essentially a continuation of the "scientific management" movement which took shape about 50 years ago. Indeed that movement toward cost-reduction or efficiency increase was even then often called "rationalization," particularly in European comment. It gave rise to much agitation about "technological unemployment" and to the cult of Technocracy in the period between World War I and World War II. It led also to the movement toward "economic and social planning." Whereas "scientific management" was directed toward rationalization of the operations of the individual plant or company, national economic planning had as its goal the rationalization of the operation of the whole economy. The point I want to make here is that this new development of electronic computers and controls and this enlarged concept of continuous process makes contributions to the objectives set out in the U. S. Employment Act of 1946.

This is simply the concept of "continuous process" applied to the economy. A very simple model of a stable free enterprise economy would show companies and individual producers buying their materials, their capital, labor, and other productive resources in free markets, turning them into finished goods which are distributed to consumers and other users at prices which fall within their purchasing power and which yield returns to the producers which enable them to maintain, improve, and expand their facilities in step with growing population and advancing techniques and provide the owners with residual profits adequate to compensate risk and provide incentive for continuing the process at full tide.

This economic process is carried on through an elaborate system of price and income relations largely arrived at through market bargains, but with participation of government at a few places where we the people have decided that such participation is needed or beneficial. These price-and-income relations have been working out so well for the last decade that we have had high and rising prosperity. The issue which automation now raises is this: Will it alter present economic relations in such ways as to disturb these favorable conditions, or will our business system be able to translate these technological improvements fully and promptly into still greater general prosperity and higher standards of living? It is evident that it will change wage income both by number of jobs, some places up and some places down, and by wage rates upgraded here and downgraded there. It will obsolete some capital equipment and make important demands for new capital equipment. It will affect unit costs for some products, but not all; prices in some markets, not in others; profits and dividends, tax yields, and public spending.

As a long-run generalization we are justified, I believe, in accepting the oft-repeated dictum of businessmen that "each successive step in mechanization has created more jobs than it destroyed, and has been followed by the necessary price and wage adjustments—and will do so in the present instance." But on so sweeping a proposition we need to study the fine print rather than just taking the bold-faced captions. While the historical and statistical record does show a consistent uptrend in general production and average real income or well-being along with technological progress, there have been severe short-run dislocations, local catastrophes, and painful readjustments for both employers and employees.

In contrast to the preponderant attitude of business executives, labor union officials have been outspokenly concerned about the economic impact of automation on the well-being of the mass of worker-consumers in the years immediately ahead. Some among them, to be sure, simply follow blind instinct and fight any labor-saving change. But top labor leadership began early to talk in terms of "the challenge of automation." This challenge they state somewhat as follows: "The guiding purpose of labor organization is to raise the level of well-being for the working masses. This demands maximum production and economic expansion. Automation as the current phase of technological and managerial development is designed to promote this end, and organized labor welcomes it and seeks to cooperate in its growth. But we believe that much study is needed by all parties if the gains are to be made as large and as steady as possible and the temporary dislocations and local burdens or losses made as small as possible and most equitably shared." With this view I find myself in accord rather than with the idea that the problem will take care of itself or be disposed of automatically by the invisible hand of free enterpise.

TECHNOLOGY IS INDIVISIBLE

Here I want to make three points: (1) that automation is only one inseparable part of the larger problem of technological progress; (2) that the application of our advancing technology to the ever-better satisfaction of human wants goes forward through a continuous flow of money relations essentially like the "continuous process" of physical production in an automated factory; (3) that the economic problems posed by this technological advance can be solved only by a combination of competitive pressure, business statesmanship, and constructive public policy.

As to the first point, there seems to be a considerable tendency to blame (or credit) automation for impacts on the economy that stem primarily from other factors in our advancing technology. This is misleading, but at the same time it is quite impossible to separate these several sources of increased productivity from one another and to measure and deal with each of them separately. We are not in fact confronted by a specific economic problem of automation but with a broad total problem of trying to capture the values of higher productivity put within our reach by scientific progress and avoiding either nonuse or misuse of these potentialities. We are all aware that the technology that gives us our present high level of productivity includes the internal combustion engine and jet propulsion; the newer metallurgy, including light metals and a great range of alloys; synthetic chemistry, with its versatile development of plastics; radiology and atomic fission. Automation is only a way of harnessing this varied and highly productive technology. It happens at the moment to have stolen the spotlight, but we shall have to deal with the problem of assimilating the total technology into our economic institutions and practices. We have not done this step by step as the successive technological changes took place, and hence the accumulated problems have become a cause of concern and possible friction. Fortunately, automation itself furnishes us some tools for tackling the first problem—both an intellectual tool in the concept of "continuous process" or integration, and a mechanical tool in the device of the electronic computer.

INTEGRATION OF THE ECONOMIC PROCESS

When businessmen or others say that technological progress is good *per se* and that it takes care of its own economic process, they invoke a simple logic of the free enterprise economy. The entrepreneur seeks profit by adopting a device for raising efficiency. This lowers cost. Price falls proportionately and thus broadens the market. This restores the number of jobs or even increases them and raises the level of living or real incomes. This comfortable formula presupposes a state of complete and perfect competition in a quite simple economic environment with great mobility of labor, both geographical and occupational. But these are not the conditions of today's industrial society, with large corporations and administered prices; with large unions and complicated term contracts covering wages, working conditions, and "security"; with complex tax structures, credit systems, and extensive government employment and procurement. The smooth and beneficent assimilation of sharp and rapid technological change has to be effectuated through intelligent and even generous policies painstakingly arrived at by administrative agencies, private and public.

It is of the essence of the automation concept and program that it integrates the several parts of a complex total process so that its successive or related steps shall mesh smoothly together without conflict, lost motion, or lost time. As an economist, I am moved to stress the fact that the operational flow of our modern industrial system is not merely a matter of the physical movement of material objects from their native state through the processes of extraction, fabrication and delivery as finished—often highly complex—products. It involves also the flow of price and income relationships that furnish purchasing power to consumers—individuals, business concerns, and government procurement agencies—as well as capital formation out of profits and savings, and finally incentives to enterprisers and to workers of all grades to prepare themselves for and apply themselves to the kinds of activity that the character of our technology makes possible and requires. How does automation affect these relationships? Are we aware of what are the components of this operational flow and are we integrating them so intelligently as to attain maximum production?

Space does not permit a systematic anaylsis of these large questions, but I shall undertake to hit a few high spots which will illustrate some of the practical problems in the integration of wage structure, profit rates, price policy, and investment practice. I shall do this in terms of short-run "foreseeable trends." My major premise is that we are now in a critical situation which might, hopefully, be described as pregnant or, apprehensively, as explosive. Since World War II technology (with infant but growing automation) has been put to full use under conditions of extraordinarly high and sustained demand, public and private. Labor, viewing this unparalleled rise in productivity, has sought to capture the largest possible share in the form of successive rounds of widespread wage increases in basic rates, escalation formulas, and fringe benefits. As the unit cost of labor went up, management sought to maintain or improve its earning position by raising prices and/or by introducing labor-saving machines and administration. The first solution of management's problem—that is, price raising—has been facilitated by our elastic monetary system, and we drifted along on a Sybaritic course of mild inflation as a way of life. The second solution of management's problem of meeting labor's wage demands has accelerated piecemeal mechanization, yesterday's infant "scientific management," today's adolescent automation.

Adolescence is always a stage of storm and stress, and this is no less true of this problem child of industry than it is of the teenage kid. The badly adjusted youth has part of his psyche ahead of other parts, and I fear that this is also true of automated industry. I strongly suspect that we have already built up at many spots a productive capacity in excess of the absorptive capacity of the forthcoming market under city and country income patterns that have been provided, and employment patterns that will result from this automated operation. We are told on impressive authority that we have not been making adequate capital provision for re-equipping industry in step with the progress of technology. This is probably true if it means making full application of electronic devices and Univac controls generally throughout our industrial plant. But we have not yet demonstrated our ability to adjust the actual market of today to the productivity of the production lines we have already "modernized." They have not yet come to full production, but as they do we see an alarming amount of unemployment appearing. Suggestions have been made that balance could be restored by lowering prices or by cutting the work week. Both processes take time and present their own difficulties. Meanwhile, the current trend is toward higher prices reflecting wage advances already negotiated.

Let us get down to cases. In 1955 several very complicated and quite novel wage contracts were concluded in two basic industries—automobiles and steel. In effect the United Automobile Workers and the United Steel workers put a punched card into the control mechanism of our national economic process that instructed it to divert income over the next few years to specific groups of workers at specified rates. Almost simultaneously the steel executives instructed the continuous-process money mechanism to channel funds to them at a higher rate than formerly. The automobile executives waited several months before they raised prices. Both of these steps were attempts to plan or rationalize the economic process to make it work better for particular groups. The unions argued that more purchasing power must be diverted to consumers if a growing product is to be sold, and that this income could be drawn off from company treasuries without harm to the industry. The companies argued that they could not pay higher wages, expand plant, and improve technology unless they could get more for their product and that their customers were able to pay these higher prices while still maintaining or enlarging the volume of sales.

I am not suggesting any judgment as to the correctness of the setting of these particular controls on the economic process, but am simply calling attention to the well-known fact that once these persons in authority in administrative areas private or public pre-set any one of these control devices on our money flow system, they—and many other groups in the economy—have to abide by the inevitable consequences of these interventions in the continuous process or else revise the setting and give more workable instructions to the mechanism.

A second illustration could be drawn from agriculture. We installed some gadgets to stimulate production under war and early postwar conditions of domestic and foreign demand. As technology improved, mechanization was accomplished, and demand fell off, we have been unable to change the settings on the control device to bring this part of the process in step with other parts. We go on with surplus plant, surplus workers, and surplus product.

A third case of very basic character in the adjustment of money flows to technological conditions and changes may be taken from the field of investment needs and tax policy. It is quite possible to invest so much in automatic installations as to entail loss rather than gain. Labor has pointed out that automation in excess of market demand may entail idleness for some workers and waste of labor resources by the nation. Shifting taxes onto or off of corporations or income classes is a means of influencing these interconnected economic processes. The point often hidden by the smoke of battle is that the optimum tempo of technological advance, investment, and productivity is not self-determined but should be relative to the price-income structure and the spending and saving patterns of the mass of the population.

We need much more research and carefully checked experimentation if we are to discover which revisions of current business practice will do most to incorporate technological improvements into economic gains. The continuous-process concept should guide such studies and the use of electronic computers should facilitate the processing of the vast array of data needed for making adequate interpretation of results under different proposals. It is arguable that a closer approach to optimum rate of adoption not merely of automatic devices but of all mechanization and other technological advances would be attained if our institutions required companies to consider and provide not only for the cost to them of new equipment or processes but also for the cost to the worker resulting from such changes. Many devices other than time guarantees for wages (for instance, severance pay or reimbursement for time lost or skills scrapped) need to be explored. The real question is how the incidental costs of social progress are to be distributed between private and public agencies.

Every time the Congress passes a money bill, every time it revises our tax structure, every time it passes a regulatory measure for price maintenance (alias "fair trade"), farm price supports (alias "parity,"), or stockpiling of copper, rubber, wool, or silver it is giving punch-card or tape instructions to some part of the continuous flow mechanism of our economy. Public policy on all these matters should be framed in the light of the fullest possible understanding of the integrated character of the price-income structure and behavior of our economy, with an eye single to promoting "maximum production, employment, and purchasing power" for the whole people, not to serve the immediate interest of any special group.

This sort of scientific and engineering rationalization of our national affairs calls for a simply stupendous amount of grass-roots data as to what is actually happening at an infinite number of spots in the economic process. This mass of data is too voluminous to be seen, classified, and evaluated by statisticians, economists, and statesmen and processed into generalizations which can guide legislators and executives, public and private, in discharging their necessary function of programming the economic process and of pre-setting the control mechanism that determine the value flows throughout the economy—and thus lead to full and efficient use of our resources or to delays, wastes, or breakdowns in the mechanism.

Fortunately, the development of the electronic computer or "mechanical brain" makes it possible to process these vast bodies of relevant data economically and accurately, thus giving an adequate and reliable base on which human judgment can be exercised as to the course which economic policy and action should follow. Business concerns can use and are using such computers to analyze market capacities and responses, to calculate optimum investment policy (such as the practicability of automating their operations) or wage policy (particularly as to pension programs or the possibilities and costs of making annual or less-than-full-year employment guarantees). The "input-output" project at Harvard University is designed to give us an empirical and analytical overview of the integrated operation of the whole economy. This sort of data processing is supplemented by the newer "expectations" and "projections" techniques superseding the methods of hunch, prejudice, and interest-group pressure which stand in the way of rationalization of our economic affairs. These methods of objective analysis of economic problems seem destined to have increasingly wide practical application in administered price-making, in the negotiation of national wage-bargains and security plans in basic industries, and in the handling of our money-and-credit system.

Of course, no amount of data and no improvement in its processing will yield *final* answers on industrial, financial, and commerical policy. They give the tools for value judgments rather than demonstrable solutions to social problems. To get maximum well-being from our rapidly advancing technology we need truly competitive institutions backed by mutual forebearance by the parties at interest. We have to depend on the mores of business above and beyond the structures set up by law. Each segment of an automated continuous process has a built-in "responsibility" or interrelation to every other part through its mechanical or electrical interconnections. Something comparable to this kind of control is to be seen in an authoritarian economy. But in a free enterprise system, human judgment is given play at most of the important points of interrelationship. Unless the responsible executives seek to integrate their operations to the prosperity of the whole economy and use the full apparatus available for gathering and processing the data relevant to policy determination our economic process will disintegrate into wasteful struggles for individual or group short-run advantage. Much of the potential benefit of technological process (of which automation is one particular expression) may be lost through failure to make our economic structure and practices equally scientific.

EDWIN G. NOURSE was born in New York State, but grew up in northern Illinois. He received his A.B. degree from Cornell University, and his Ph.D. from the University of Chicago. After fifteen years as a college teacher of Economics at the Wharton School of the University of Pennsylvania and at state institutions in South Dakota, Arkansas, and Iowa, he turned to research work with the Brookings Institution in Washington. He was active in agricultural, labor, and industrial studies, particularly in the field of price policies and income relations. He served for some years as Vice President of the Brookings Institution. He is a past president of the American Economic Association and of the American Farm Economic Association, and for several years was Chairman of the Social Science Research Council. At various times he served as delegate from the United States to the Biennial Assembly of the International Institute of Agriculture at Rome, and as American member of the Mixed Committee on Nutrition of the League of Nations. After the Employment Act of 1946 was passed, Mr. Nourse was appointed Chairman of the Council of Economic Advisers in the Executive Office of the President. He served in that position from August 1946 to November 1949. From 1950 to 1953, he held a fellowship of the John Simon Guggenheim Memorial Foundation, and in 1956-57 was a Visiting Scholar for Phi Beta Kappa. He has published numerous books, including *America's Capacity to Produce, Price Making in A Democracy, and Economics in the Public Service.*

CHAPTER 17

MEETING AUTOMATION FULL-ON

CLEDO BRUNETTI, *Executive Assistant to the Vice-President,
Food Machinery & Chemical Corp., San Jose, Calif.*

Automation as all technical developments have done creates new technological and social problems as it brings with it a new look to the industrial, labor, and particularly the educational picture.

Historically despite the benefits derived from increased use of machinery, the relationship between man and machine has been at times an uneasy one, as early Industrial Revolution history will attest. It was feared that the machine would usurp man's own prerogatives, yet this has never happened. Man created the machine, and like a wistful puppy it is dependent on man for guidance, nourishment and repair. The machine cannot think alone. It "thinks" only by selection of a number of choices man has designed into it. Nothing in the world of electronic marvels presents any other possibility for the machine.

Fifty years ago Henry Ford introduced mass production in the automobile industry. Certainly those concerned about the effect of mechanization on jobs could have viewed that event with serious alarm. But an understanding of what benefits it could bring not only to industry but to labor, the customer and the national economy would have spared many the unnecessary concern and worry which did take place. Yet in Henry Ford's days it was easier to see how automation would cut the cost of products and make cars available to more people.

We face a situation not at all different from the one of Henry Ford's days. The outlook is not only encouraging, it is very good. We are not starting out as Columbus did with a hope based on meager data and a strong heart. We have enough past experience to have seen and to know that our greatest hopes for progress and happiness lie in making maximum use of the technology available to us.

If we were to take away or to have limited our progress in modern oil refining, business machines, and automatic screw machines, we would have a problem on our hands that would make any unemployment we have today seem like a grand utopia.

We faced problems but lived through well the era where machines stepped in to do the manual tasks and we created more *manual* jobs. Today the emphasis is on equipment to guide the operation of the machines and these equipments are creating many more new jobs requiring more *brains* than brawn, as well as more manual jobs. We are moving up on the ladder of intellectual development and our men and women in the ranks of labor are moving up to meet the challenge of this development.

May I repeat this important concept—when we introduced machines to do manual tasks many years ago we created more manual jobs—the equipments and machines we are using now to do mental jobs are creating more mental jobs. I am pleasantly surprised whenever I hear anyone, even some of the engineers and scientists of my profession, talk about machines without human participation. We appreciate the compliment, but may I set things straight from the start—there is no development on the way, planned or even anticipated, that will make it possible for industry to operate without workers, and hard workers.

For though machines can work and do what we design them to do they cannot make the type of decisions man can do and in many cases they cannot compete in cost with human labor. In this highly mechanized era it may surprise people to know that a human operator over a period of years can be much cheaper and much more flexible to changing industrial production requirements than a complex piece of machinery.

It is machine and man that will bring the progress in the many, many years to come—for our descendants and for many generations thereafter.

WHAT IS AUTOMATION

Automation, a newly-coined word, has created in some minds the impression that this is not a process of natural growth. This "newness" interpretation of the word is confusing for the word seeks to set end limits on a continuous growth process. Automation cannot be said to have begun on any certain date, nor can it be said that it will end at any definite time. Automation is in truth but a phase of our continuing technological advance. It is just more of a kind of stuff which has been taking work out of work—creating more and better jobs—and raising our level of living with each improvement since the invention of the wheel.

Put very simply, automation is the continuing improvement in the tools we use and have been using. Automation, in effect, had its start in the days of the Neanderthal man when that early ancestor learned that the animals of the forest could be subdued more easily, more effectively with a club than with his hands. The desire to lessen work, to do things more quickly, effectively, and easily, produced the wheel in time. Ultimately, man learned to make combinations of wheels, shafts, and other parts into *machines* capable of even more complex operations. To the machine was added cheaper, more efficient, more abundant *power*. Then came programming telling the machine what to do, and finally control to check the quality and quantity of machine output. Put them all together—machine, power, programming and control—and you have automation.

OUR DOMESTIC NEEDS

Domestically there are sound economic reasons for continued automation. A rapidly increasing population, a greatly stepped-up demand for consumer goods and services, and the prospect of a proportionately smaller labor force to produce them make further technological development mandatory. Accepted social trends such as shorter working hours, longer in-school time for youngsters, and better retirement plans releasing more workers at 65 will serve as natural controls on the size of the labor force.

From government and other figures we see that our outlook in the U. S. for the next five years is about as follows: Our present population of around 168,000,000 will increase about 7 per cent. Our labor force of slightly over 68,000,000 will increase about 6 per cent. Our Gross National Product, the total of all goods and services produced and sold, can be expected to increase 20 per cent or better.

If production does not keep step with demand, there will be a greater demand for the proportionately fewer goods and services available, and inflation will be the end result; things will cost more. A cut in the purchasing power of the dollar for you, for me, for everybody, means a lessened standard of living. Increased productivity will be the only way of controlling any such inflationary pressure. The continuous increase in natural productivity has been accomplished, not by working longer, but by constantly inventing better machinery to supplement human energy with mechanical power.

OUR INTERNATIONAL PICTURE

While we concentrate on our domestic picture we must also think on the international level as well. We have one way of life; other nations have theirs. If we cherish our way of life, and we do, we must be prepared to support strong leadership with a strong economy.

One of our world neighbors, Russia, has made astonishing strides in the past decade or two. Let us look at the facts.

Nation's Business reports that in 1938 the Soviet machine tool industry was producing 1,800 machine tools a year of 100 different types; in 1940, 65,000 of 500 different types; a probable 260,000 of perhaps 3,000 types in 1955. It also points out that our own armament industry absorbs 14% of our steel, while in the USSR over 30% has been going into armament since 1938.

In the area of technical training where the long range effects really will count, we are turning out something like 27,000 engineers a year, the USSR 50,000; we train 50,000 technicians, the USSR 1,600,000.

The choice today is not in our hands. Russian engineers, scientists, and technicians are every bit as good as we are, and under this program are bending every effort to pull abreast of and past the United States. This new technology is not confined to the United States alone. It is available to any country to use. We know that many other countries are either applying it now or planning to apply it aggressively. Remember the race for superiority in atomic weapons. The race for superiority in productive capacity started 25 years ago and the United States is no longer gaining the fastest.

WHAT HAS AUTOMATION BROUGHT US

What has automation brought us in this technologically geared country of ours? Those industries that have adopted mechanized means of increasing their output have shown striking growth and the reasons for this growth are apparent. These industries have succeeded in reducing the cost of their products through technological advance, have successfully met competition, have enlarged their markets by increasing consumer purchasing power through higher wages and reduced prices, and have added to their payrolls. Facts demonstrate clearly that reduced costs, and a resultant increased demand for goods and services, have continually meant a better business and increased manpower requirements.

Mechanization and technological progress have resulted in the creation of vast new industries such as the automobile industry, now employing over 900,000 workers; the aircraft industry with its employment of nearly 800,000, and the telephone industry with about 700,000 on its payrolls. Automation, in one form or another, has been responsible for the vast nationwide network of dial telephones, and the vast numbers of new jobs created in such organizations as the American Telephone and Telegraph Company. Installation of dial equipment, far from throwing thousands of telephone operators out of work, has actually meant a three-fourths increase in the number of telephone operators in the past ten years.

Today, because of mechanization, the average American family enjoys a standard of living over 30 per cent higher than in 1940. Home ownership is steadily increasing, with nearly 30 million families now owning their own homes compared with half this number in 1940. Nearly 40 million families, or three quarters of the U. S. total, own their own automobiles and over 10 per cent are two-car families. And yet we have untold numbers of new products which will find their way into our homes in the years to come. For while 98 per cent of U. S. homes are supplied with electricity and radio, only 68 per cent have telephones, 56 per cent have vacuum cleaners, 5 per cent clothes drier and only 3 per cent have air conditioners. We want and need many more products of many kinds.

One gets a very revealing glimpse into the workings of our economy when the pattern of income distribution is examined. Disposable income is already around $260 billion. The U. S. has long been "middle class." Now it's getting to be upper "middle class." The number of families in the lowest income groups has been and is falling in a steep decline. The number moving into what might be called the upper middle group is doubling and tripling.

To glean another view of what mechanization has done for us let us take a look at the past and compare it with our present situation. Our present labor force is 68,000,000. Suppose we were to trace back to what it might have been without any mechanization. Our country would not have turned out to be the "land of opportunity" and we would find a United States with a population closer to 40,000,000 than 168,000,000. Our economy would be one of farming, fishing, hunting, and hand manufacture of furniture and home necessities and trinkets. Including the principal home worker we would have about 14,000,000 jobs. This leads to the sober realization that we owe some 54,000,000 of our present jobs to mechanization.

I have long noticed in our travel toward a better and better world that mechanization or automation is only another phase in the history of human progress. Consider for a moment that less than 500 years ago a few thousand Indians lived in this vast country of ours. They had no coal, no petroleum, no metals beyond chunks of pure copper. They had a precarious food supply, flimsy wigwams for homes, mystical medicine, and chronic warfare. It is a startling thought that all our machinery, our homes, bridges, roads, buildings, all the products we have today were in the ground beneath the feet of those Indians. Man's mind alone created the products. The same land now supports 168 million people at the highest standard of living ever achieved in history.

COMPETITION FORCES AUTOMATION

In order to establish a sound reference for considering the reason for and effects of automation let us re-examine some fundamentals of running a business. These apply to the cobbler doing business in a small shop as well as to the largest of companies. A manager, cobbler or Divisional General Manager must use every resource to keep up the volume of products sold by his establishment. For every item or product he has on the market there are competitors by the score. His most serious and difficult problem is how to keep customers coming in the door.

He does this first by offering the product at the most attractive price. To this he adds the best quality he can put in the product at the price the customer can afford to pay. This will make his product more attractive and he hopes to sell more by doing so.

Next, to keep up his volume and hopefully increase it he adds new lines of products, usually of the type he can sell using his present selling methods. The cobbler adds shoelaces and shoe polish. Larger organizations who can sponsor research try to develop new products, for example, color television instead of black and white.

This would be adequate, except the day when an organization had a monopoly on a new product is a matter of history. Today even when a company comes out with a new appliance its competitors immediately announce they will have a similar and better product soon and expand every effort to get it out in the shortest time possible.

So we still find price the controlling item in most of our sales. Management is ever skating on the thin line of cost of its products. To keep his prices as low as possible our manager introduces whatever methods or practices he can to cut down costs. If he doesn't, his sales drop and his ability to keep people on his company's payroll drops with it. His reponsibility is to sell all the product he can and in doing so he creates work so people can have jobs.

It is not a question of choosing or not to devise and adopt new and better ways to cut costs. Business survival makes it imperative to find new ways to reduce costs by increasing the productive efficiency of each employee. Business has no choice. If it refuses, it will soon find its markets invaded by a more progressive competitor.

Automation, in its true essence, starts with marketing. If marketing shows that by reducing prices he can raise the volume of products sold, reach the fellow with fewer dollars to spend but the same desire for the product as present customers, the manager goes to work on costs. Searching for new cost methods he looks into Automation.

AUTOMATIC EQUIPMENT IN THE ELECTRONICS INDUSTRY

One of the subjects which is of most concern to us today stems from the statements that automation is moving in on us rapidly, that it will soon be dominating our production picture and unless we do something fast we may plunge into a difficult economic situation. This is based on a feeling that electronic computers will displace entire office loads of people or that we have machines that can turn out 1000 radio sets a day, now requiring 200 people, with only 2 people, a reduction in workers of 100 fold. It has been somewhat more than amazing to those of us who are developing better tools for production to see what glories have been piled on our shoulders as to what our new developments can do.

The mechanized part accounts for a small portion of the total working force. This is hardly cutting the labor force down by a factor of 100. This improvement in manufacturing was accomplished only after an enormous number of man years of work had been put into the development of the printed circuit and mechanized assembly techniques by industry, government and other research and development organizations.

What is the overall effect of this type of automatic equipment on jobs? First, since the lead time of such investment is measured in terms of one or two years, it permits a gradual redistribution of labor to be effected within the industry before the full impact of the particular new automatic machine is felt in production. Second, this investment creates jobs at the machine-using plant, at the plants supplying components for the radio sets and at the plant which developed and produced the machine. These act to set up a purchasing potential which contributes to absorbing an increased quantity of goods when they are finally produced.

AUTOMATION IS FOR SMALL BUSINESS AS WELL AS FOR BIG BUSINESS

Capital investment of small, medium and large business for the past few years has shown that the small and medium business spending collectively averages more than big business. In 1956 for example small business spending for automatic handling, automatic assembly and automatic control equipment averaged approximately $225,000 and for medium size businesses $290,000. Big business spending for that year was estimated at $475,000 per plant. Because automation plays a major role in the modernization of plants and because small business in the aggregate will contribute a healthy slice of the total being spent for new equipment, there can be no question but that small business is rapidly adopting automation methods. The individual totals may not be as large as those invested by big companies but the men who run small businesses are just as keen and their bankers, you can be sure, are backing their judgment. While the sums appear to be large it should be pointed out that these expenditures are actually a very small fraction of the total capital structure of these businesses. Thus industry is demonstrating an optimistic but cautious approach.

AUTOMATION WILL NOT TAKE THE COUNTRY OVERNIGHT

The word automation brings to some visions of the push-button factories springing up with unattended machines and the product gliding throughout the manufacturing plant in one uninterrupted flow untouched by human hands. Let us not fool ourselves as to the magnitude of man's accomplishments, past and present. We are a long way from the push-button factory. We have learned that automation can not be accomplished by the mere addition of electronics to our present tools. Automation requires the understanding of new principles, the development of new processes, and the application of new materials. All this requires much of research and many, many man-years of hard work to develop and apply.

The evolution of automation will continue to be a gradual process. Automation will not run wild because it has the following built-in feedback system to regulate it from precipitating destruction of any specific source of mass employment.

1. It takes many man-years of work to develop automation equipment.

2. Over-mechanization of a plant is not good management or even economic. It can, on the contrary, be very costly and inefficient.

3. Much depends on the free and unpredictable customer's willingness to permit his wants to be standardized long enough to liquidate the cost of the automatic equipment.

4. Automation requires that industry have a continuing market for the product because once the plant is tooled up for a given item, the expense of making modifications to the product is large.

5. The production of goods involves many basic functions other than processing, though we instinctively associate it alone with the automatic factory. Such functions as design for automation, the handling and inventory of raw materials and components, the storage of inventory of finished assemblies or parts

to be used for maintenance, and the servicing problem also become essential to the whole. Before a push-button factory becomes practical each of these functions must be automatized and integrated into the over-all to a high degree of perfection.

EFFECT OF AUTOMATION ON OVER-ALL EMPLOYMENT LEVEL

The notable advances in employment in those segments of our economy most affected by automation (manufacturing jobs) do not confirm the story that automation will permanently displace large numbers of people. For instance, employment in the automobile industry has grown from 460,000 in 1939 to more than 900,000. Employment in chemicals and allied products has grown from 400,000 in 1939 to more than 800,000. In electrical machinery employment stood at 335,000 in 1939 and has skyrocketed to some 1,200,000. The aircraft and parts industry, employing 64,000 in 1939, has jumped to over 800,000. Employment in steel at the blast furnaces, steel works, and rolling mills has grown from 435,000 in 1939 to over 650,000.

Though our plants have been mechanizing intensively for years and years, excluding war peaks, when the increase was higher, we have actually increased the number of jobs in the past twenty years at the rate of 1,200,000 every year.

To see what using the capacity of tomorrow's machines to compute today's unemployment does to our figures, let us consider the gross national product of our country. It represents a very real measure of our workload, stating in dollar terms the total productive output of goods and services. In these terms our workload in 1955 exceeded 370 billion dollars. This work was done with a labor force of 65 million people. This means that each person in the labor force produced about $5,700 worth of goods or services to contribute to the gross national product of the nation. In 1940 we had a gross national product of 101 billion dollars and a labor force of 54 million people, each member of the labor force producing about $1,900 worth of goods or services. Let us go back to that time and assume that some forward-looking exponent of automation had described the machinery which we have available today which enables us easily to produce $5,700 per member of the labor force. Noting this effect on the 1940 workload, we would have predicted a displacement of about 36 million people in 1955.

Conversely, suppose we had frozen our tools in 1940 design and attempted to turn out 1955's workload. We would have needed a labor force of 195 million people, three times the available labor force in the United States at that time.

An analogous situation exists when we attempt to consider today's workload in terms of tomorrow's machinery. We can easily show that the tremendous potential capacity of automation, if fully applied to today's 400 billion dollar workload, would result in unemployment. However, if we look forward and apply the capacity of tomorrow's machinery to tomorrow's workload, we may find that the situation is not nearly as serious as we have been led to believe. We may even see a labor shortage. In the next ten years our gross national product will increase by some $160 billion. The vast fields of electronics, chemical, paper, ceramics, aviation and machine industries, and the service industries, the banking, insurance, department stores, and others will continue to grow and add jobs.

Although we have 68 million people employed and may add another ten million to the labor force in the next ten years, we should not concern ourselves about lack of jobs for the future; for automation, if allowed to grow in its normal way, will absorb all these people in better and better jobs. I believe that automation will add 15,000,000 new jobs in the next ten years.

NEED FOR RETRAINING

What must we do to meet this new look on business and labor? We must immediately begin a training activity within our labor force to meet automation full-on. This is nothing novel; it is only a reflection of the dynamic nature of our society which requires training and retraining as we progress through life. This is true not only of our technical, engineering, and scientific personnel but of all our labor force. One can find very few situations in present-day industrial life in which a laborer has gone to work as an apprentice in industry and continued to practice without changing the skills which he learned during his apprenticeship. Labor has always met this challenge.

Some examples of new jobs being created by automation are: machine operators, machine supervisors, electronic test operators, machine maintenance mechanics, electromechanical technicians, component materials handlers, product designers, research engineers and scientists, methods engineers, machine service staff, key punch operators, computer programmers, instrument mechanics, and control equipment technicians.

In addition to the above types of technical jobs, automation will open more opportunities for administration, clerical, procurement, wholesalers or distributors personnel, warehousing and transportation personnel, retail sales personnel. These are but a sample of the jobs created.

What reasonable plan can we follow in setting up this very necessary retraining program? It is plain economic sense that ways must be found to do this. Short-term dislocations will occur but can be minimized by advanced retraining. Industry and labor unions should join forces to prepare for the changes automation is bringing. I should like to recommend the following course of action: 1. The new jobs created by automation should be defined. 2. The worker must be acquainted with the nature of the new jobs. This is a function of both the industry and the labor unions. This information should be clearly presented to the employees, so that they can, perhaps for the first time, see what automation will mean to them. 3. Having been acquainted with the available new jobs the employee must provide the initiative to obtain the basic skills necessary to accomplish the job of his selection. 4. The employee may learn these basic skills from a public or vocational trade school of a type already in existence but too few available. Our community facilities need to be expanded and re-oriented to accomplish this task. 5. Industry must continue and in fact increase training in the operation, use, or support of the specific automatic machines which it installs. 6. Procedures must be set up by industry to properly evaluate the new skills which the worker now brings to his new job and reward him for his efforts.

American labor has never in the past failed to rise to the occasion, to take on any new machine dreamed up by engineers and introduced by management. The unfailing adaptability of labor to change has been demonstrated again and again in the past. Why should we now depreciate labor's known adaptability and tremendous capacity to take on and master new machines and new techniques, to acquire new skills?

Experience has shown that as the level of general education has risen, there has been a corresponding rise in the demand for skilled jobs and a rejection of menial jobs.

Change has its hardships, but it has its nice points too. Man is a creature of change; the greatest hardship is that of no change, of boredom. As Ann Morrow Lindbergh said in her refreshing book, *Gift From the Sea*: "One resents any change even though one knows that transformation is part of the process of life." Let us strive to keep it in balance, but let us not resist change, particularly when we know that it is necessary for survival and that in the long run we shall all be better off for it. Let us healthily face the job of redirecting our efforts, retraining our people, and looking optimistically at the laws of nature. We must continue to grow for survival and for happiness.

Selected Bibliography

R. W. Bolz, "Industry Plans for More Automatic Operation," *Automation*, II, 3 (1955), 41-47.

J. F. Dewhurst & Associates, *America's Needs and Resources, A New Survey* (New York: Twentieth Century Fund, 1955).

S. Lee Everett, "The Engineer and Automation in the Process Industries," *Electrical Engineering*, LXXIV,5, (1955), 371-373.

Herbert Harris, "Russia's Gaining on Us," *Nation's Business*, XLI, 9 (1953), 25-27, 74-77.

"Incomes: Everybody Gains—Mostly the Man in the Middle," *Business Week*, (May 28, 1955), 134-135, 138, 140, 142.

Henry Lefer, "Scientists Shortage Threatens Defense," *Aviation Week*, LXII, 12 (1955), 31-34, 37-38, 41, 45-46, 50.

Stanford Research Institute, *Symposium on Automatic Production of Electronic Equipment; Proceedings*, (April 19-20, 1954), 6-7.

Ira Wolfert, "What's Behind This Word 'Automation'?" *Reader's Digest*, LXVI, (May, 1955), 43-48.

CLEDO BRUNETTI is executive assistant to Executive Vice President for Engineering and Ordinance at Food Machinery & Chemical Corporation, San Jose, California. He has received numerous awards from Army, Navy, War Department, OSRD, Department of Defense and is a Fellow of the IRE and AIEE and is also listed in several Who's Who's.

For seven years he was a civilian member of the Panel on (Electronic) Components of the Research and Development Board. He organized and headed the Board's Subpanel on Packaged Subassemblies. Since 1953 he has been and continues to be a consultant to the Assistant Secretary of Defense (Research and Engineering) . He is also a member of the advisory group on reliability of electronic equipment, and has also been chairman of the Advisory Group on assemblies and assembly techniques since 1955. He earned his Ph.D. at the University of Minnesota in Electrical Engineering. He remained there from 1932 until 1941 to teach electrical engineering and to direct graduate research in electronics. From 1941-49 he was with the National Bureau of Standards in Washington, D. C. Before coming to his present position, he was also an associate director with the Stanford Research Institute, Stanford, Calif., and then director of engineering research and development for General Mills, Inc.

CHAPTER 18

AUTOMATION: WEAL OR WOE

EDWARD R. MURROW, *Vice-President,*
Columbia Broadcasting System
and FRED W. FRIENDLY, *Producer,*
Columbia Broadcasting System

(Editor's Note: On June 9, 1957, SEE IT NOW, edited and produced by the partnership of Edward R. Murrow and Fred W. Friendly, gave the nation a ninety minute report on automation over CBS Television. It was called Automation: Weal or Woe. The section of the script which is reproduced here reveals the mixed feelings of rank-and-file labor and the view of one of its leaders, Walter P. Reuther, UAW president and vice-president of the AFL-CIO, on the new technology.)

A VETERAN STEEL WORKER: The steel industry as a whole used to be twelve hours, only in the sheet mill, and they were eight—the open hearths were twelve hours—the labor just twelve hours—the galvanizing was twelve hours. Well, they called it twelve. They worked eleven hours in the day time and thirteen hours at night, but they just called it twelve hours, and as far as wages is concerned, oh, the wages when I went to work, if we made three dollars—three dollars and a half a day, we were wonderful. That was a good job. Those same jobs today, working under the same conditions, is worth twenty bucks— twenty dollars right on it. We—we—we worked, oh, out of the ordinary I would say and I don't see the way that they make sheet iron today how we ever stood it, but it didn't seem to bother us. We'd even go home so tired that we didn't know which way home was, and sit down, and get a little bit to eat and run around all night, and start right in the next morning.

The same thing over day in and day out. It didn't make any difference to us. We got used to it. We got hardened to it.

Q. **Do you think that the future of the man who works in the factory on the production line is doomed because of machinery?**

SAME STEELWORKER: Not completely, no. No. Brains— I don't think—I think we—we'll always have to have some men to do some jobs, but it would be an awful long time before machinery will eventually place—replace man. I read the scientific papers and read them and follow that line up and see that—what it says. Look at the—look at the propellors—the jet propellors, look at 'em. They're goin' to the stratosphere as far as they can go,—8, 9, 10—I don't know how many thousand miles they can travel today. All right. I just—when—when I was a boy that— that was foolishness for anybody to talk that way—you'd never thought of it but if you'll read your papers, as I said, you'll find out that there's nothing impossible.

(**Members of Local 6 of the Bakery Union, Philadelphia**).

THE CHAIRMAN: But they never get to the humanitarian side of people who are let go by automation—what their particular problems are—what confronts them when they're let go of their jobs when they've accumulated fifteen to twenty years of seniority—when they have gone beyond the age limit of forty. According to some of the statistics, you're just as well dead at the age of forty. We don't know where it's going to end. The floor is open for any questions that may come from the floor.

FIRST UNION MEMBER: As you know, I have been up there in the Packing Room for sixteen years. I have five children. Now, if I am displaced out of my position, what is going to happen to me? I know we should stick together and I for one am going to do the same as I've been doing in the past. I am not going to take any additional work upon my shoulders, and I believe that should stand for the rest of the membership in this building today.

SECOND UNION MEMBER: I was there for fourteen years and I'm forty-four years old, and as far as I know, I understand no place else was going to hire me. The only place I can go to work is a pick and a shovel.

THIRD UNION MEMBER: Automation! Automation! These meat heads up here, our business representatives, have done nothing about it, and they are meat heads. We're being pushed out of jobs and thrown out of jobs because of them. They knew this was comin' and they've done nothing to help us; yet they say they're lookin' out for our welfare. Are they lookin' out for our welfare? They're only lookin' out for their own pocketbook—every one of 'em.

FOURTH UNION MEMBER: I know that there's a lot of dissatisfaction because of the layoffs and what can the union do. With a problem like this, I don't know what we can do and while it was mentioned earlier that maybe it would be a good idea to take the people out on strike, if that were the answer we would be the first ones on the picket line, but that isn't the answer. Maybe some of you have some answer to it. Very frankly, we don't.

FIFTH UNION MEMBER: I hear a lot of talk goin' around —goin' around up the shop, but when you come down here you don't have a damn thing to say. Why don't you get up and say it now? A lot of you guys are doing two or three different jobs instead of refusing to do but your own job. That's one of the reasons they don't need as many men as they do, but now that it's hurtin' you guys, youse are hollerin' like hell. Why don't you get up there and say it now?

SIXTH UNION MEMBER: I—Mr. President, I disagree with the brother that called our representatives "meat heads." I'm a member of the sweet dough department, and at one . . .

VOICE: You're a meat head too.

SIXTH UNION MEMBER (continuing): . . . and at one time we had plenty of work until the company eliminated all hand work and got a machine and a faster machine, and a belt, and now all they're puttin' out are cinnamon buns, and they're just cinnamon bunnin' the public to death.

SEVENTH UNION MEMBER: We'll never get no place if we're going to work nine, ten, eleven hours a day just to help the company out. They're not lookin' out for us, so why should we worry about more than eight hours' work?

EIGHTH UNION MEMBER: Not an eight or nine or ten or eleven hour day and go home eleven and a quarter hours so you can be back within the twelve-hour intermission, but about a six-hour day and there is no—no more authority in this country than our Vice-President, Mr. Nixon, who said if that would be the answer to everybody working, then keep them working a thirty-hour week.

FOURTH UNION MEMBER: (Second Appearance): And maybe a shorter work week would alleviate the problem but that too wouldn't be the answer. The important thing, whenever there is mechanization or automation, is that there must be a corresponding increase in business to take up the slack of work.

(A meeting of some U.A.W. Shop Stewards in Detroit).

FIRST UNION MEMBER: People have been displaced with machines. People have been displaced because of jobs leaving the plant and we are not prepared to meet the change because we are not notified. I think that as responsible people, we should be notified by management from model to model what do they anticipate, what operations are going to be eliminated, how many people will be displaced, and why is it important that we should know? We should know because we can, with an organized training program, allow people to train for the new type of machines and for the new type of work coming into these plants.

SECOND UNION MEMBER: Now, none of us, and I don't think our union, kicks about the introduction of automation. We welcome it because unless this country automates, it's going to be in rough shape in—in competition with other countries. The problem for us is who pays the cost of automation. The U.A.W. is for automation. We're for progress, but we are not for the working people, we are the least able to afford it, to pay the total cost of automation, and in terms of human misery and suffering, that is exactly what is going on in Detroit and in the country. For example, we have eighty-five—eighty-nine thousand unemployed workers in Detroit now and the industry's going full speed. What happens when we have a letdown? What happens when we have more automation next year? How are these people going to live? It's difficult to move elsewhere. It's difficult to find a job especially if you're an older seniority employee. That means the U.A.W. no matter how good it is, how strong it is, or how much money we have in the strike fund, can't solve the problem of automation alone. Thank you, Brother Chairman.

THIRD UNION MEMBER: You think over here in our machine shop they've cut down, and they have, because I remember the time when thirty-one hundred people worked in there and they didn't get a hundred motors an hour, and this included truck motors as well as passenger motors. But today, they can run a hundred and twenty-five motors per hour in there and they have less than nine hundred people in that machine shop of Chrysler—Chrysler-Jefferson.

It only means one thing—that the motors are needed and they're being produced but the men that produced them are not needed to the degree that they were before. But the fellows by the thousands who were displaced, it's not automatically living for them, and it gets to be worse as time goes on unless the union, the government does something about it.

(In an interview with Walter Reuther).

Q. What has automation done and what is it going to do to the automobile industry?

REUTHER: Well, I think I can say, that automation has already had some impact upon the employment opportunities in the automotive industry. Based upon the best figures that we can get, we've lost, roughly, about a hundred and fifty thousand jobs during the last nine years, but I think we need to understand that automation really is just in its infancy; that we've just begun this whole mechanization of the manufacturing and assembly processes, and that essentially we're sort of standing on the threshold of the second Industrial Revolution. I have great confidence in the future. I personally believe that if we—if we work at the problem, I know that we can solve the problems. What I'm frightened of is that maybe people will believe that we can just coast. There are certain industry people who have got the notion that—that automation is not the beginning of the second phase of the Industrial Revolution. They say it's just an extension of the old technology. Now, that is not true. That's really running away from reality.

Several years ago, when I went through the Ford-Cleveland engine plant where they had fully automated the production of engines, and I looked over the acres and acres of automated machinery, I was asked by a management person how I liked the situation, and I told him I was very much impressed, and he says, "Well, you won't be able to collect dues from all of these automated machines," and I said, "You know, that is not what is bothering me. What is bothering me is, how are you going to sell cars to all of these machines?" And you know, we can make great progress in the production of automated equipment, electronics and all that, but we're still going to have to make the consumers the old-fashioned way.

Q. But what's going to happen to the men who are displaced by the machines?

REUTHER: Well, this is the big problem and I don't believe that we can truly measure the impact of automation upon the displacement of labor yet because it's too early, but certainly, based upon what has already happened, this will become an increasingly serious problem unless we begin to project plans into the future. Now, many things have got to be done. Obviously, we've got to re-train workers so that they can be absorbed at new skills as automation changes the character and the composition and the requirements of each job, but all of this thing will require some advanced planning, and one of the things that concerns us at the present time is that there is no place where we can get a total look as to what is happening to the American economy in terms of the long pull. General Motors knows about the automotive industry. General Electric knows about the electrical industry. The Standard Oil Company knows about petroleum.

DuPont knows about chemicals but no one—there is no place now in the United States where anyone has pulled together all of the threads of this problem and tried to weave it into a total pattern, and what we've been suggesting is that free labor and free management, in cooperation with free government, ought to create some sort of a technological clearing house where there could be a central place where we could assemble all of the data—what is happening today—what we can project for tomorrow—what we can think about beyond tomorrow, so that as a—as a rational process we can begin to know the facts and then begin to plan as free people to meet the problems and to realize the promise of the greater abundance that these machines will make possible.

Q. Well then, you're not suggesting that the government should decide but that it should be a cooperative venture between government, labor and management?

REUTHER: Yes. The government would essentially provide the mechanics by which the voluntary groups using those mechanics would make voluntary decisions. When the most complicated tool was a hoe, it didn't matter whether the person who owned that hoe used it responsibly or irresponsibly. It had no impact upon society. Maybe the family of the person who owned that hoe wouldn't eat as well, but as the technology and the tools of production become complex and more productive, the ownership of those tools takes on broad responsibilities. Labor, since its decisions have an impact upon the use of those tools, labor takes on new responsibilities.

Q. Well do you think the four-day week is just over the horizon?

REUTHER: I think the four-day week will be with us much quicker than we realize, because I believe that the impact of automation and atomic energy and these other new technologies is going to come much faster, and you cannot measure the future by the standards of the past, because this whole process is going to be accelerated a great deal, and that's why I believe that— that we will get the four-day week long before we can use it intelligently unless we begin to work hard now on how can people use their new leisure creatively and constructively. That's the problem, I think, that needs a great deal of attention, because that's the area—this again, you see, is this lag between the —in the social sciences as compared to the physical sciences. We always make more progress in working with machines than we do with men.

As we automate production, the individual worker gets further and further removed in the production process from the ultimate end product. When a worker makes a Cadillac—but he doesn't have any feeling that he—he helped create that— his little pieces—and we've got to find a way in—in the leisure hours of people to give them an opportunity for personal creative expression, and this is, I think, the great challenge. We can feed and clothe and house and take care of man's material needs, but having done that, we ought to find a way to enable people to grow into better—better human beings by enabling them to grow and mature and develop culturally and spiritually and intellectually.

Q. More hospitals, more schools, more reception facilities?

REUTHER: These are the things we need to put greater and greater—I think we've got to rearrange our priorities. We've got to somehow get our values in sharper and clearer focus so that we know precisely really what are we trying to do? I happen to believe that if we're just in a race with the Communists to see who can achieve the greatest material prosperity just in terms of bath tubs and plumbing and radios and T.V.,

I don't think we have any assurance that the Russians can't do a job equally if not better.

There's got to be behind this great material prosperity a kind of a sense of moral purpose, because power without morality is power without purpose, and it's in these—terms of these intangible basic human values that I think the free world has to maintain its margin of superiority over the Communists, and I believe that here again, that if we can approach this in terms of economics of abundance instead of the economics of scarcity, we can solve our problems, but the tragedy is too often people always think in terms of dividing up scarcity when we ought to be thinking about how we can cooperate to create and share abundance.

EDWARD R. MURROW, Columbia Broadcasting System Vice-President and broadcaster extraordinary, and FRED W. FRIENDLY, co-producer with Murrow of the TV "See It Now" series, first got together in 1948 when the Petrillo ban on musical recording inspired Friendly, then running his own radio package show agency, to see Murrow about narrating a series of recorded speeches of famous statesmen and the historic events in which they participated. It resulted in the three album "I Can Hear It Now" series. Shortly thereafter, Murrow and Friendly teamed up on radio with "Hear It Now" and on November 18, 1951 they launched their history making "See It Now" venture on television.

After graduation from business college, Friendly became a writer of news programs for a small radio station, then at the age of 21 went in for himself producing and selling his own radio shows. During the war he was an Army correspondent in the Pacific area, and then he moved on to Europe for the fighting climax there. His exploits and accounts of battle earned him the Legion of Merit and the Soldier's Medal. As co-producer of the "See It Now" series, he handles the filmed portions of the program as well as reporting, editing and continuity chores.

Born in Greensboro, N. C., Murrow's early life was quite typical of any American lad. He went to the usual grade and high schools and then on to Washington State College, from which he graduated Phi Beta Kappa in 1930, with majors in speech and history. He took the job as president of the National Student Federation that same year and traveled extensively in America and Europe. This led to his next position in charge of foreign offices with the Institute of International Education in 1932. In 1935 he joined CBS as Director of Talks and Education and two years later he was in Europe, where from 1938 until 1945 he headed the roster of news correspondents which reported the war and the London blitz. His program "This Is London" became the listening post for the free world and made him world famous. His postwar news coverage has continued to bring into sharp focus events and dramas of current history when and where they happen.

CHAPTER 19

AUTOMATION, EMPLOYMENT AND
ECONOMIC STABILITY

W. S. BUCKINGHAM, JR., *Professor of Economics and Industrial
Management, School of Industrial Management,
Georgia Institute of Technology*

Since World War II some spectacular discoveries in the fields
of electronics and communications have permitted the manufacture of various types of automatic computing machinery. These
machines are capable of translating a large body of previously-
developed, theoretical, economic and business principles into
practical significance. Called electronic computers, they are
capable of processing data with almost unbelievable speed. When
information is fed into them, usually on tapes, they can perform
a series of logical operations and can choose among several previously anticipated courses of action based on built-in criteria.
They even adjust automatically for errors. The operation of
these computers to solve scientific, commercial or industrial
problems is very often referred to as automation.

Also in the last few years a number of automatic or semi-
automatic machines have been constructed to supplement conventional assembly line operations in factories. These machines perform hundreds of individual mechanical functions without direct
human intervention. The operation of these machines is likewise
commonly called automation.

Finally, scientists, computer manufacturers and science fiction
writers have shown, hypothetically at least, how the administrative and manufacturing processes of an enterprise could be integrated into a single, silent, automatic monster which could grind
out an endless chain of products without a man in sight. This
awesome picture has charged the imaginations of some and struck

terror in the hearts of others. The possibility of such developments is also called automation.

In addition to this definitional confusion many speculations,
hypotheses and fragments of theories concerning the broad economic and social implications of automation are currently being
expounded. In this flood of verbage there is no shortage of
imagination but there is a notable lack of the kind of critical
thought and careful documentation which yields quantitative,
scientifically accurate results. There is a great need to collect,
sift, classify and evaluate the empirical evidence which alone can
test these generalizations.

It is not the purpose of this presentation to provide any conclusive, concrete facts or to try to verify any particular arguments
by amassing evidence. Rather the main purpose here is to try to
establish a frame of reference within which the results of subsequent empirical investigations can be logically fitted so as to
determine the probable impact of automation on employment
and economic stability. Some facts will be used here to illustrate
problems which require study but this statement will seek to
achieve its main purpose by presenting (1) a definition of automation based on four basic principles which underlie all of the
various popular concepts, (2) an estimation of the probable
scope and speed of automation in the future, (3) a classification
of eight major types of direct effects of automation, and (4) an
evaluation of the impact these effects on five principal tests of
the performance of an economic system with particular emphasis
on the maintenance of full employment and economic stability.

PRINCIPLES AND DEFINITION

The variety of popular uses of the term automation necessitates some definition which is both precise and relevant for
analysis. Such a definition can best be derived from an examination of the major principles which underlie most if not all of
the popular concepts of automation. There are four such
major principles—mechanization, feedback, continuous process
and rationalization.

Mechanization means the use of machines to perform work.
Sometimes mechanization substitutes machinery for human or
animal muscle. The steam engine did this. Sometimes mechanization substitutes machinery for brainwork at the lower, routine
levels. The electronic computer does this. Because of the power,

compactness or speed of machine operation mechanization usually permits tasks to be performed which could never be done by human labor alone no matter how much labor was used or how well the enterprise was organized and managed. Mechanization increases wealth and reduces drudgery in the long run but in the short run it may cause hardships to workers whose skills are rendered obsolete, diluted by a further specialization or whose jobs are abolished altogether.

Feedback is the second principle inherent in automation. This is a concept of control whereby the input of machines is regulated by the machine's own output so that the output meets the conditions of a predetermined objective. As in a simple, thermostatically-controlled heating system, the conditions created by the output automatically control, in turn, the amount of input and hence the performance of the machine. When controlled by the feedback principle, machines start and stop themselves and regulate quality and quantity of output automatically.

Continuous flow or process is the third principle of automation. This concept is of increasing importance because it is spreading from many individual production processes to the business enterprise itself and on to the entire economy. Mass production, increasing interdependence, and now automation all embody this principle which is leading to a concept of the business enterprise as an endless process. Business for the most part has ceased being an operation that can be started and stopped with small loss. The regulation of a constant flow of goods has become a major concern of management.

This continuous process idea has changed the function of management. The man of daring and imagination who relied on hunch supported by experience has become a technological casualty. The shrewd bargain has given way to the carefully calculated risk. The increasing size and complexity of business enterprises precludes the top executives from having knowledge of the details of the firm's operations. Decisions must be made by groups who rely on reports from the sales, production, accounting and other departments. Top executives today are forced to view their functions as consisting of planning, controlling and coordinating the firm's operations and harmonizing the interests of the firm with those of employees, investors, suppliers and customers. Because of the high degree of interdependence in the economy the decisions of these executives intimately affect the lives of millions of people.

Rationalization—the fourth principle of automation—means the application of reason to the solution of problems or to the search of knowledge. In a production system it means that the entire process from the raw material to the final product is carefully analyzed so that every operation can be designed to contribute in the most efficient way to the achievement of clearly enunciated goals of the enterprise.

Actually rationalistic philosophy is nothing new, having become an important force in the world with the Renaissance. However, the scientific, rationalist philosophy takes on numerous new implications when it can be implemented by modern electronic machinery. The rise of electronic computers has led to a fascination with the possibility that super-rationalism in the business and scientific spheres might spill over and transform society into an exact mechanism in which all elements of chance, risk, capriciousness and free will—as well as all spiritual values—would be eliminated. Although this kind of speculation is highly dubious, nevertheless it is one logical extension of this fourth principle of automation.

Following these four principles—mechanization, feedback, continuous process and rationalization—automation can be given a definition precise enough to be useful for logical analysis. It can be said to be any continuous and integrated operation of a rationalized production system which uses electronic or other equipment to regulate and coordinate the quantity and quality of production.

AUTOMATION IN THE FUTURE

For the purpose of determining the extent to which automation can be applied to productive processes, industries can be divided into three groups. The first includes those industries in which production can be reduced to a continuous flow process. Oil refining, flour milling, and chemical production are illustrations of industries in which automation has made, and should continue to make, significant progress. In other industries it is possible to revamp the productive mechanism in such a way as to convert it from a series of unit operations into a single endless process. While some industries utilize processes which are not conducive to automation, new methods of production may be conceived which are more acceptable.

A second class includes industries in which some automation

is possible, but full or nearly complete automation is not likely. Indeed, it is possible that some industries may have automatic machines applied to seventy-five per cent of their operations, yet the cost of making the plant completely automatic would more than offset the savings achieved from the use of partial application of automatic machines. In this category would be found industries which require substantial information handling and accounting functions but in which the method of production or the nature of the product is not adaptable to continuous flow techniques. Such industries would include transportation, large-scale retailing, and the manufacture of certain nonstandardized consumer products like furniture.

The third group into which all industries may be classified includes those in which no significant application of automation seems likely because of the highly individualistic nature of the product, the need for personal services, the advantages of small scale units or vast space requirements. These would include agriculture, mining, professional fields, and most construction and retailing.

Other limitations on the scope and speed of automation are more temporary but are nevertheless significant at the present time. These include (1) the high initial cost of equipment which for the time being at least prevents all but the larger firms from using it, (2) the shortage of highly trained operators and analyzers and (3) the time required to analyze the problems, reduce them to equations, program the computers and translate the answers into useful data. The solution to the problem of rethinking through the entire production process is likely to come slowly because of the tremendous mental inertia which is confronted in such cases.

EFFECTS OF AUTOMATION

Following the principles and definition of automation already derived, the direct consequences of applying automation to a productive system can be classified as follows:

1. Many direct production jobs are abolished.

2. A smaller number of newer jobs requiring different, and mostly higher, skills are created. These new jobs include equipment maintenance and design, systems analysis, programming and engineering.

3. The requirements of some of the remaining jobs are

raised. For example, the integration of several formerly separate processes and the enhanced value of the capital investment increase the need for comprehension and farsightedness on the part of management. Also greatly decreases inventories and more rapid change-over times create tensions which require more alertness and stamina.

4. Production in aggregate and per man hour is enormously increased.

5. The production of new and better goods of more standardized quality becomes possible. However, there may be a loss of variety. Many different models are possible from combining a few standardized processes in different ways but, as in automobiles, the final products still look pretty much all alike.

6. There is an increase in the quantity and accuracy of information and the speed with which it is obtained. Management can thus have a clearer picture of its overall operation and by knowing the consequences of alternative courses of action it can act more rationally.

7. In most cases a more efficient use is made of all the components of production—labor, capital, natural resources and management. In a few cases high operating speeds waste materials but even here the loss is usually justified by saving other resources including even time which is a valuable component of production.

8. A continuous pace is often set at which the plant must be operated.

ECONOMIC IMPACT

In order to determine the economic impact of the above eight major effects of automation they should be evaluated in terms of the performance tests of a properly-working economic system. The criteria of an economic system's performance, or the goals which an economic system should seek to maximize, can be classified as follows: (1) the level of employment of all resources; (2) the stability of employment of all resources: (3) the satisfaction of consumers' desires, i.e. a pattern of resource and product allocation which always satisfies the more urgent requirements first; (4) the efficiency of production, i.e. output divided by cost in human effort, physical recourses and lost opportunities; (5) progressiveness, i.e. the rate of increase of productivity. Of course, these goals cannot all be increased simultaneously.

For example, maximizing short run aggregate living standards (number 3 above) requires a more or less equal distribution of income if it is assumed (and it cannot be proved otherwise) that different people have the same basic needs and the same capacities of enjoyment. Now equal income distribution, and hence maximum human satisfaction, is partially inconsistent with progressiveness (number 5 above) because some inequalities of income are necessary to provide the incentive to increase productivity and hence long run living standards. Consequently some optimum combination of these five goals—particularly some compromise between short and long run living standards—should be sought.

The first goal—full employment—is now generally accepted as both an economic and a political necessity. The enormous costs of unemployment—particularly of labor—have been well-documented. Human resources depreciate with time rather than use and they depreciate at an accelerated rate when they are unemployed because of the decline of knowledge, skills and morale. The main economic cost of unemployment is in production that is permanently lost.

However, the social costs of unemployment far exceed the economic costs since unemployment also contributes in large measure to crime, disease, family disintegration, race and religious prejudice, suicide and war.

IMPACT ON LABOR

It is on the employment of labor that automation has its greatest impact. A recent Ph.D. dissertation by David G. Osborn[1] at the University of Chicago revealed that in twelve cases of automation ranging from chocolate refining to railroad traffic-control the reduction in employee requirements ranged from 13 per cent to 92 per cent with an average reduction in employment of 63.4 per cent.

In the oil refining industry employment has fallen from 147,000 to 137,000 in the last seven years although output rose 22 per cent. The Federal Reserve Index shows that production in mining and manufacturing was about the same at the end of 1954 as at the beginning but total employment in these indus-

---

1. David G. Osborn, *Geographical Features of the Automation of Industry* (Chicago: Research Paper No. 30, Univ. of Chicago Press, 1953).

tries was down by almost a million. It is often said that such declines will be offset by increases in employment in the most dynamic sectors of the economy but even in the electrical machinery industry itself employment remained constant at about 1,100,000 from 1952 to 1954.

It is true that there have been no mass layoffs, from automation but this is apparently because automation has proceeded slowly enough so far to allow normal turnover to disguise the displacement. The worker displaced is not fired. He is the one who is not hired.

Another rather subtle form of displacement is in the so-called "hidden unemployment" of downgrading. It is true that automation creates a demand for new skills of a higher order and no doubt there will be a long run upgrading of the labor force. However, because automation renders many skills obsolete and dilutes other skills by a further division of labor, and since the new skills require extensive training and education, workers may not be able to move easily into the new jobs. When they cannot they are often downgraded in work even though their pay is not reduced.

There is further evidence of this lack of upward labor mobility in the critical shortage of engineers and other highly trained specialists.

A recent National Science Foundation study shows that out of the upper 25 per cent of high school students about half are unable to go to college and another 13 per cent drop out before finishing college. Thus nearly two-thirds of those with the greatest potential for scientific leadership never receive a college education through to the Ph.D.

Industrial location is affected by automation and this, in turn, affects employment. There could be a shift in labor oriented industries away from low labor cost regions for two reasons. First, the smaller labor force reduces the savings from lower wages and second, there is a smaller wage differential between the skilled worker of different regions than between the unskilled and it is the more highly skilled workers who are likely to be retained if automation is introduced. For example, the new Corn Products plant at Corpus Christi, Texas was located in an area which normally would not supply a large, skilled labor force. However, since automation reduced the importance of a large labor supply this plant could be located closer to its markets and its sources of raw materials and fuel.

Since automation will be limited to industries which now employ only about 25 per cent of the labor force and because automation creates many new jobs for which the necessary education and training will delay the entry of young people into the labor force, there would appear to be no reason to fear long run, mass unemployment. However, there is no automatic regulator in the economic system that guarantees full employment and the great advantages of automation can be insured only if there is a continued expansion. As automation advances in our basic industries the American economy becomes like a rocket which must continue to accelerate or else fall from the sky. This leads to the next criterion of successful economic performance—the necessity for economic stability.

Here also the long run outlook is good but the short run poses problems. In general there is no more reason to expect a recurrence of the depression conditions of the nineteen thirties than there is to expect another epidemic of smallpox. In both cases the causes are well known and the remedies are effective if they are applied. However, automation makes the need for vigilance all the more imperative because it has unstabilizing effects in the short run just like the original Industrial Revolution had. By greatly increasing the fixed costs of the plant and setting a continuous pace at which it must be operated the adverse consequences of shutdowns are magnified.

Unfortunately the very increases in efficiency and technological progressiveness which automation brings are a potential threat to continued stability. The abundance of production itself which increases living standards also frees people from spending all of their incomes unless they so desire. Whenever basic necessities can be secured by most people with only a part of their incomes, full employment becomes precarious because prosperity is then sustained by that portion of total spending which is dependent on confidence rather than on physical needs. A prosperous economy is always potentially unstable in the sense that small changes in expectations can have magnified effects.

The costs of instability, however, are large for everybody. The businessman must maintain expensive inventories and hedge against price changes, if possible, or else take great risks of loss. The cost to the worker, however, is the greatest of all from a personal standpoint because he lives from day to day and thus he suffers first, and most acutely, when his income falls. It is clearly the responsibility of businessmen and government—since they are the basic economic decision-makers of the country—to insure as high and stable a level of production and employment as possible.

The third criterion of economic performance—maximum satisfaction of consumers' desires—seems to be well met by automation. The great increase in output and improvement in quality of goods is bound to raise living standards if full employment is maintained. There are a few danger spots even here however. A wealthy economy like the United States must continue to be reasonably equalitarian because its prosperity depends on mass purchasing power. If the benefits of automation are not shared with workers in the form of productivity wage increases and with consumer in the form of lower prices these mass markets will be threatened.

There is an additional need for maintaining a high level of consumption. This is because automation does not seem likely to create the great waves of primary and secondary investment that earlier technological developments did. The automobile, for example, stimulated vast investments in the oil, rubber, highway and construction sectors of the economy. If the electronics industry does not call forth such secondary investment, consumption will need to rise to fill the gap.

The fourth and fifth criteria of economic performance—efficiency and progressiveness—are both well met by automation. By its very nature automation increases productivity and accelerates technological progress. In the University of Chicago study referred to above, productivity increases in twelve cases of automation ranged from 14 per cent to 1,320 per cent in a case of office automation with the average for all cases being 382 per cent. Space requirements alone were reduced from 12 per cent for printed circuit fabrication to 94 per cent for lard rendering with an average for all cases being 59 per cent.

Whatever the short run maladjustments and conflicts may be, automation favors the long run improvement of economic well-being. America's high living standards are not due to any monopoly of knowledge, brain power or industriousness. They are due largely to the enormous amount of capital equipment which both sides of industry—management and labor—have with which to work. This capital increases efficiency and automation accelerates the process.

Although automation rides the wave of the future, it is understandable why workers and consumers should be concerned. Our leaders of industry, who are men of vision, are also men of wealth and position who can afford to take the long run view. But the rest of us live in the short run, unfortunately, and that is where the potential dangers lie. A high degree of public responsibility from the leaders of industry, labor and government will be required if the mistakes of the first Industrial Revolution are to be avoided.

WALTER S. BUCKINGHAM, JR., Professor of Economics and Industrial Management at Georgia Institute of Technology, was born in Florida, received his B.S. degree from the Georgia Institute of Technology (1948), M.S. (1949), M.A. from Indiana University (1950), and Ph.D. (1951). He is also Research Economist at the State Engineering Experiment Station and Management Consultant and Director of Executive Courses at the Georgia Institute. He served 3½ years in the U.S. Navy during World War II, and is the author of such works as *The Theoretical Economic Systems* (1958).

Selected Bibliography

G. B. Baldwin & G. P. Schultz, "The Effect of Automation on Industrial Relations," *Monthly Labor Review* (June, 1955).

Solomon Barkin, "Automation, Productivity and Industrial Relations," *Proceedings,* the Seventh Annual Meeting, Industrial Relations Research Association (New York, 1955).

Yale Brozen, "The Economics of Automation," *American Economic Review* (May, 1957).

John Diebold, "The New Industrial Revolution," *Nation* (September 19, 26, October 3, 1953).

Peter F. Drucker, "The Promise of Automation, *Harper's* (April, 1955).

Paul Einzig, *The Economic Consequences of Automation* (New York: W. W. Norton, 1957).

E. M. Hugh-Jones, Ed., *Automation in Theory and Practice* (Oxford: Blackwell, 1956).

"How Automation Affects Employment—A Survey," *Management Review* (March, 1956).

O'Mahoney, Diebold, Campbell and Buckingham, *The Challenge of Automation* (Washington, D.C.: Public Affairs, 1955).

Frederick Pollock, *The Economic and Social Consequences of Automation* (Oxford: Basil Blackwell, 1957).

Pyke Magnus, *Automation: Its Purpose and Future* (New York: Philosophical Library, 1957).

Ted F. Silvey, "The Impact of Automation and the Workers," *Free Labour World* (October, 1955).

Report of the Subcommittee on Economic Stabilization to the Joint Committee on the Economic Report, Congress of the U.S., on *Automation and Technological Change* (Washington, D.C.: U.S. Government Printing Office, October, 1955).

CHAPTER 20

AUTOMATION AND RESPONSIBILITY

WILLIAM W. BARTON, *President,*
Barnes Co., Rockford, Ill.

My company's primary products are special machine tools. Our customers are mainly the mass producing manufacturers of the country. We process, either "for them" or "with them," their parts. We suggest methods of manufacture, quote the cost of constructing the machines necessary to accomplish these requirements and estimate the rate of production that the machines will produce. Our company has been in the machine tool field since 1872 and in the special machinery field since 1924. As such, I should say that we have always been active in promoting and developing automation.

Automation is a new word for a very old process. The trend of automation, I suppose, had its origin as far back almost as the invention of the wheel. Sometimes I wonder if the Athenian legislature may not have inquired into the economic implications of the advent of the harnessing of the energy of the wind. Did they inquire into the fate of all the galley slaves that the sail put out of work? I doubt if those replaced workers worried about losing their jobs as much as we see some of labor worrying today about the loss of theirs.

But remember that the history of the effort of man is to harness and utilize more and more energy and to replace by mechanical technology or automation the jobs of the galley slave so as to free him and his progeny from manual labor.

I could spend a week detailing examples of automation in various fields, such as the office, the telephone exchange, the kitchen, or others, or in exploration in greater detail examples of the machines and installations of our Company.

All of this would add up to the conclusion that automation always has been with us and is the backbone of our economic progress. This conclusion is based upon the knowledge that, since the greater the total capital investment per worker, the greater the productivity of any society—and the more a society has to divide, the higher the standard of living. The degree of this advance is the measurement of society's economic progress.

I, therefore, inquire as to the definition of automation.

Harder, of Ford, who has been given credit for coining the word automation, originally gave it a very narrow meaning, defining it as "handling (of) parts between successive production operations."

Peter F. Drucker,[1] in his recent articles in Harper's Magazine which attracted so much attention, stated that

"Economic progress might be defined as the process of continually obtaining more productivity for less money. The means to achieve this is INNOVATION—the improvement in living standards is the result of innovation."

He then goes on to state that innovation may be "technological" or "non-technological" and that the innovations that have had the greatest impact on our economy have been the non-technological, such as our changes in distribution—our development of the new concepts of business organization—the new basic management tools such as controls of budgets, cost accounting, and production scheduling. He defined automation "as the use of machines to run machines" and proceeded to explore the philosophical foundations thereof.[2]

Norbert Wiener, the MIT mathematician, predicted that automation would lead to "the human use of human beings" and declared that "the automatic machine was the precise economic equivalent of slave labor."[3]

Many others have incorrectly defined or spoken of automation as the "second industrial revolution."

---

1. P. F. Drucker, "Promise of Automation," *Harper's* (March, 1955), 31.
2. Drucker, "Promise of Automation," *Ibid.* (April, 1955), 41.
3. "Business Week Reports to Readers on Automation," *Business Week* (October 1, 1955), 78.

Carroll W. Boyce, in his article in the September copy of *Factory*,[4] stated that

"To most people, automation is just about anything that spells technological progress (or) if it means more production with less work, it's automation."

The article proceeded to outline what automation would mean to our economy. It was followed by comments by a dozen or so well-known men among whom was Senator Joseph C. O'Mahoney.

The October 1955 issue of *Business Week* in an article on the subject finding so many divergent definitions, stated that it was "fool-hardy to try to add still another."[5] The article then proceeded to analyze some of the features and areas of application.

Personally I prefer to think of automation in the larger sense —as an innovation created by man to increase his production; TECHNOCRACY, if you will.

AUTOMATION'S APPLICATION

In a broad sense, it can be stated that any feat that hands and body can perform can be duplicated automatically, given enough time and money.

Too often recently those misguided individuals who fear the term automation and some of those who welcome it have left the impression that the trend is spontaneous and new.

Analyzing the application of automation under the impetus of the ever faster pace to today's society—and studying the means with which it has been put to use in the past, we find one underlying fact—that automation does not just *grow*—nor is it, nor will it ever be applied overnight to every activity of man. Were this possible, then automation should be greatly reverenced, for we should suddenly have achieved that long looked for Utopia.

Automation's applications are a studied piece-meal process, created only after its *need* is recognized and measured and then only after its cost is weighed against that *need*.

---

4. C. W. Boyce, "What Automation Means to America," *Factory Management and Maintenance*, CXIII (September, 1955), 84.

5. "Business Week Reports to Readers on Automation," *Business Week* (October 1, 1955), 78.

Its creation is not a product of a moment or of one individual. Generally it is a result of years of effort, of trial and error, and the product of many creative minds employed to answer the question of how.

Sometimes the need is recognized and known to exist; yet, the end product to satisfy that need can only be made commercially by automation.

Successful automation could be referred to as the "accumulative correction of errors."

In our narrow field of the application of automation we never produce a machine or appliance that is not obsolete before it is shipped. Every machine we build can be improved upon and often is by our competition or customer. Every repeat order for a piece of automation produced by us can be bettered. The decision as to whether we try to do so or not is economic, a measure of the selling price, the cost, and the degree of success of the previous model.

AUTOMATION'S PRODUCTS AND BY-PRODUCTS

The product of automation can only result in good for society as a whole—and it is gratifying to see that almost every article on the subject, including those written by the labor leaders, express this thought.

It is true that in the past many attacks have been made upon certain applications of automation, but never for long if the innovation truly freed man's hands and produced more goods with less work.

On the other hand, it is equally as axiomatic that the introduction of automation has and will create dislocations in the labor force—these may affect a few or many; they may be departmental wide, plant wide, or community wide.

It is also true that while upon the whole the result of automation will upgrade the vast majority of labor and make available more and more leisure time, that there will be those cases where because of age or lack of will or mental inability, some will be downgraded.

But, when a radio poll in Detroit showed that listeners feared automation next to Russia instead of welcoming it as the basis of the well-being of the city, the magnitude of the misapprehension and false fear of a large portion of our society is better understood.

This type of reaction is similar to the reaction of a large section of our society towards atomic energy. Only recently are the people of our country awakening to the human benefits of atomic energy, and then, only after a great national effort to publicize the "atoms for peace" movement.

RESPONSIBILITY CREATED BY AUTOMATION

Accepting as facts the broad concepts and without exploring in more detail other causes and effects, I should like to be so bold as to add a few of my personal thoughts on the responsibility that the three major sections of our economy—industry, labor and government should take towards automation.

All three of these sections of our society must undertake to support and encourage the orderly growth of more and ever more automation. They must constantly publicize and educate all of society of the good that its greater productivity will create.

Industry through enlightened management and ownership needs to recognize that the increased production (that is the increase in the wealth produced) must be divided between labor and owners in a fair manner. The division of this increase cannot be *only* between higher wages and more dividends. Some of the increase must be used to support the retired and displaced worker. More and more of it must be plowed back into capital. If automation's trend is to continue, industry must constantly have faith in the future and spend more for development, expansion and plant improvement.

Labor must not be short-sighted and demand so large a share of the added wealth that it will stifle the ability of industry to place more capital per worker at labor's disposal.

Government at the local, state and national levels must not neglect their obligation to enforce a fair division of our created wealth and must not spend their share ineffectively or wastefully.

If there was ever a proper place for the encouragement of automation, it is in government for its benefits there can only result in the good of the society as a whole.

Industry, through education within its own limits and divisions, can improve the humanitarian relation toward its workers, can seek to encourage better means of lightening the burden of the displaced worker, and can through education and training assist in the upgrading of its labor. It is, of course, true that many units of industry are making great progress along these lines but many others are not. Labor relations' policies of today are a vast improvement over those of only a few years back. Both labor and government should recognize and differentiate between those units of industry which are progressive and those that are not and encourage the progressive units.

Labor, and particularly those divisions that are organized, have in many instances of late years better recognized their share of responsibility, but much improvement can be asked for yet. It is not enough any longer for the Unions to demand for their members only higher wages and more fringe benefits, but they must seek to find better means to tax their membership for the aid of the technologically displaced members. This may sound like heresy to some labor leaders but it has been a long established and accomplished fact by others—much more progress can be achieved, however, in this field.

There should grow within labor a better knowledge of the "harm to all" that the encouragement of "feather bedding" does, for if automation is *good for all,* then inversely "feather bedding" is *bad for all.*

Nor will the Guaranteed Annual Wage, in my opinion, prove to be the blessing that some of its authors think it will be, and large sections of labor can never fall under its protective wing.

Labor must also recognize and educate its members that it is the duty of all workers to deliver an honest day's work and that a "dole" of any kind from any source is a harmful thing, a temporary measure to assist in cases of hardship; that it must not be abused and that every effort humanly possible should be exerted to get off and back to gainful employment.

Government also on all levels can be mindful of these factors and should not establish allowances for unemployment either so low as to be of no help or so high as to encourage idleness. Much better control over this aspect of the problem can be done at local and state levels than is now being done.

Labor by better co-operation with industry can accomplish much more than is now being done to improve the education and upgrading of its members. If by no other means than through a publicity campaign on its part, they can explain the good of automation and encourage their members to prepare themselves for better and more responsible jobs.

Government responsibility being for "THE GOOD OF ALL" must keep the confidence of the governed high—must be mindful of the good of the whole even at the expense of the few. This does not mean that government has to accomplish everything by legislation, particularly at the national level. Much can be accomplished better by the orderly assembling and evaluation of facts then armed with the facts give proper emphasis and publicity to the conclusions. Many correcting measures can then be taken at local and state levels with better and more equitable results. Such results can assist in shortening the suffering in distressed areas.

A recognition on the part of all sections of our society from the "individual" to the "congress of the people" must realize that our society must be kept dynamic and its produced wealth per capita kept ever increasing.

Automation must be encouraged—means must be found to force its growth, by more liberal tax and depreciation policies, by better aid to our small and growing progressive concerns, and by assistance of the displaced worker without destruction of his will or need to work.

WILLIAM W. BARTON has been President of the W. F. and John Barnes Company in Rockford, Illinois for more than twenty years. A graduate of the United States Military Academy, he served in the Army Field Artillery for a number of years before becoming associated with the Chelsea Fibre Mills in New York. Shortly thereafter he joined the Barnes Company and began his training and five years later administration in the field of special machine tools. Prior to World War II, through his foresight, Rockford Ordnance Plant was built and operated until 1953. This division was cited many times for its production of war equipment. Also during this period he was a member of a United States Strategic Bombing Survey Group in Europe. His interest in automation has aided Barnes in progressing with the new trends in the field. The Company has not only expanded in size but has broadened its interests to include developing, designing and building specialized mechanical, hydraulic and electrical equipment of various sorts as well as food processing and canning machinery. Barton's detailed knowledge and understanding of machine tools have been a major factor in establishing their special automotive equipment. During recent years he has become keenly interested in the atomic energy field and aided in the production of a linear accelerator. One of these machines is being utilized by the Barnes Company, under the name of Midwest Irradiation Center.

CHAPTER 21

QUALITY IN AN AUTOMATED ECONOMY

JOHN T. RETTALIATA, *President*
Illinois Institute of Technology

Automation has been called progress in action. The distinguishing characteristic in automation is the emphasis on a continuous flow of work as contrasted to stop-and-go methods that retard the progress of even the most advanced conventional forms of mass production. Machines pass parts to one another, give orders to one another, and inspect their own product.

To some, the process is a new industrial revolution; to others, it represents evolution, an extension of mechanization. Whatever the viewpoint, it must be apparent to all that automation will be employed to an increasing extent in the years ahead.

All signs point to a future economy of almost limitless proportions. The nation's population is expected to double to well in excess of 300 million persons in the next 50 years; the gross national product, it is predicted, will double to more than $800 billion in the next 25 years.

Obviously, conventional methods of operation cannot be relied upon to attain the kind of economy being prophesied. Its realization will require utilization of, and further advances in, all of the techniques of automation, controls, electronic computers, motion study, materials handling and others, in addition to some yet to be devised.

And, of necessity, automation will be employed increasingly for reasons of more economical production and to compensate for a relative decline, compared with national output, in the effective labor force.

A recent issue of Instruments and Automation (February 1957) carried the results of research conducted by James R.

Bright of the Graduate School of Business Administration of Harvard University. Mr. Bright investigated 13 highly automated manufacturing plants and, among other things, questioned them on their primary motives in automating.[1] He found that the four principal motives, listed in the order of frequency, were (1) reduced direct labor content per unit output; (2) increased capacity; (3) reduced indirect labor, and (4) higher quality of product. Incidentally, he found that none of the plants had experienced a cut back in the total labor force as a result of automating.

The incentive to automate is that of profit, a motive which, properly conceived and applied, is legitimate and constructive in our competitive, free-enterprise economic system.

Automation can lower production costs through higher productivity of the individual worker and through better production and business control. Automation can, and almost always does, yield a better quality of manufactured product or processed item. The combination of these characteristics leads to a more favorable relationship between quality and costs from the consumer's standpoint, which improves profit and, at the same time, advances the standard of living.

Thus, the significance of automation is its worth in terms of economics and people, the latter both as consumers and employees. To be worthwhile at all, automation must produce results in terms of increased production, lower costs, better products, greater safety and other benefits.

In creating the greatest satisfaction for the least expenditure of human effort, automation is a phase in the historic process through which American technology has gone. Our technology has given us the highest standard of living ever known. It has lightened our physical burdens while reducing our hours of labor.

With little more than 6 per cent of the earth's population, and about the same percentage of its land area, we enjoy an abundance greater perhaps than all of the rest of the world combined.

In the plant automation starts with the facilities for processing, handling, and control, all integrated to produce the desired results. The key element is proper integration . . . the organization of the operation into a complete whole. The full potentiali-

---

1. J. R. Bright, "Myth of Automation," *Instruments and Automation* (February, 1957), 249.

ties of automation cannot be reached if we think only in terms of the individual machine, or even several machines. The entire process of manufacture from raw material to consumer, must be geared to automation.

Automatic production differs from manual efforts in many ways, of course. It relieves the men and women on the assembly lines of most of the heavy work that is present in many operations, enabling them to devote their entire attention to the more vital aspects of the operation.

Better quality of inspection . . . the key to quality production . . . can be performed on an economic basis as the element of human fatigue, often the result of sheer boredom, is eliminated. Nor are machines subject to distractions that sometimes affect the individual.

There is another important difference between the machine and people. Machines cannot be laid off if sales slip. The huge investment in an automated production line requires that all phases of the business . . . raw material procurement, manufacture, warehousing, sales, and distribution . . . be scheduled for the most effective and economic use of the production facility. In a sense the entire company must be automated.

These goals are attainable only through top management understanding and program coordination. New management problems are introduced. Plant communications take on much greater importance; proper employee education and training are vital; increased emphasis on long-range planning becomes necessary; cost elements and job descriptions must be revised.

These are functions which cannot be delegated to a machine. They call into play another quality . . . the human mind. Quality control, like other developments in industrial progress, must originate in the minds of men. The devices they create and apply are merely new steps in the long train of human effort, which began with the invention of the wheel, and which substitute other forms of power for human muscle.

Industry faces a challenging task. Population increases and rising incomes will swell the demand for goods and services, and the facilities of production, distribution, transportation, and communications must be increased to meet these needs.

Bureau of Census figures indicate that in 1965 there will be more than 190 million Americans. There are about 172 million today.

New families will need new homes, and everything that goes

into them. There will be need for more hospitals, stores, churches, schools, water, food, electricity, and many other things necessary to sustain life and advance further the standard of living.

These are some of the challenges that demand intensified research and development in the years ahead. It will be necessary to bring into play forces which do not now exist. These will be in the form of further advances in technology brought about by new developments and the useful application of knowledge yet undiscovered.

This emphasizes the importance of well-educated men and women who will be capable of conceiving and developing the tools industry needed to produce the things necessary for the satisfaction of the public's ever-increasing demand for better living.

In particular, it will be the scientist and engineer to whom we shall have to look to accomplish this. But, even at today's levels of production, we are experiencing serious shortages in these fields, and we can expect to continue to do so for years to come.

In a period of spiraling technology, we have witnessed drastic reductions in the number of scientists and engineers being graduated by the nation's colleges and universities. In 1950, the total of graduates in these fields was nearly 110,000. Five years later the total had been cut in half, amounting to less than 54,000.* Only recently have we started upward again.

We must have more engineers, and we must have engineers possessing the quality of superior education. New scientific discoveries, fast moving technological developments in automation and other fields, and the increasing complexity of our social and industrial structures demand more engineers with more education.

We are entering an era in which the undergraduate program must be modified in order to produce the well-rounded engineer. There is too much to learn in too short a period.

As a result, steps are being taken in the nation's colleges and universities to prepare engineers who are well grounded in fundamentals and possess the ability for creative thinking. More attention will be devoted to the sciences and the humanities, with

---

* Figure based on statement by James R. Bright at Armour Research Foundation of Illinois Institute of Technology sponsored Automation Conference, Chicago, April 14-15, 1956.

less emphasis on application. Specialization will move into the graduate area.

For many years the bachelor's degree has been the accepted badge of admission to the engineering profession. But it is becoming more difficult at the undergraduate level to prepare a student adequately, and it is my belief that it will not be too long before a graduate degree will be the minimum requirement in engineering, as is the case in the sciences.

Since scientists and engineers will exert a major influence in the development of the greatly expanded industrial economy of tomorrow, they must be prepared to accept corresponding responsibilities.

The first of these, and perhaps the most important, is that of leadership.

There have been changes not only in our processes and products, but in our organizational structures, in growing corporate complexity, in methods of financing and selling, in the relations of business with labor and government, in concepts of business ethics, and in the laws under which business and industry operate.

The changes will affect everything the engineer does. Upon him, in increasing measure, business and industry will place greater reliance. He will move more and more into the broader spheres of management as a whole.

It is the engineer's knowledge and skills, his foresight, judgment, and decisions that weigh heavily in the successful outcome of the aim of the industrial venture, which, of course, is earning a profit.

From a fractional representation a few decades ago, about 40 per cent of present-day industrial management is engineer-educated, according to a Columbia University survey. The engineer is replacing the lawyer and financier in top industrial posts of the nation.

If the engineer is to be counted upon more and more to share the helm, and to have a larger sphere of influence in this high-paced, dynamic economy, two things must happen: (1) The ranks of the engineering profession must be continually reinforced and replenished. (2) The educational process must encompass vastly more than the imparting of scientific knowledge and technological skills.

The task of making the engineer increasingly useful to society devolves primarily, perhaps, upon the engineering colleges. We who represent the engineering educational institutions know that the embryo engineer must be exposed to economics . . . history . . . psychology . . . political science . . . law and literature.

The young engineer must be made aware of the cost factors in production, the labor factor and market ability. He must be made conscious of advertisers, suppliers, and customers . . . all of this beyond a solid grounding in the basic principles of his chosen engineering field. It is a tremendous responsibility, this preparation of the engineer for a world of new and vital comprehensions.

As I have indicated, the problem of engineering education is keenly recognized by the technological colleges of the nation. In my own institution, we have steadily been effecting changes and extensions aimed at greatly broadening the base of engineering education and building into the engineer the potentialities of widest usefulness in our industrial civilization.

The task of the technological institutions, however, is not simply turning out a finer product. We must educate enough of them . . . the right ones . . . and educate them superlatively well.

The challenge before us is to intensify greatly the effort to interest more young men and women in the profession, to give them compelling motivations for entering the field. We can use, and would welcome, a great many more women engineers, and there are attractive opportunities for them in many fields.

I would call your attention to the striking fact that only six of every 10 of the top 5 per cent of high school graduates go to college.

A major effort must be made to bridge the gap between high school and college . . . to eliminate or reduce the tragic loss of potential scientific and engineering talent through the failure of great numbers of college caliber high school graduates to continue their formal education at the higher level.

There are other calls upon the American engineer which do not permit discussing. One, however, is too vital to omit. That is the responsibility of conserving our natural resources.

In America we rejoice at our good fortune. Our abundance is great compared with that of other nations. But we are paying an enormous price for it.

And we are paying for it in terms of a commodity much more valuable than dollars. We are paying for it through expenditure of our natural resources which the voracious appetite of an expanding economy is depleting at a rapid rate.

We have been living in a period of raw material luxury. We now face ever-increasing periods of poverty of natural resources. Conservation, substitution, and replacement must be the necessary and essential ingredients of our future economy.

We must cope with the doubly-difficult assignment of achieving a greatly augmented economy at the same time that the resources required for such an economy are diminishing.

Automation, despite its proved performance and tremendous progress in recent years, is in its infancy. Anything which relieves man of drudgery, while giving him better quality products at the same or lower cost, is fundamentally good. Its use will expand as its benefits become more widely known.

JOHN T. RETTALIATA, an internationally recognized authority on steam and gas turbines and jet propulsion, was named president of the Illinois Institute of Technology on February 4, 1952 at the age of 40. He also is president of two important organizations affiliated with the Institute: Armour Research Foundation of Illinois Institute of Technology, one of the nation's largest independent research organizations, and the Institute of Gas Technology, the utility gas industry's own educational and research facility. A native of Baltimore, he received his bachelor of engineering degree (1932) and his doctor of engineering degree (1936) from Johns Hopkins University. In 1956, he was awarded the honorary degree of doctor of engineering by Michigan College of Mining and Technology. He joined Illinois Institute of Technology in 1945 as director of the department of mechanical engineering after nearly 10 years of engineering development for the Allis-Chalmers Manufacturing Company in Milwaukee. He became dean of engineering at Illinois Tech in 1948, and vice president in charge of academic affairs in 1950. During World War II, he visited Europe at the invitation of the Navy, where, in 1943, he studied British developments in jet propulsion; in 1944 he was named to the subcommittee on turbines established by the committee on power plants for aircraft of the National Advisory Committee for Aeronautics; and in 1945, he investigated the steam turbines developed by Germany for hydrogen-peroxide submarine operation. He holds directorships in 13 corporations.

Selected Bibliography

N. H. Ceaglske, *Automatic Process Control for Chemical Engineers* (New York: John Wiley, 1956).

S. Lilley, *Automation and Social Progress* (New York: International Publishers, 1957).

E. M. Grabbe, *Automation in Business and Industry* (New York: John Wiley, 1957).

Samuel E. Rusinoff, *Automation in Practice* (American Technical Society, 1957).

John Diebold, *Automation, The Advent of the Automatic Factory* (New York: D. Van Nostrand, 1952).

Carl Dreher, *Automation: What It Is, How It Works, and Who Can Use It* (New York: W. W. Norton, 1957).

Paul Einzig, *Economic Consequences of Automation* (New York: W. W. Norton, 1957).

Gilbert H. Fett, *Feedback Control System* (New York: Prentice-Hall, 1954).

Millard H. La Joy, *Industrial Automatic Controls* (New York: Prentice-Hall, 1954).

American Management Association, *Keeping Pace with Automation* (American Management Association, 1956).

H. S. Levin, *Office Work and Automation* (New York: John Wiley, 1956).

L. R. Bittel, *Practical Automation* (New York: McGraw Hill, 1958).

Floyd E. Nixon, *Principles of Automatic Controls* (New York: Prentice-Hall, 1953).

E. M. Hugh-Hones, *Push-Button World: Automation Today* (Norman, Okla.: University of Oklahoma Press, 1956).

C. R. Walker, *Toward the Automatic Factory* (New Haven, Conn.: Yale University Press, 1957).

CHAPTER 22

AUTOMATION WILL BEAR WATCHING*

One highly gratifying thing which appeared throughout the hearings was the evidence that all elements in the American economy accept and welcome progress, change, and increasing productivity. Not a single witness raised a voice in opposition to automation and advancing technology. The fact that representatives of organized labor are watchful lest the material gains of automation become the sole objective, without recognizing the individual hardships that may be caused by job losses and skill displacements, ought not to be turned into a charge that labor, as such, is obstructive to new developments. Both organized labor and management are apparently aware of and intent upon seeing that these human elements are not disregarded.

However much we may welcome the fruits of advancing technology—however optimistic one may be that the problems of adjustment will not be serious—no one dare overlook or deny the fact that many individuals will suffer personal, mental, and physical hardships as the adjustments go forward. The middle-aged worker particularly, who may find his skills rendered obsolete overnight or his job abolished as his work is turned over to a machine, has every right to expect that industry, his union, and society will recognize his plight and assist in his retraining, or his relocation if necessary.

The trend toward automation will bear watching to make sure that it does not add to troublesome pockets of local unemployment. When we are told, for instance, that automation in Detroit means unemployment in South Bend, Ind.—when we know that

* This chapter is an edited summary of The Report of the Subcommittee on Economic Stabilization to the Joint Committee on the Economic Report, Congress of the United States, on *Automation and Technological Change* (Washington, D. C.: Government Printing Office, 1955).

such progressive steps as the dieselization of the railroads are partly responsible for persistent unemployment in such localities as Altoona, Pa.—when it appears that automation, by speeding obsolescence of northern cotton mills, contributed to a major shift in the location of that industry—it is imperative that industry itself, with the sympathetic support of labor, must develop specific and concrete programs to ease the problems of adjustment. To the extent that those directly involved fail or are unable to cope with the problem, the Federal Government may find it expedient and desirable to assist local people to find solutions to these problems rather than risk their spreading to larger areas of the economy.

These hearings will not have been in vain if, in arranging for them and hearing the many helpful witnesses, a feeling of social consciousness about the problem has been stimulated. It is easy for those in business who are absorbed by cost reduction to forget that automatic production, if it means fewer and fewer jobs and a disregard of human costs and hardships, will in the end be damaging to the foundations of our free society. While most industrialists, by their willingness to consider these problems have demonstrated understanding of the social responsibility of free business, the subcommittee, unfortunately, found evidence that some of those busy in advancing the technical side of labor-saving machines are still apparently unaware of the overall significance which their activities have to the economy.

The best and by far the most important single recommendation which the subcommittee can give is that the private and public sectors of the nation do everything possible to assure the maintenance of a good, healthy, dynamic, and prospering economy, so that those who lose out at one place as a consequence of progressive technology will have no difficulty in finding a demand for their services elsewhere in the economy.

The subcommittee recommends that the spirit and objectives of the Employment Act of 1946 continue to be a safeguard and that the statutes be given active instrumentation and support by the executive agencies, the Congress, and the people as a whole.

The subcommittee also recommends that industry, and management for its part, must be prepared to accept the human costs of displacement and retraining as charges against the savings from the introduction of automation. In saying this, the subcommittee is not unmindful of—and was, indeed, gratified by—the extent to which enlightened management is already aware of and accepting responsibility in this respect. Nevertheless, by careful planning and scheduling, the adjustments of workers and due recognition should be given to the timing of investment and technological changes with an eye on the state of general business and the needs for increased employment.

Organized labor should continue to recognize that an improved level of living for all cannot be achieved by a blind defense of the status quo. The education of its members, of management, community leaders, and Government officials, such as has been provided by these hearings, is an important function of union responsibility as well.

Section IV

Automation and Society

CHAPTER 23

AUTOMATION AND EDUCATION

WILLIAM E. DRAKE, *Professor History and Philosophy of Education, University of Texas*

It is not possible to provide even a partial answer to the full effects of automation on education in the United States in the years just ahead; yet there is enough evidence already in hand to indicate that the changes will be of a revolutionary character.

THE MEANING OF AUTOMATION

In its most fundamental sense automation calls for not only the machines and power which characterized the first Industrial Revolution, but the programming and control which characterize the new. In programming, you tell the factory what to do, and, in control, you build the particular mechanisms into the machine that are needed to accomplish the desired purposes. The resulting effect is a complete integration of formerly separate units of production into a unified process, thus creating a "flow production" from raw material to finished product, including packing. In short, the word "automation" includes "all technical developments that make automatic development more possible."[1] Note, however, that automation is a novelty; it is the latest development of the application of science to daily life. And this process of industrialization, combined with the ever-increasing power of the forces of mass communication, has resulted in transforming all forms of social relationships. In turn, education in

---

1. England, Department of Scientific and Industrial Research, AUTOMATION (London: Her Majesty's Stationery Office, 1956).

our technological society is forced to face squarely these changes, since it has to not only utilize the educational principles and methods which time has proved to be conducive to the formation of good men, but also modify them in order to answer the needs flowing from the individual and social consequences of automation.[2]

AUTOMATION AND THE EDUCATIVE PROCESS

Glaringly enough, then, the implications of an automated society for the school are obvious. The complicated nature of an automated economy calls, for instance, for designing engineers in large numbers for the entire plant as well as product. It calls for the retraining of all workers and the upgrading of those of exceptional talent. Mastery of basic skills acquires increased significance because of the need for research in all areas and the high degree of specialization demanded. Automation calls for a high degree of integration in the arts and the sciences, of philosophy and mundane affairs. More discipline, more knowledge, a changed attitude, a rethinking through process, and a life time of reorientation, all are a part of the new era. A new type of executive, and a high level professionalized teacher are a part of the new order of business.

The first Industrial Revolution was the underlying force in bringing about the major educational developments of the past century. Industrial developments in Germany, England, the United States, Japan, and the USSR, called for a program of mass education. Along with the education of the people, there was a reform of higher education with major emphasis upon the applied sciences, individualism, specialization, and the practical arts. In a sense the whole movement was anti-intellectual, for it was both anti-rational and anti-humanistic. In education, the em-

---

2. UNESCO, *Education in a Technological Society* (Paris: UNESCO, 1952): C. D. Dobinson, "The Impact of Automation on Education," *International Review of Education*, III (1957), 385-398; International Labour Office, Report of the Director-General: Part I: *Automation and Other Technological Developments. Labour and Social Implications* (Geneva: International Labour Office, 1957); UNESCO, "Social Consequences of Automation," *International Social Science Bulletin*, X, 1 (1958); G. S. Counts, "Education and the Technological Revolution," *Teachers College Record*, LIX, 6 (March, 1958), 309-318.

phasis was upon quantity rather than quality, and further pointed up by the fact that freedom was defined in terms of an opportunity to make money rather than freedom to know.

There was much that was positive in the educational reforms of the past century. In the economic field, and by virtue of the development of the applied sciences, education made possible the mass production of foods and material goods, the like of which the world has never known. The United States became a land of abundance and the envy of the world. This was the end result of the development of the colleges of agriculture and mechanic arts, the great technological and professional schools, and the great schools of science in the universities. The foundations of this program were rooted in the development of a graded elementary school, a public high school, and the development of programs for the mass training of teachers. Of primary significance was the large program of research carried on in the graduate schools of the institutions of higher learning. Here there was significant advancement of knowledge in the social sciences and the humanities as well as in the physical and biological sciences.

On the negative side, while there was much application of knowledge to the further production of goods, there was far too little application of knowledge to the solution of or an attack upon the social problems created by or related to the Industrial Revolution. The teacher was politely told to stay away from those matters pertaining to religion, politics, race, sex, and economic interests, unless they dealt with something that happened a generation or so ago. The teacher was to teach good citizenship, but, at the same time, never take sides. A crisis situation may be said to have prevailed in the periods of the two world wars and in the early fifties now known as the Era of McCarthyism.

It does not seem possible that the social, economic, and political conditions which so marked the first Industrial Revolution can continue to prevail in the Era of Automation, primarily because (a) of the overall need for brains rather than brawn, (b) the rationalization of industry calling for a rationalized society, and (c) to consume the products produced by automation people must have high consumer purchasing power. It is conceivable, however, that many will not be able to return to their jobs when business picks up, that "millions of workers may suffer this kind of loss year after year even while statistics indicate that productivity, gross income, and the like are high enough to justify calling the over-all economic condition healthy."[3]

Present experience with the automating process has demonstrated the need for an upgrading of our educational program in all areas and at all levels, although those needs most immediately observable are in the fields of mathematics and the sciences. There is a need for better brains, for everybody to know more, for catching on quicker, for more intelligent planning, for more effective and more intelligent teaching, for a life time education. The principles involved relate to all aspects of production—cost studies, reorganizing the productive process, rethinking the product, organizing for a major change, designing, and redesigning. It could, almost, be said that not only do we need a new attitude, we need a new mind.

Foundationally, our greatest deficiency seems to be the lack of a sense of mathematical or rational mindedness; yet, the increasing complexity of not only the machines to be created and operated, but of the conditions of everyday life, demand a higher order of skill and social adjustment. The problem of integrating man and the machine to the larger productive end and the evolutionary increase in size and complexity, have made automation necessary in the area of information as well as production. Only through the method of mathematical analysis can such needs be met.

One has only to observe the complicated nature of some of the present automating processes, such as the electronic computer, to appreciate the relation of education to automation at all levels. One of the serious obstructions to our full development is the lack of an adequate supply of technically able young men. Here the problem is not so much one for the colleges but the high schools, the elementary schools, and the parents. The new approach demanded for machine designing is one of function, of means, rather than end or product. Such a development indicates a pronounced preference for the young man or woman who has had the necessary kind of formal training at the college level. Under these conditions there will be a definite preference for book learning over the knowledge that has been gained by working on the old time job, be it a tool and dye maker or an expert arc welder.

The increased emphasis upon and need for book learning

---

3. Warner Bloomberg, *The Age of Automation* (New York: League for Industrial Democracy 1955), 15.

places upon the school a far greater and more extended responsibility than has been true in any period of our past history. To meet this responsibility the school, at all levels, will need to operate with a higher degree of efficiency, freedom, and professional responsibility than has ever been true. There is a need for change in content, methods, materials, skills to be learned, and in basic educational theory. We shall need more and better books as well as a wider use of tape and film and work experience. Since the emphasis will be on higher knowledge, skills and insights, and since the machine will determine the output, the attitude toward work will be one of joy and satisfaction rather than slave labor.

Of necessity, there will be a shift toward an emphasis on the personality of the worker.[4] What kind of education will best enable the individual to fit into an automated industry and society? This applies to management as well as to the engineer, to the teacher as well as to the student or office worker. The issue of leadership is especially critical because of the challenge for world influence coming from the USSR.

New educational responsibilities seem to fall in four areas: (1) the training of a new type of executive and social leader, (2) the education of specialists—teachers, scientists, engineers, economists, political scientists, journalists, and research workers of all types, (3) the training and retraining of workers for creation, installation, and maintenance of automated machines, and (4) moral and ethical foundations, meaning and purpose of the new age.[5] The increased need for original ideas calls for more emphasis on creative education and research as well as a closer relation between industry and education. Also, as has been stated by one of the leading scholars in the field of automation:

> If we are to attack the roots of the problem of human resistance to technological change, considerable attention must be given to the upgrading of our educational system so that young people entering the labor forces will possess the abilities necessary to adjust themselves to industrial change.[6]

---

4. Harold H. Punke, "Social and Educational Problems of Automation and Longevity," *The Journal of Educational Sociology*, 30 (May, 1957), 401.
5. Cecil Rockover, "New Responsibilities Vested in Our Colleges," *Peabody Journal of Education*, 34 (May, 1956), 26-30.
6. John Diebold, *Automation* (New York: D. Van Nostrand Company, 1952), 131.

Probably the most serious and important of all educational problems confronting the schools in the new age lies in the area of meaning, purpose, and value. The need of the world that is passing was well expressed by John Dewey in his *Democracy and Education,* but, since World War I, the heated controversy between those of traditional philosophic mind and the pragmatists has continued, often at heated point. The crisis presented by the new age is one which demands that we, both groups, break through our present conceptual barriers in education to new grounds of thought and action in curriculum building, administration, teaching, and learning. To follow a behavior theory because of its parallel relation to circuits and feedback in the automated machine is the kind of mechanism that spells out, not only the doom of freedom, but the destruction of the human race.

AUTOMATION AND SCHOOL ADMINISTRATION

One can anticipate vast and significant changes in school administration in the age that lies ahead. In view of the higher professional level at which the schools must operate, both the political and business roles of the administrator will diminish in significance. On the other hand, in addition to being a leader in the socio-cultural sense, the administrator must be able to lead in creating the best possible conditions for teachers to teach and carry on research, and for pupils to learn.

Financing the public school must be at a level sufficient to enable the school to compete with industry and the other professions for the necessary brain power. To achieve this end the federal government of necessity must carry equal financial responsibility with the states and local governments. Mass education for all will be carried through the public community college. Beyond this point scholarships must be provided for all who are lacking in funds, but who have the genius to do outstanding work, for in the age of automation a premium will be placed on brain power.

From the standpoint of function, the most crucial of all issues in education is related to the already dangerous shortage in quality level teachers. The economic law of supply and demand having failed to work with teachers' salaries because of the fact that teaching has been involved in the art of politics rather than the logic of economics, must bring about the growth of a strong and unified profession of teaching. Teachers must participate effectively in policy making at all levels, and in cooperation with public officials, laymen, and administrators.

Maximum utilization of the school plant must be achieved by operating a twelve months school, six days a week. Each child and each teacher is to be on duty, however, only nine months out of each year, or for three quarters. The school must be re-organized so as to serve all the varied adult interests as well as those of the elementary, secondary, and junior college level. Now that adequate funds are available, much attention must be given to the problem of counseling and to individual differences and to the problem of culture lag, of conformity, of human values, and of leisure time. It should be recognized that the training of youth in the techniques of the industrial revolution is of no greater importance than its effects, demands and challenges.

It is clearly evident that the new era demands a more intel-lectual type of instruction in both the elementary and secondary school. The wide range of new skills demanded calls for a new approach. "Our young folks will have to be given opportunities for education on a much higher plane than now available. More and more money will have to be set aside by the government and industry for this purpose."[7] The crucial nature of the issue is such that every possible effort must be made to utilize all pos-sible human resources.

It is imperative that both the elementary and secondary school provide the kind of scientific orientation that will enable each child to think and create in terms both material and social. At the elementary level, there is too much of birds and not enough of sputniks. According to Professor Glenn G. Blough, President, National Science Teachers Association, science teaching in the elementary school continues to be of the "age of the bustle and the blacksmith." The fact of the matter is that the great majority of teachers in the elementary schools of the nation have little or no qualifications for the teaching of science and mathematics in knowledge, interest, or attitude. The fault is not that of the teach-ers, but rather that of the public, in not demanding and being willing to pay for the qualifications needed. The same holds true for the human sciences. The situation is so bad that in one of our states only one per cent of the elementary teachers has a four years degree. Furthermore, the situation is aggravated not only

---

7. E. M. Hugh-Jones, *The Push-Button World* (Norman: University of Oklahoma Press, 1956), 70-72.

by the fact that the great majority of elementary teachers are women, but that there is an enormous turnover in the staff.

That the curriculum issue is not merely a matter of science and mathematics is indicated by the fact that the principles behind automation are not basically scientific. This is illustrated by the contrast between science as knowledge and the problem of management. Scientists more often than not have a problem of communicating with their fellow men. The scientifically oriented program of the schools must be undergirded with a quality well-rounded humanized program based on the human sciences. One can expect an understanding relationship between the scientists and management, but such does not necessarily hold for management and labor. Also, it needs to be recognized that the man who was responsible for the digital computer (Babbage) could not write a coherent report of it.

The problem of the curriculum at the high school level involves much more than a high degree of specialization. The need is equally as great for a sound program of general education built on the quality program of the elementary school. New types of programs are needed to enable high school youth to arrive at a higher level of intellectual maturity, to acquire more skill in logical analysis, to develop a well disciplined social and scientific imagination, to acquire greater occupational efficiency, and to exemplify positive growth in human relations and value judgments.[8] The solid subjects of the secondary school, including science, mathematics, language, and the human sciences are a must for all secondary school youth who are able to profit from their study.

The pressing needs of the automation age are a major challenge to the secondary school. Among the needs that must be met are the development of more social insight and responsibility, a far wiser use of leisure time, more adequate and extensive science laboratories and equipment, a reduction of the gap between the knowledge of the scientist and common knowledge, more efficient and effective guidance programs, upgrading of vocational education, a tightening up on the solid subjects, and a careful re-examination of the total administrative program.

---

8. T. Eugene Holtzclaw, "Automation and the Curriculum," *The Clearing House*, 31 (March, 1957), 426.

AUTOMATION AND HIGHER EDUCATION

One of the most disturbing aspects of our national situation lies in the increasing shortage of qualified teachers at the higher educational level. President A. McLemore of Mississippi College is reported as saying that

"As a consequence (of more attractive remuneration in business and industry), the colleges have been deprived of a large proportion of their most talented personnel. The very wellsprings of America's future leadership are being dried up at their source."[9]

The significance of this statement can be more fully appreciated by emphasizing the point that there is serious doubt whether or not there are enough high level brains to fill the job demands even if we get them all through the universities, but we cannot begin to get the job done without well qualified institutions of higher learning. The importance of the task is indicated by the fact that in 1900 there were 250 factory workers to one engineer whereas in 1956 the ratio was only fifty to one.

There are those who have seen the solution to the problem of the shortage of quality university teachers in closed-circuit television. This is an interesting point of view, for it reflects the extreme mechanization of our age. While there are certain areas of knowledge which lend themselves best to the television medium, the essence of good teaching at any level is a matter of the quality of the human relation existing between teacher and pupil. This the television medium destroys far above everything else. Television not only standardizes, but it likewise puts a premium on showmanship. Here we have probably our best illustration of the potential vices and virtues of automation. Closed circuit television can be used to produce a canned, dehumanized educational product or to strengthen and supplement our human resources. It can never be a genuine substitute.

The job of the university is an all-around responsibility. Automation calls not only for the training of scientists, engineers, and managers, but for a higher degree of competency in philosophy, politics, law, medicine, social work, and all other areas of spe-

---

9. *Phi Delta Kappan,* XXXIX (January, 1958), 184.

cialization. There is the problem of locating individuals of special abilities as well as securing a wholesome balance between the various areas of specialization. There is the ever increasing need for a higher university output in teaching as well as in research. The need for a closer working relation between the university and the community is clearly indicated. Even in the field of language there is need for technological improvement.

An illustration of the new and larger responsibility to be imposed upon the university is to be found in the effect of automation on economic theory. It is possible that there will be considerable effect on the old law of supply and demand. Eventually, through planned output and through greater management responsibility, a stabilized production pattern can be developed. The secret lies in the area of production cost, but this does not necessarily imply a lowering of prices. Problems of foreign exchange, the business cycle, raw materials, wage scales, and undeveloped countries, all call for continued research and investigation at the higher educational level. This is not to minimize the overall significance of continued research in the area of electronics.

It is important also to note that the needs of the new era call for an extensive development of the community college. This is necessary not only to relieve the university of the cost and problem of carrying those incapable of carrying on a high level of quality work, but to provide a terminal program of instruction for all types of technicians and maintenance workers. In addition to an elaborate well equipped laboratory set up, the community college is to have a widely developed program of field service working in close contact with industry, as well as with community agencies.

Also, there is a deeper and more penetrating issue confronting higher education, that of basic function and philosophy. A reexamination of basic purpose is in order for neither the orthodox religion heritage nor the scientific individualism of the past century is adequate for our day.[10] Confusion, chaos, and indifference, must be supplanted by a sense of social direction and felt need.

10. Arnold S. Nash, *The University and the Modern World* (New York: The Macmillan Company, 1944), 98.

AUTOMATION AND ADULT EDUCATION

It is immediately evident that there must be considerable expansion of the adult education program to meet the changing conditions and demands of the new age. Evening schools, operated as a part of the public school program, are destined to reach a size equal to that of the day school. In order to carry on this program the present number of teachers will need to be more than doubled. The program needs to be designed especially for the retraining of both office and factory workers, for new careers, especially in areas such as recreation and government service.

In addition to the establishment of a broad and extensive evening school program, there must be a decided change in the use made of the medium of television. Instead of the adolescent level of cheap entertainment which so characterizes the modern T.V., we must have a solid program of adult education on the national level. This should be directed at the upgrading of the cultural level of the people, especially to the end of achieving a higher level of social intelligence. General knowledge in all areas, science, mathematics, language, history, economics, political science, music should be given special consideration.

To achieve this end it is necessary that the corporation make a considerably larger contribution to the financing of the educational program. In 1936, the corporations were contributing only $36,000,000.00, or one-third of one per cent of their income; $226,000,000.00 in 1945, or one and one-fourth per cent, and in 1954, only $314,000,000.00, or eighty-five hundredths of one per cent. When it is known that corporations should give up to five per cent of their income, the deficiency becomes very apparent. Only twenty and five tenths per cent of the alumni of colleges made educational contributions in 1956, while many of the 500 largest corporations have never given anything at all.

AUTOMATION AND HUMAN VALUES

The basic research of Norbert Wiener and others shows a parallel relationship between the operation of living individuals and the newer communications machines, but this is so only in a psychological and not in a philosophical sense. It should be remembered that, while man has created the machine, the machine **cannot create man.**

The field of cybernetics has done much to stimulate educational research in such psychological areas as memory, thinking, and learning, but when we turn to the areas of perception and conception the issue is not so clear. The idea that perception is nothing more than electrical impulses transformed into patterns is of the kind of mechanism that has been discarded even by the physicists. Along with scientific analysis, experimentation, and logical problem solving we must continue to rely on intuition and insight.

Possibly, the most significant educational question raised by automation is whether or not we can bring about deep rooted and vast changes in the quality and structure of our society. It needs to be emphasized that professional philosophers like George Orwell and Aldous Huxley may prove to be more than scientific fiction writers unless we take them seriously. Our danger is not only of being controlled by the machine but of being moulded into one.

Since the eighteenth century man has passed through a cycle, from religious tyranny through rationalism to scientific individualistic mechanism. Are we now to have a few people commanding the labor of many slaves, a push button age where man is but the joker in the deck? Sound logic tells us that the most important problems of our age are of a social and philosophical character. To illustrate, we know that the power of wealth without intelligent mercy is more destructive than an atom bomb, that leisure without culture is disastrous, and that high income and low culture result in terminal decadence. Also, there is the question as to what will be the effect on our people of the passing of the gospel of labor and of our increasing involvement in cheap programs of commercialized entertainment.

One truth stands out from the reports that have been made in the field of automation and that is the imperfections of our present knowledge of economic and social forces as compared with our knowledge of technical possibilities. Thomas A. Edison once remarked that "the dynamo of our God-given ingenuity is running ahead of our equally God-given humanity."[11] The effects of our push button civilization upon our political, social, economic, and social ways of living are so near at hand that there is a dire need for extensive immediate research. Automation has

---

11. Dwayne Orton, "Automation—Bugaboo or Boon," *Junior College Journal*, XXVI (May, 1956), 494.

been defined as a barbarous word, but as to whether or not such is the end result of the movement depends upon man.

We have been characterized as a bunch of dreamy poets who have carried our idealism to a point where we can no longer bear to face a hard material fact when we meet it.[12] This is illustrated by our use of the products of science while at the same time failing to act scientifically. We accept the philosophy of science in relation to the object but refuse to accept it in relation to our fellow man. Stated otherwise, we seek to enjoy the power of science without accepting the responsibility. Put in terms of automation, the new machine can dignify the role of labor, provide more leisure time for thought and cultural uplift, and bring about an economy of abundance, but it cannot determine the quality of the relations that are to exist between man and his fellow man. This is the role of education, for to become morally free, man must be intellectually free.

At the present time there is no evidence that the American people are willing to translate their concepts of truth, goodness, justice, mercy, and beauty into operational terms that will produce a social correlate comparable to the genius which has produced the automated machine. Proof that we have done so will come only when we have poured into the upgrading of the educational program the same kind of effort and money that we have poured into the military since the beginning of the cold war, and when we have adopted for education the same kind of methodology that we have adopted for the automated machine, the methodology of social engineering.

EDUCATION AS SOCIAL ENGINEERING

It is unfortunate that at a time when the spirit of liberalism, the essence of the democratic process, was of greatest need, so many social philosophers of eminence, including Jacques Maritain, Reinhold Niebuhr, Arnold Toynbee, and Karl Mannheim, should have exemplified so little faith in the potentialities of the future.[13] It is understandable, however, that such should be the case for present conditions are indicative of social failure insofar

---

12. John Fisher, "The Editor's Easy Chair," *Harper's Magazine*, 216 (January, 1958), 12.

13. For a detailed discussion of this point read Frankel, *The Case for Modern Man* (New York: Harper and Brothers, 1956), Chapters 4-7.

as the hopes of liberalism are concerned. Yet, the failure is not that of liberalism, but of the people to become liberal. This is true especially insofar as the methodology of social engineering is concerned.

The method of using machines to control themselves is awesome in terms of the potentiality of relief from both physical and mental drudgery, but such potentiality is not to be realized until we apply the same rational principle of design to social living. Until we are willing to make this application there can be no true design for education.

The essence of the principle of automation has been with us throughout the entire history of western civilization. It is the central thread of all the great utopias, such as Plato's *Republic*, Augustine's *City of God*, and Moore's *Utopia*. It is the same principle which led to the designing of the first machine, and the rise of the modern industrial corporation. With the coming of automation it is imperative that design be applied to our economy and to the development of our communities else we wind up destroying ourselves both mentally and physically.[14] The fact that the communist world has accepted the principle of social design, even though brutally and despotically applied, is the essence of the real challenge with which communism confronts us.

All of what has been said calls for a revolution in education in the United States. Antiquated notions of teacher training and teacher education must be supplanted by the idea of the teacher as a social engineer. Undoubtedly, the direct producers of goods have been far more effective in their chosen task than have the indirect producers—teachers, lawyers, priests, et al. Can this not be attributed to the fact that the principle of design, of engineering, has been applied to the production of goods, whereas the indirect producer has continued to rely upon a verbalistic laissez-faire approach to his problem?[15] It is evident that this is something which the teacher cannot bring about in and of himself, and that the determining factor rests in the source of community power. How to make those who have this power realize the significance of this issue is a responsibility which all who have any insight into the nature of our problem must assume.

---

14. Richard Neutra, *Survival Through Design* (New York: Oxford University Press, 1954), 91.

15. William E. Drake, "Education as Social Engineering," *Educational Theory*, III (April, 1953), 151.

Top level minds, are there enough? Not if judged by past level performance, but there is much evidence that men do not work at anything like their high level organic potentiality. We know that the IQ of performance is related to our culture background, family, community and social order, and that through education, if properly financed and motivated, the culture background of the child can be improved. A careful analysis of conditions across the western world today indicates that the challenge of automation can be met, whether of scientists, technologists, or technicians, if there is sufficient will in the soul of the social order to do so.[16] We need to remember, as Aristotle once said, that the purpose of leisure is not merely to play games, "for that would imply that amusement is the purpose of life," but rather to provide the opportunity for an improvement of the human mind and soul.

Selected Bibliography

Books

Warner Bloomberg, *The Age of Automation* (New York: League for Industrial Democracy, 1955). The book is written out of a deep concern for the human aspects of automation.

John Diebold, *The Advent of the Automatic Factory* (New York: D. Van Nostrand Company, 1952). This is one of the best of the studies covering the general field.

George Friedman, *Industrial Society* (Glencoe, Illinois: The Free Press, Publisher, 1955). Very critical toward the social implications of the movement.

E. M. Hugh-Jones, *The Push-Button World* (Norman: University of Oklahoma Press, 1956). By and large a dramatization of what may be expected in tomorrow's world.

Robert Jungk, *Tomorrow Is Already Here* (New York: Simon and Shuster, 1954). This is a disturbing book especially because of the implications of mechanization.

Magnus Pyke, *Automation: Its Purpose and Future*. (New York: Philosophical Library, 1957). A challenging study from the standpoint of science and logic and their social implications.

Frederick Pollock, *Automation, A Study of Its Economic and*

---

16. C. H. Dobinson, "The Impact of Automation on Education," *International Review of Education*, III (No. 4, 1957), 394.

Social Consequences. (New York: Frederick A. Praeger, Publishers, 1957). Thorough and philosophical in treatment. Contains an excellent bibliography.

Norbert Wiener, *Cybernetics or Control and Communication in the Animal and the Machine* (New York: John Wiley & Sons, Inc., 1948.) The standard and most authoritative study in the field.

Magazine Articles

B. P. Brodinsky, "Automation," *The Nation's Schools,* 58 (August, 1956), 35-38.

John H. Fischer, "Automation, Implications for Education," *The School Executive,* 76 (December, 1956), 64-66.

T. Eugene Holtzclaw, "Automation and the Curriculum," *The Clearing House,* 36 (March, 1957), 426-428.

Maurice P. Moffatt and Stephen G. Rich, "Implications of Automation for Education," *The Journal of Education Sociology,* 20 (February, 1957), 268-274.

Dwayne Orton, "Automation—Bugaboo or Boon," *Junior College Journal,* XXVI (May, 1956), 493-497.

Harold H. Punke, "Social and Educational Problems of Automation and Longevity," *The Journal of Education Sociology,* 30 (May, 1957), 398-404.

Harold S. Spielman, "A Challenge to Educators," *Science Education,* 39 (March 1955), 102-140.

Eric A. Walker, "Automation and Education," *Junior College Journal,* XXVI (May, 1956), 498-503.

WILLIAM B. DRAKE received his B.A. from the University of North Carolina (1924) and M.A. (1928) and Ph.D. (1930) from the same institution. He served as Principal, Murphy School, Orange County, Northy Carolina (1924-1926), Superintendent of Schools, Columbia, North Carolina (1926-1928), teaching fellow, University of North Carolina (1928-1930), Assistant Professor of Education, Pennsylvania State College (1930-1939), Associate Professor of Education in the University of Missouri (1939-1944), and as Professor of Education and Head, Department of History and Philosophy of Education (1944-1947); since 1957 he has been Professor of History and Philosophy of Education in the University of Texas. During several summers, he has been Visiting Professor in the Universities of North

Carolina, Illinois and Texas. Among his several other academic appointments, he was the National Secretary Treasurer of the Philosophy of Education Society and President of the Pennsylvania Federation of Teachers (1936-1939). He was awarded the Alphonso Smith research prize (University of North Carolina, 1930), and is a member of the Editorial Board of *Educational Theory*. He is the author of *The American School in Transition* (1955), and co-author of: *The Sociological Foundations of Education* (1942), *Significant Aspects of American Life and Postwar Education* (1944), and *Teaching World Affairs in American Schools* (1956).

CHAPTER 24

AUTOMATION'S IMPACT ON CAPITAL AND LABOR MARKETS

YALE BROZEN, *Professor of Economics,*
The School of Business, University of Chicago

Automation is conjuring up visions of a society in which human beings are as obsolete as horses. Machines run machines, repair machines, program production, and, in some science fiction fantasies, run governments and rule men. Union leaders collect no dues and businesses lack customers because, presumably, the production-line worker will no longer be required.

To place this development in perspective, we must realize that, although automation may be an engineering revolution, in the economic sense it is nothing more than a continuation of an evolution which has been going on for centuries. Beginning with a world in which men could barely capture enough to keep themselves alive, we went through a *first* industrial revolution that ended about 3000 B.C. It raised productivity sufficiently so that societies could afford priestly classes, an aristocracy, armies, and a bureaucracy. We even reached the stage of the Greek democracies. These were democracies of the few, however, resting on a large class of slaves.

The second industrial revolution (the one normally thought of as *the* industrial revolution) abolished slavery. It became cheaper to use mechanical power than human muscles, and created the base on which mass democracy now rests. The economic gains of the first industrial revolution went to the few, those of the second have gone to the many; perhaps the gains of the automation revolution will carry us on from a mass democracy to a mass aristocracy, provided we do not sabotage it by creating more incentives to dissave and to not-produce than we already have.

In the past century, we quintupled the productivity of the average man and tripled his annual income. If we can do the same in the next century, average family income can go to $18,000 a year instead of the present $6,000. The common man will become a university educated, world traveler with a summer place in the country, enjoying such leisure time activities as sailing and concert going; able to call on superior medical services to maintain his health, eating exotic foods from the far corners of the world in fine restaurants, and living in a home equipped with beautiful furniture and paintings. Essentially, this is what automation promises.

AUTOMATION AND UNEMPLOYMENT

But what about the unemployment that is threatening? Walter Reuther testified before the Joint Congressional Committee on the Economic Report that he was alarmed over the effect of automation on labor requirements. He cited a Ford official's statement that automation reduces direct labor requirements by 25 to 30%. He told about the new automatic engine plant in Cleveland which turns out 154 engine blocks per hour with 41 men. This, he compared with the 117 men required under former methods.

What are the probable employment effects? Those who talk about the unemployment caused by technological change usually say something like the following. In 1954, unemployment hit a peak of 3,700,000. If productivity had not increased between 1953 and 1954, private industry would have needed 730,000 more workers to turn out the goods it did. If productivity had not increased, then, there would have been less than 3,000,000 unemployed.

One way of reducing this absurdity is to see the number of workers required to turn out 1954 production if we had stopped increasing productivity in, let us say, 1957. In that case, we would have needed 8.1 million, obviously an impossibility.

The Federal Reserve Bank of Minneapolis, in its May 31, 1955, *Monthly Review,* said that unemployment was higher at that time than it would have been with the current expansion of output because "the installation of new equipment has raised output per worker in many plants." It implied that the rise in productivity was responsible for unemployment. But this was simply not the case. Neither output nor employment would have been as large as it was if productivity had not increased. Real wage rates were raised at a rapid clip in 1953. As a result, it became uneconomic to use marginal facilities. Workers were laid off in late 1953 and early 1954. As more efficient productive facilities became available during 1955 and 1956 through new capital formation, high-priced workers were re-hired. The new equipment was sufficiently productive so that employers could afford to hire workers whose wage rates were too high to permit using them on old equipment without suffering losses. Expansion of output and employment occurred because of the increasing availability of more productive equipment. If productivity had not increased, unemployment would be greater now, than it is.

We must recognize that there is no such thing as technological unemployment. If there were, then the number of unemployed should increase with the rate of technological change. Yet, we find that periods of large unemployment have been periods of slow technical change. Periods of most rapid change have been associated with minimal amounts of unemployment.

There may be technological displacement but not unemployment. We seem to always recognize this *displacement,* but seldom do we pay much attention to the *absorption* of workers which occurs because of technological change. If we were to measure the number of jobs created by technological change, as well as the number of workers displaced, we would find that, at any given wage level, more jobs have been created than have been eliminated.[1] The number of jobs available now would be far

---

1. S. C. Gilfillan estimates, on the basis of a sample of 120 most important inventions, that only 33% of inventions are labor-saving. Even labor-saving inventions do not necessarily displace labor. By decreasing the cost of a product, the rate of demand is expanded frequently to the point where large numbers of workers are *absorbed* because of the labor-saving invention, not displaced. This occurred, for example, in the automobile industry. A recent example is provided by Admiral Corporation, which has automated its circuit production for TV sets. As a consequence, it has been able, in the past year, to cut its 24-inch console model receiver from $500 to $340. The increased sales have resulted in absorption of workers. No displacement occurred despite such changes as the elimination of 425 hand-soldered connections from each television receiver.

Of the balance of the inventions examined which were not labor saving,

larger than the available work force, despite—or rather because of—all the new techniques that have become available in recent decades, but for the job destroying effects of the great increase in real wage rates which have occurred in past decades.

Employment levels, other things being equal, are a function of the wage rate. From the beginning of the third quarter of 1952 to the third quarter of 1953, for example, real wage rates were raised by 6%. If there had been no increase in productivity, we would have had 12 million unemployed in early 1954 instead of 3,700,000. The elasticity of demand for labor, according to Senator Paul Douglas' studies, is between minus three and minus four. A 6% increase in wage rates, then, would cause about a 20% decline in employment, other things being equal.

Fortunately, other things were not equal. Productivity rose by about 4%. Consequently, the net unemployment causing rise in wage rate was less than 2%. As a result, the decrease in man-hours worked amounted to only 5% instead of 20%. The rise in productivity *prevented* unemployment. It did not cause it.

If increases in productivity were to reduce employment, we would expect the industries which have had the greatest rise in productivity to have had the greatest decrease in employment. Those which had the least increase in productivity should show the least decline in employment. Yet the opposite has been the case. From 1909 to 1937, Solomon Fabricant found employment negatively correlated with unit labor requirements. In industries in which labor requirements per unit of product were cut most drastically, above average increases in employment occurred. In industries in which unit labor requirements were cut least, employment increased least or declined.[2]

If the correlation has any causal significance in the relationship of productivity to employment, we can expect employment to rise in automating industries. It is argued, however, that automation is a new kind of technological change. We cannot judge its consequences, according to this argument, by what has occurred in the past.

Even if we grant that automation is different from other kinds

---

Gilfillan found that 45% were new developments of consumer goods. The remaining 22% were capital-saving inventions. Both of these varieties, as will be shown later in this paper, act to increase the demand for labor and create additional jobs at any given wage level.

2. *Employment in Manufacturing, 1899-1937* (New York, National Bureau of Economic Research, 1942).

of technological change, we should not blind ourselves to history to the point of saying it is something new. Automation is an old technique in some industries.

Essentially, what automation will do will be to turn machine operators into machine tenders and servers. This has already occurred in the textile industry, to name one example. Walking into the loom room of a modern mill, your first impression is that of a vast space filled with busy machinery and no people in sight. Yet employment in textile mill products is above one million.

An automatic, card programmed loom was devised by Jacquard over 150 years ago. An automatic flour mill was built in 1784. Automatic silk looms were designed by Jacques de Vaucanson in 1741. Early steam engines had automatic governors.

The chemical and petroleum refining industries are continuous process, automatically controlled industries with many years of experience behind them. They, too, provide employment on a large scale, amounting to more than 600,000 for the two industries.

The effect of automation has been to reduce the *relative* number of machine operators required, but it has greatly increased the number of maintenance men, engineers, office employees, and other non-machine operators required. This again, is nothing new or peculiar to automated industries. In 1899, less than 7% of the labor force of the manufacturing industry consisted of persons other than production workers. Today, over 20% of the employees in manufacturing industries are not production workers. Taking only the period since 1939, we have seen production workers in manufacturing rise by 55% while other workers increased by nearly 95%.

TECHNOLOGICAL DISPLACEMENT

Suppose we grant that there is no such thing as technological unemployment, and that neither automation nor increasing productivity is new. Does it not remain that automation is coming with a rush since we have developed electronic controls and feedback and self-adjusting systems? Does this not mean that we will displace people on a scale not previously encountered? A professor from M.I.T. has said that factories will be fully automated in 10 years. Will this not affect 16 million manufacturing employees and require adjustments of an unparalleled magnitude?

Suppose we argue that all the $26 billion worth of equipment purchased annually (the 1956 level of purchases) in the United States will be used solely for labor saving purposes. How much displacement will be caused?

I have examined a number of new developments in order to get a measure of what to expect. To give you a few examples, the Commonwealth Edison Company of Chicago installed an IBM computor in 1956. Its cost was about $1,000,000 and it reduced the billing force requirements from 470 clerks to 270 clerks. This was directly offset only by three maintenance men and an applications engineer after the machine was first installed. Average investment in equipment per employee displaced was $5,000, aside from an investment of about $10,000 per displaced man in preparation and transition costs.

The new Fairless Steel Works outside Philadelphia was built to produce about 300 tons of steel per man-year with an investment in equipment of $90,000 per man. At that time, the steel industry produced only 160 tons per man-year. To save one man, then, required a $100,000 investment.

In the automobile industry, the Cross Transfer Machine, which costs $2 million, reduces operating man-power requirements from 33 to 9 men. A rise in maintenance labor is required. Net saving in man-power amounts to approximately 22 men. To accomplish this, $90,000 of equipment must be purchased per man saved.

The average new equipment purchase required per man year reduction in labor requirements seems to run about $35,000. On this basis, if the $26 billion of new equipment put in place in 1956 is the prevailing level of the future and is strictly for the purpose of saving man-power, approximately 700,000 men per year will be displaced from present jobs. An independent estimate, made by other means by R. L. Meier of the University of Chicago, arrived at a figure of 300,000 men per year as the outside limit of displacement.

Actually, surveys of the capital programs indicate that half of the new equipment currently purchased is intended for expanding capacity, the other half for modernization and replacement of old equipment. On this basis, about 300,000 people are likely to be displaced. While this is not an alarming number, how can we handle the problem?

Commonwealth Edison's approach is indicative of what can be done. Despite the 200 man reduction in billing force requirements, not a single person was laid off. People were transferred to other departments of the company to fill vacancies. Most firms face a voluntary quit rate of 20 per 100 employees per year. In *manufacturing alone,* then 3 million people per year must be hired to replace voluntary quits. The 300,000 men per year displaced by automation in *all* industries, then, can usually be transferred to jobs in the same company vacated by voluntary quits. Even if they must find jobs elsewhere, there are a large number constantly becoming available through quits and, in addition, through *expansion.* Our only real problem is the older worker who usually finds it difficult to obtain a new job. Many more of these men, however, are displaced by changes in tastes and business failures than by automation. This is not a problem whose proportions are importantly affected by these new techniques.

AUTOMATION AND WAGE RATES

Actually, the most alarming potentialities of automation are in the field of wage rates. Although automation of the type resulting from new inventions in recent years will raise the *average* productivity of the work force, wage rates depend on *marginal* productivity. This may be decreased, although not necessarily.[3] If an industry introducing newly-invented automatic techniques drains large amounts of capital from other industries, and releases workers, the capital-labor ratio in other industries will fall. Marginal productivity and wage rates will decline under these circumstances despite the rise in average productivity and national income.

For example, despite the increase in national income, the marginal productivity and the wage rate of labor dropped from $80 to $75 per week (measured in 1956 dollars) because of automation of the variety assumed here. If the wage rate did drop, the employment would (from 65 to 55 million under the assumed conditions) .

This effect occurs only under the narrow assumptions made

---

3. See Y. Brozen, "The Economies of Automation," *American Economic Review,* May 1957, for a complete discussion of the circumstances which determine whether marginal productivity will increase or decrease.

above, however. With the quantity of capital growing more rapidly than population, automation in a few industries would have to absorb the excess of current savings over that needed to provide for a larger population, and in addition, capital from other industries to cause this to happen. Also, some types of automation are capital saving, and other capital saving inventions and innovations normally occur constantly. Capital using automation would have to absorb this capital saving as well as current saving and capital from other industries to bring about the consequences described above.

With a constant supply of labor, weekly marginal productivity over a four-year span seems likely to rise from $80 to $90 (1956 dollars). Of course, in a four-year period, the labor supply will rise by about three million. Despite the large rise in the marginal productivity curve, then, the marginal productivity of the actual labor supply will rise somewhat less, perhaps only to $85, because of the population increase.[4]

AUTOMATION AND CAPITAL REQUIREMENTS

The capital required to automate industry to the levels possible with available techniques amounts to a staggering sum. Electronic computers involving investments of a million dollars and more are common. Price tags on automatic grinders run $300,000, automatic machine production lines for electronic components cost $700,000 as against a $90,000 investment for conventional techniques, and transfer machines often cost two million dollars and up. Over-all, the investment necessary to automate the manufacturing industry alone would run on the order of one-half trillion dollars.

Our net supply of savings in excess of that required to equip the numbers added each year to our growing work force will be on the order of 10 to 15 billion dollars annually for the next few decades. If this entire amount were used to automate only the manufacturing industries (which constitute less than a fourth

---

4. The effect of the innovations introduced during the late eighteenth and early nineteenth century seems to have been of this character. Average productivity rose rapidly during the industrial revolution, but marginal productivity (and wage rates) lagged. This was partly because of the capital using character of the innovations of the period and partly because of the rapid population increase. Fortunately, a very high rate of capital accumulation prevented marginal productivity from dropping.

of all industry, measured by volume of employment), at least thirty years would pass before automation to the extent possible with present techniques would be complete. Some of the capital will go for the modernization and automation of industries other than manufacturing. Automated railroad yards, for example, have been running between 5 and 34 million dollars per yard. It is very improbable that manufacturing industries will achieve the level of automation envisioned as possible in less than half a century. Forecasts, put forward by some writers several years ago, of completely automatic factories displacing all others in ten years, are nonsensical.

When considering the amounts of capital required to build modern plants, it becomes obvious that the net investment required to modernize industry, assuming no population growth, will run in excess of two trillion dollars. The 108 new paper mills which were put into operation in 1956 used $25,000 of plant and equipment per worker. Even these did not apply all the capital which would have been economic if more savings had been available. One of the mills, for example, used $36,000 per man and, as a consequence, was able to produce 17 pounds more paper per man-hour than the average for all 108 mills. As old mills are replaced, less than $5,000 per man will become available through depreciation and conversion of the capital embodied in old plant and equipment. This means that a net investment of at least $20,000 per man will be absorbed in the paper industry.

A modern steel plant uses $110,000 of capital per man as compared to a present average investment in plant and equipment of about $10,000 in the steel industry. Between 1952 and 1955, average investment per worker in electric power plants rose by $24,000. Half of the capacity now in place in the industry will be replaced in the next two decades which will result in a further rise in investment of at least $50,000 per employee. Adding up these requirements for all industry, the estimate of a two trillion dollar requirement for modernization appears modest.

In view of the slow pace at which savings are being accumulated and made available for this purpose, the demand for funds for modernization will be strong for many decades. The only apparent restrictions which seem likely are either an adverse movement in the tax structure or an overpricing of labor to the point where it will absorb so much of the revenues from prospective investments that they are not worth undertaking.

The probable consequences of automation together with the usual amount of new investment and capital saving inventions and the expected population increase consist of the following:

1) National income will increase in the next four years by about 15 per cent.

2) Wage rates will rise by about 8 per cent (assuming they increase only to the extent compatible with maintenance of full employment), but wage income will rise by 12 to 13 per cent (because of the 4 per cent increase in labor supply).

3) Wage payments will be a proportionately smaller share of national income.

4) Returns per unit of capital will rise 5 or 10 per cent. Total return to owners of capital (including the return to government from the share of capital's earnings confiscated through tax measures) will rise 20 per cent (because of the larger supply of capital).

5) Payments to those receiving the returns on capital (including the government) will become a proportionately larger share of the national income.

Automation will produce consequences whose order of magnitude is much smaller than those to which we have adjusted in the past. If any special measures are desired because of fears of automation, they should be shaped in the direction of making the economy more flexible and adaptable and toward increasing the rate of saving and investment. While the objectives of the guaranteed annual wage, which seems to be the United Auto Workers' method of meeting the threat of automation, are desirable, it is not the means which will accomplish the objective. Turning variable labor costs into fixed costs introduces additional rigidities into our economy rather than making it more flexible. Insofar as it introduces additional risks in saving and investing, it will reduce the rate of increase in capital which will be forthcoming at any given rate of return. This will slow the rate at which the marginal productivity of labor moves upward and, consequently, the rate at which we absorb the unemployed at any given wage level, or alternatively, the rate at which the wage level moves upward. Insofar as capital is actually consumed supporting laid off workers (the higher the level of unemployment compensation, measured as a percentage of take home pay, the longer people will remain unemployed), less capital is left to purchase equipment which will raise the marginal productivity of workers.

Although technological change (including automation) generally reduces the man-hour requirements for turning out a unit of product, it does not reduce total employment. If product demand is sufficiently elastic, reductions in costs and prices expand the rate of demand and increase employment despite reduced labor inputs per unit of product. If demand is inelastic, reduced prices do not expand the rate of demand sufficiently to maintain employment in the directly affected industries. But the reduced prices leave more income in people's hands. This makes it possible for them to buy more of other products. The displaced laborers are absorbed, then, in the industries to which people turn when they have more money left to spend. That is why manufacturing industries have expanded employment as farm labor requirements have dropped. If manufacturing employment drops because of automation, then education, research, entertainment, medical, and leisure time industries will expand as people turn their spending to the tertiary group.

Automation is different in its results only to the extent that it is turning machine operating labor into machine maintenance and machine tending labor. To the extent that this type of labor is less routine, more creative, it is tending to put interest back into jobs. Also, insofar as this type of labor requires more skill and judgment, and less muscle, it is likely to improve the demand for older workers in the long run and make displacement of them from other causes a less awkward problem.

APPENDIX

A. Technological displacement and technological unemployment.

It may seem paradoxical to admit that an increased rate of technological change may result in a larger number of employees losing their jobs each year, and yet argue that there is no such things as technological unemployment (and that the number of jobs may even be increasing and unemployment decreasing because of technological change). The paradox can be explained in the following way.

In a normal year, the quit, layoff, and discharge rate generally runs about four per 100 employees per month.[5] This means that, out of an employed work force of 50,000,000 wage and salary employees, about 25,000,000 will quit, be laid off, or discharged each year. If the average duration of unemployment (before a new job is found or the worker is recalled) is one-eighth of a year, average volume of unemployment will be $3\frac{1}{8}$ million.

Let us suppose that an acceleration of the rate of technological change causes an additional 2,000,000 workers to lose their jobs each year. (As was pointed out above, this is far larger than the actual number of workers likely to be displaced.) At the same time, the acceleration of change will create a large number of new job openings producing new products, increasing production of products whose cost has decreased and whose demand is elastic, and increasing production of products whose demand expands because of income freed from the purchase of items whose cost has decreased and whose demand is inelastic. As a consequence of the greater frequency of job openings, the average duration of unemployment falls to, let us say, one-ninth of a year. Despite a rise in number of quits, layoffs, and discharges to 27,000,000 a year because of accelerated change, average unemployment will fall to 3 million from $3\frac{1}{8}$ million. Displacement rates may increase, then, and unemployment simultaneously decrease because of technological change.

B. Capital and marginal productivity of labor.

In examining wage rates around the world, one is struck by the correlation of the quantity of capital used per worker and wage rates. In Brazil, for example, manufacturing wage rates are approximately $600 a year (for a forty-hour week) and capital used per manufacturing employee is around $1,700. In the United States, wage rates are around $4,000 a year and capital employed per manufacturing employee is about $13,000. Other countries show approximately the same sort of correlation. It seems that every extra $1,000 of capital adds about $300 a year to the wage income of workers under the circumstances of the technology practiced in today's world.

An extra $1,000 of capital adds about $40 to the net national income in the United States. Twenty dollars of this goes to local

---

5. *Economic Report of the President* (Washington: U. S. Government Printing Office, 1955), p. 159.

government (in real estate taxes), fifty dollars to the federal government (corporate income tax), and $300 to wage earners. Those who save and invest the $1,000 get about $60 return in the form of interest, dividends, or rights to retained earnings. It seems that labor gets the lion's share of the extra income produced by extra capital.

C. Automation.

The effects of automation differ depending on the forces which bring it into being. Much of the automation we are getting is the simple consequence of a rising stock of capital. It is adaptation to changing circumstances rather than invention which brings this type about.

As capital increases, wage rates rise. As they rise, it pays the individual firm to shift to capital intensive techniques which were not previously economic. From the point of view of the firm, the wage rise makes the shift economic and forces it. From the point of view of the economy, it is the rise in the quantity of capital which makes it possible.[6]

The second force which may operate to bring automation to an industry is technological change in other industries of a type which raises the marginal productivity of labor. As labor becomes more expensive, technological adaptation will be forced on other industries similar to that described above.

An illustration of the influence of the combined forces of an increasing supply of capital and rising marginal productivity of labor in other industries is provided by the Fairless steel works. The tremendous quantity of capital invested per man would not be possible or economic in a society possessing less of this resource than our own. Nor could this investment be justified except on the basis of the high cost of labor (high marginal productivity in competing uses). As a matter of fact, the investment probably would not be economic at the wage rates prevailing at the time the plant was built. Only because labor rates were expected to rise 25% or more in the coming decade was it economic to use such large amounts of capital per worker at the time the plant was built.

---

6. See Y. Brozen, "Determinants of the Direction of Technological Change," *American Economic Review*, May 1953, for an analysis of this point.

The third force causing automation in an industry may be capital saving inventions applied in other industries which cause them to release capital. Its price will drop and cause firms to shift to automated methods of production for the same reasons as in the first case described in the appendix.

Wage rates or employment will rise. The proportionate share of national income going to labor will increase and that going to property owners will decrease.

The fourth force leading to automation may be new inventions directly applicable to this industry. New inventions may have any one of several effects. They may save only labor (in the technical coefficient sense).

How this affects the economy depends on several conditions. If the demand facing the industry is sufficiently elastic, its total labor requirements will increase. It will use a larger amount of capital in the aggregate. This means that under static assumptions it will absorb capital from other industries. The marginal productivity of labor in other industries will drop, which will be reflected back into this industry. Consequently, wage rates will drop or unemployment will occur, despite increased employment in this industry.

If it uses less labor than formerly, it may still use more capital. The consequences for wage rates or employment, under static assumptions, will be the same.

If it uses less of both labor and capital, because the invention is both labor and capital saving in the technical coefficient sense and demand is inelastic, the consequences will depend on the ratio of released capital to released labor (assuming the capital invested in old equipment can be salvaged.[7] If the released capital has a higher ratio to released labor than the average prevailing in other industries, marginal productivity of labor and wage rates will rise. If it has a lower ratio, these quantities will fall (under static assumptions).

An example of the type of automation which is labor saving only is provided by the development of an automatic method for machining cylinder heads. The old method required 6 machines costing $240,000 and 6 operators. The new method

---

7. See Y. Brozen, "Adapting to Technological Change," *University of Chicago Journal of Business,* April 1951, and "Welfare Theory, Technological Change, and Public Utility Investment," *Land Economics,* February and May 1951.

requires one operator and an automatic machine costing $230,000 for the same output.

A new automatic method for crankpin grinding saves labor but requires a larger investment for the same ouput. Five hand operated grinders costing $201,000 can be replaced with an automatic grinder costing $280,000. Here, the consequence is clearly one of absorbing capital relative to labor in proportions which reduce marginal productivity of labor in the economy (under static assumptions), regardless of whether the industry uses more or less labor.

Another type of automation is illustrated by the addition of newly invented automatic controls of existing machines. A $250 automatic control added to a $5,000 riveting machine, for example, increases net output by 10% by reducing the number of rejects.

The effects here will depend on the same sort of considerations as those mentioned in the first instance of a new automation invention. If labor is absorbed in quantities relative to capital absorbed larger than the average prevailing in other industries, marginal productivity and wage rates will rise.

Other types of automation can be cited, but these are sufficient to illustrate the analysis.[8] On balance, it seems likely that automation will be capital using, particularly in view of the difficulties of salvaging capital from old equipment, except when the change takes place largely by adding automatic controls to old equipment. This would imply that rates of saving and investment should be increased to avoid depressing wage rates or employment. In terms of tax policy, this objective may be accomplished by reducing corporate income taxes and upper bracket income taxes.

Selected Bibliography

"Automation—Bogeyman or Bonanza?" *Steel*, Vol. 140 (October 15, 1956), 109-116.

Discusses the effect of automation on productivity and marketing. Also provides many examples of automation and its consequences.

---

8. See Y. Brozen, "The Economics of Automation," *American Economic Review*, May 1957, for an analytical discussion of various types of automation.

Automation and Job Trends (Chicago: Council for Technological Advancement, 1955).

Beginning with an analysis of automation, the origin of the idea from its inception is traced, together with its impact on jobs, education, and our economy. The place of automation in the American production system is explained.

Automation, Operations Research, and Business Planning (Chicago: University of Chicago, 1958).

Papers dealing with past experience with automation and forecasts of marketing, engineering, and research problems associated with automation. Also, the use of operations research and game theory in dealing with automation problems is described.

Report of the Joint Committee on the Economic Report, *Automation and Technological Change,* Senate Report No. 1308 (Washington: U. S. Government Printing Office, 1956).

The findings of the Subcommittee on Economic Stabilization flowing from extensive hearings on the nature and implications of automation are presented. These findings effectively summarize the implications of nine days of testimony on automation and its effects from 25 representatives of business, labor, government, engineering colleges, and other national groups.

Y. Brozen, "The Economics of Automation," *American Economic Review,* XLVII (May, 1957), 339-385.

A technical discussion of several different types of automation classified by causes and by capital and labor demand effects.

E. Schiff, *The Primary Employment Effects of Productivity Gains* (Chicago: Council for Technological Advancement, 1954).

Analyzes the effect of reduced unit labor requirements on employment. The analysis considers the direct consequences under various demand situations in the industry in which the change occurs. It also points out the effect on employment in other industries.

Proceedings of Second Annual Conference on Automation for Senior Officers (Chicago: University of Chicago, 1957).

A collection of papers by outstanding authorities dealing with business applications of computers, capital budgeting, personnel, and other problems involved in planning the introduction of automation in business enterprises, and automation policies of business firms.

Y. Brozen, "Studies of Technological Change," *Southern Economic Journal,* XVII (April, 1951), 438-450.

Reviews various studies of technological change under four

categories: (1) historical and descriptive studies, (2) employment effect studies, (3) studies dealing with sociological and political consequences, and (4) those which deal with determinants of the direction and rate of change.

Ira Wolfert, "What's Behind this Word 'Automation'?" *Readers' Digest,* (May, 1955), 43-48.

A concise article on the uses of automation, with examples, told in an easy-to-read style. An optimistic viewpoint of its promise of more jobs, better working conditions, and a higher standard of living is taken by the author of the future of an automated industry.

YALE BROZEN, Professor of Economics in the School of Business of the University of Chicago, was trained as a chemical engineer at the Massachusetts Institute of Technology. He studied economics at the University of Chicago (where he received his Ph.D. degree). He served on the faculties of the University of Florida, Illinois Institute of Technology, University of Minnesota, Northwestern University, and Escola de Sociologia e Politica de Sao Paulo (Brazil); in addition, he was the training administrator of the U. S. Army Signal Corps Radar Training program during World War II, consultant to the Social Service Research Council's Committee on the Social Implications of Atomic Energy and Technological Change and to the National Science Foundation, etc. He is also Director of the Air Research and Development Command Program and teaches in the Executive Program in the School of Business of the University of Chicago, and is also the Director of Research in the Transportation Center of Northwestern University. His *Workbook for Economics* has been used widely in the United States and abroad.

CHAPTER 25

AUTOMATION AND LEISURE

WILLIAM A. FAUNCE, *Assistant Professor, Department of Sociology and Anthropology and The Labor and Industrial Relations Center, Michigan State University*

Twentieth Century Western Civilization is confronted by at least one problem unique in human history. Man's historical preoccupation, working for the means of livelihood, is increasingly involving a diminishing portion of his time and energy. The problem of extracting from the earth the means of subsistence in an economy of scarcity has been replaced by a new set of problems in an economy of machine-made abundance. One such problem is the proper utilization of larger and larger amounts of available leisure time.

Many societies in the past have supported leisure classes of varying sizes with varying amounts of time free from productive activities. Leisure today is no longer the privilege of the few, however, but the prerogative of many. Never before has the question of how leisure time is to be used assumed such importance.

This new leisure as well as our economic abundance is machine-made, the product of the increasing mechanization of production technology. Spurred by the necessity of increasing productive efficiency during World War II, American industry appears to have entered another and perhaps unprecedented era of rapid technological change in production facilities. The term automation describes one direction which these changes have taken. In this chapter we will be concerned with the effect of automation upon amount of available leisure time, the effects of increased leisure upon patterns of use of leisure time, and the impact of an increasingly leisure-oriented society upon various American culture patterns.

AUTOMATION AND WORKING HOURS

The number of hours and days worked per week during any period tend to be regarded as natural and immutable. The five day, forty hour work week is currently so regarded. It has only been a decade or so, however, that this amount of working time has been the standard work week in this country. In the past century the average number of hours worked per week by non-agricultural workers has decreased from about sixty-five to about forty hours and this decline has been fairly steady, averaging about three hours per decade. As Seligman has noted,[1] it is the process of reduction of working time which appears to be "natural and immutable."

The increasing productivity of our economy is what has made this reduction in working hours possible. For the economy as a whole, output per man-hour has been rising by two to three percent per year over the past fifty years. In the non-governmental sector of our economy, output per man-hour has approximately doubled over the past twenty-five years. The benefits of this increased productivity have been distributed between income and leisure on roughly a 60-40 basis, 60 percent going into greater income and 40 percent for more leisure time. Whether this particular way of cutting the "productivity pie" will continue or not is difficult to predict. Some of the factors affecting the likelihood that automation will produce a further increase in available leisure time may be considered, however.

Examples of dramatic increases in output resulting from automation are legion. A chemical company has recently opened a magnesium mill which is capable of producing more magnesium sheet and plate than the total previous national capacity. A musical record company has installed new automatic machines with which four men turn out eight times as many records as 250 men had previously produced. An automobile engine part which was once produced at the rate of thirty-eight per hour by five men and two machines is now produced by one man at one machine at the rate of 750 per hour. One of the most highly automated automobile engine plants now produces twice as many engines with one-tenth of the manpower of a conventional

---

1. Daniel Seligman, "The Four Day Work Week: How Soon?", *Fortune*, (July, 1954), 81.

plant. While hundreds of such instances might be cited, they are probably exceptional cases and automation has resulted in comparatively minor productivity increases in many plants and offices in which equipment of this kind has been introduced. However, the net effect of the development of rapid transfer machines, electronic inspection devices, and computer technology has undoubtedly been an increase in output per man-hour in American industry and has probably accelerated the rate of productivity increase.

The extent to which this increased productivity will be translated into shorter hours of work will probably be determined in large part by the amount of pressure brought to bear by trade unions and labor generally for reduced hours. The logic of cost reduction would, in most instances, dictate a management preference for longer hours and fewer workers when the point is reached at which a decrease in hours worked no longer increases productivity. This may be especially true where there are pressures to maintain continuous operation of expensive automated equipment in order to meet amortization deadlines. Thus far, a further reduction in the work week has not become an immediate collective bargaining objective of American trade unions generally. A number of unions, including the large, pattern-setting U.A.W., however, are committed to the shorter work week as a goal for collective bargaining and it seems probable that, with continued productivity increases, union pressure for decreased hours in other industries will follow.

Since the various industries within our economy are differentially susceptible to further mechanization, the possibility for decreasing hours of work may vary from industry to industry. Reductions in hours worked per week have occurred at varying rates in different industries in the past and will undoubtedly continue to do so. Because of the rate of change in production techniques and the likelihood that there will be greater pressure exerted to reduce the work week, it is probable that working hours of hourly rated employees in mass production industries will be among the first to be affected by automation. It is with these workers that this chapter is primarily concerned.

Relatively greater pressure to decrease the work week for production line workers may develop for a number of reasons. Automation appears to have already decreased the number of job opportunities available to such workers in some industries and, barring any great increases in product demand, seems likely to do so in others. Comparing the two year averages for 1947-48 and 1953-54 in the automobile industry, production workers increased 7.6 percent, non-production workers increased 19.7 percent, while total production increased 68.9 percent. Semi-skilled operatives and kindred workers actually constituted a smaller proportion of the work force in 1956 than they did in 1947. In a recent study by the author, a majority of management representatives as well as of workers and union leadership in the automobile industry in Detroit expressed the opinion that automation would result in at least some short term displacement of workers. To the extent that automation results in reduced job opportunities, even in the absence of large scale displacement of workers, it is likely that there will be pressure exerted by unions to decrease hours in order to spread available jobs among a larger number of workers.

Changing standards regarding adequacy of style of life among working class families may also affect preferences regarding the distribution of income and leisure from increased productivity. Amount of income desired tends to be normatively regulated and there is some evidence that once a standard of living is established there may be less pressure to raise income than to maintain it at a customary level.[2] Demands for new goods and services are constantly being created by advertising campaigns and, more subtly, through models provided by the mass media and the consumption of many of these goods and services is dependent upon *both* greater income and more leisure. Many leisure activities which were once symbols of upper class status are now within the means of working class families and it is unlikely that the demand for more leisure time in which to enjoy these activities will slacken.

In connection with the study in the automobile industry previously mentioned, a sample of workers from an automated plant were asked whether they would prefer a shorter work week, increased wages, or longer vacations if any of these alternatives were made possible by automation. Almost three-fourths indicated a preference for a shorter work week over either more pay or longer vacations. One apparent reason for this preference

2. Bakke has noted, for example, that among working class individuals one is regarded as successful when he has attained the standard of living customary among his associates. E. Wight Bakke, *The Unemployed Worker* (New Haven: Yale University Press, 1940), 20.

is the very nature of production line work. Jobs of this kind have characteristically required little exercise of skill, responsibility, or initiative and have offered little variety in the types of tasks performed. Automation does not appear to have significantly altered these characteristics of production line work. Machine operators in the automated plant, while having somewhat more responsibility in that there is a larger machinery investment per worker, do not need appreciably greater skill and many feel that they have an even less important part in the total work process than had been the case in plants using conventional machining techniques. Neither does the change from actually operating a machine to pushing a button or watching a panel of lights or gauges offer much more relief from the monotony of repetitive operations on the job. In short, automation has not altered the fact that most production line jobs do not produce the kind of occupational involvement or identification necessary to make work a satisfying experience. If work is not a satisfying experience, it would seem reasonable to expect that there would be greater pressure to decrease working hours to allow more time for what may be perceived as more important and more satisfying activities.

There has not yet been sufficient research dealing with the factors affecting the preference of workers for greater leisure or more income to accurately anticipate the preferred distribution of the benefits of increased productivity from automation. It seems likely, however, that increasing productivity will continue to result in an increase in available leisure time. A majority of a sample of both union and management leadership in the automobile industry interviewed in 1957 expressed the opinion that there would be a reduction of working hours resulting from automation within five to ten years. It has been estimated that if the rate of productivity continues to increase at two or three percent per year and assuming a continuation of the 60-40 ratio of distribution of increased productivity between income and leisure, a four day work week may be feasible throughout the non-farm sector of our economy in twenty-five years.[3] For the manufacturing industries with which we are principally concerned, a shorter work week may be possible much sooner. If a change of this type does occur, the nature of leisure activities may be affected. There is evidence from previous research that

3. Daniel Seligman, *op. cit.*, 114.

the proportion of time spent in various activities increases or decreases depending upon amount of leisure time available.[4]

PATTERNS OF USE OF INCREASED LEISURE

A number of dimensions of leisure activities may be considered in discussing the use of leisure time. Leisure may be recuperative in the sense that time is spent relaxing from the job completed and preparing for the job forthcoming or it may be actively spent in the sense of physical or emotional involvement in an activity. Leisure time may be used creatively or non-creatively. It may be self-oriented or may be other or service-oriented. It may be spent in the company of others or in solitary pursuits. It may be spent as a spectator or as a participant in various activities. Leisure time may serve as relief from boredom or as escape from involvement.

Automation may affect patterns of use of leisure time either by increasing the amount of time available for such activities or by changing the nature of the work experience. The combination of decreased physical effort required by automated jobs and decreased working hours would make possible a decrease in the proportion of time spent in recuperation from work and permit more active involvement in leisure pursuits. Since recuperative time is likely to be non-creative there would be at least the possibility for more creative use of leisure with increased time available. Production line workers desiring creative outlets would necessarily seek such experience in leisure activities because of the essentially non-creative character of work in either automated or conventional plants. Passive, recuperative time being primarily self-oriented, there would also be the possibility of an increase in service-oriented activities given more leisure time.

One of the effects of automation upon some types of production line jobs has been the social isolation of workers because of increased distance between work stations and increased attention required by the job.[5] Social isolation on the job may result in a larger proportion of leisure time being spent in activities involving others. This may be especially true for production line work-

---

4. See, for example, George Lundberg, et al., *Leisure* (New York: Columbia University Press, 1934), 123.

5. W. A. Faunce, "Automation in the Automobile Industry: Some Consequences for In-Plant Social Structure," *American Sociological Review*, XXIII (August, 1958).

ers whose occupational roles do not encourage "self esteem testing" on the job so that recognition of success is more likely to be sought from colleagues in leisure activities. For this reason it is also likely that, given sufficient leisure time to acquire skills adequate to insure some measure of success in these activities, workers may spend a larger proportion of leisure time as participants in activities rather than as spectators.[6] Finally, it follows from the preceding discussion of the nature of automated jobs and the relation of work and leisure that for production line workers, leisure activities are more apt to function as relief from boredom than escape from involvement. For any occupational group in which work is seen as a means rather than an end in itself, leisure is less likely to represent freedom from involvement than it is freedom to become involved.

There has been almost no research measuring *directly* the effects of increased leisure upon patterns of use of leisure time. There have been a number of studies in which workers in mass production industries have been asked how they *might* use increased leisure if it became available. The accompanying table is an example of the findings from one such study. The activities listed are those indicated as things a group of workers from an automated automobile engine plant would like to do if they had either longer vacations or a shorter work week. Research involving actual situations where working hours are reduced for workers in automated plants would be necessary to test hypotheses regarding the impact of automation upon leisure activities. The results of this study suggest, however, that automated production line workers may spend a larger proportion of increased leisure time as active participants with others in activities that are potentially creative and service oriented. The data reported in Table I will be referred to in the concluding section of this chapter where some possible consequences of increased leisure for American culture are considered.

---

6. A recent study by Alfred Clarke suggests that our current concern with "spectatoritis" in American leisure patterns may be unwarranted. Alfred C. Clarke, "The Use of Leisure and Its Relation to Levels of Occupational Prestige," *American Sociological Review*, (June, 1956), 304-305.

TABLE I

PROPOSED USE OF INCREASED LEISURE TIME

		Percent of Workers Listing Activity N=125
1.	Work around the house	96.8
2.	Spend more time with family	76.8
3.	Travel	53.6
4.	Go to ballgames, fights, hockey games, etc.	48.8
5.	Fishing and hunting	42.4
6.	Other hobbies	25.6
7.	Engage in some form of athletics (bowling, golf, baseball, etc.)	24.8
8.	Read more	24.8
9.	Go back to school or learn a trade	19.2
10.	Be more active in school boards, P.T.A., boy scouts, etc.	17.6
11.	Get another part time job	16.8
12.	Join more social clubs	15.2
13.	Engage in more political action work	12.8
14.	Rest, relax, loaf, etc.	11.2
15.	Swimming, boating	4.8
16.	Work on car	2.4
17.	Church activities	1.6

SOME BROADER IMPLICATIONS OF INCREASED LEISURE

The possibility that automation may increase the amount of leisure time available to large segments of the population of industrial communities has perhaps broader implications than any other of its potential consequences. In the event of a shortening of the work week there would be an obvious need for expansion of community recreational facilities. Another and more important example of the effects of greater leisure upon community facilities may be its implications for the schools. The data in Table I suggest that almost twenty percent of the workers interviewed would, with more leisure time, go back to school or learn a trade. If anywhere near twenty percent of just automobile workers decided to go back to school, existing facilities in many communities would be inadequate to meet increased enrollments. As the application of automated machinery to production in other industries becomes more common and if a shorter work week results, the problem would become even more acute.

A more important long range problem posed by increased leisure for the schools, however, may be the necessity of including within the curriculum training for the creative use of leisure time. While there is already a growing emphasis in American education upon training for citizenship in the community and a de-emphasis of vocational training, it may be important for curriculum planners to recognize that the citizen in the automated industrial community may have an increasing amount of leisure time at his disposal. Automation may require that there be more adequate provision made for training in certain kinds of technical skills as well, but, in the long run, the primary responsibility of the schools may well become that of instilling certain kinds of values and interests which permit the creative use of leisure and, in general, the teaching not of vocational but of leisure skills.

Increased leisure produced by automation may also have an affect upon the role of the local union in community power structure. If automation reduces the size of the work force in the automated plant, the number of grievance proceedings and other day to day union-management relations as well as the number of most other functions usually performed by the local union may also be expected to decrease. It is the testimony of history, however, that power once held is only reluctantly relinquished and, that, while institutional *structure* may be slow to change, the *function* of various structural units may vary more readily. If it becomes no longer necessary for the local union to perform its various traditional functions at the plant level, it may well turn with increasing interest to participation in community affairs. The local union, or in larger communities, the regional councils representing various locals, may, for example, become increasingly involved in political action at the community level. Approximately thirteen percent of the automobile workers interviewed indicated that with more leisure time they would be interested in engaging in more political action work. If this figure is at all representative of the general interest of industrial workers in becoming active participants in political affairs, local union leadership would have little difficulty in finding volunteer workers in sufficient number to wage an effective political campaign in most communities. Increasing involvement of local unions in community politics would be consistent with the increasing interest of the A.F.L.-C.I.O. in political and social action at the state and national levels and might in some communities, produce significant changes in community power structure.

Agencies outside the local community would also be affected by a shortening of the work week. The proportion of workers indicating a desire to spend more time hunting and fishing suggests that state conservation department activities may need to be considerably expanded. It may also be necessary for highway and expressway planners to take into account the effects of changing technology upon working hours. A system of roads designed to accommodate a peak traffic load composed primarily of work trips may be inadequate to handle the traffic which may possibly enter and leave industrial areas on weekends if a shorter work week results from automation. Well over half of the workers interviewed indicated that with more leisure they would more often engage in activities that would necessarily take them out of the city.

It seems probable that a further reduction in working hours might also produce changes in patterns of family relationships in American society. Seventy-eight percent of the workers interviewed in this study indicated that with greater leisure they would spend more time with their families. Ninety-eight percent of the workers indicated that they would, with more leisure, spend more time working around the house. Leisure which need not be used as recuperative time may be spent by the industrial worker in more active participation in family activities. With the worker in the home for longer periods of time, he may take a more active role in the socialization of his children, a function currently performed very largely by the mother in American families. The current process of transition from a patriarchal to an equalitarian authority structure in the American family might also be affected by an increasing involvement of the father in family affairs.

The number of possible consequences of an increasingly leisure oriented society for American culture which might be discussed is limited only by the span of years considered and the scope of one's imagination. The examples considered above are only a few of the potential effects of increased leisure suggested by the data presented in Table I.

There is one broad question of values regarding leisure not yet considered. It has been assumed that the choice of alternative benefits from increasing productivity will be between increased personal income and increased leisure. There is a third possibility. An increased national income might be used to provide services not adequately performed at present. While it seems obvious that we are becoming an increasingly leisure oriented society, it is not nearly so apparent that we *should* become so. The increased national product resulting from a continuation of the present pattern of working hours plus increased productivity from automation might be used to provide funds for hospitals, schools, and other service agencies but, perhaps most importantly of all, for research in the social, physical, and life sciences.

The "age of atoms and automation" is nearly upon us and the magnitude of the decisions occasioned by its approach becomes increasingly apparent. The conflicting values inherent in the possibility of alternative cultural orientations toward leisure or service require the careful consideration of decision-makers in American society.

Bibliography

C. D. Burns, *Leisure in the Modern World* (New York: The Century Co., 1932). A discussion of the impact of increasing leisure upon a variety of social customs and standards.

A. C. Clarke, "The Use of Leisure and Its Relation to Levels of Occupational Prestige," *American Sociological Review,* XXI (June, 1956), 301-307. A report of research dealing with social class differences in patterns of use of leisure time.

J. Diebold, "Automation and Jobs: The effect on the Worker" *Nation* (October 3, 1953), 271-272. An analysis of the relationship between the nature of automated jobs and significance of leisure activities.

P. F. Drucker, "America's Next Twenty Years: The Coming Labor Shortage," *Harper's Magazine* (March, 1955), 27-32. A discussion of the relationship of increasing productivity, population increase, and working hours.

N. Hurley, "Automation's Impact on Education," *Instruments and Automation,* XXIX (January, 1956), 57-58. Implications of automation for training in the humanities, methods of research, and use of leisure time are discussed.

G. Lundberg, et al. *Leisure* (New York: Columbia University Press, 1934). A report of a detailed study of use of leisure time in a suburban community.

R. S. Lynd and H. M. Lynd, *Middletown* (New York: Harcourt, Brace and Co., 1929), 225-312 and *Middletown in Transition* (New York: Harcourt, Brace and Co., 1937), 242-294. The indicated sections of these two works provide an interesting contrast between patterns of leisure activities at two points in the history of a middlesized, midwestern community.

M. H. and S. E. Neumeyer, *Leisure and Recreation* (New York: A. S. Barnes, 1949). A discussion of factors producing increased leisure, its social and psychological effects, and various problems in the use of leisure time.

"Recreation in the Age of Automation," *The Annals,* CCCXIII (September, 1957). An entire issue devoted to varying analyses of aspects of leisure in contemporary American society.

D. Riesman and W. Bloomberg, Jr., "Work and Leisure: Fusion or Polarity?," in C. M. Arensberg, et al., *Research in Industrial Human Relations* (New York: Harper and Brothers, 1957), 69-85. An analysis of the relation of work and leisure and a discussion of the function of leisure in American Society.

D. Riesman, *The Lonely Crowd* (New Haven: Yale University Press, 1950). Chapters XVI and XVII especially contain an insightful analysis of the changing function and character of contemporary American leisure.

D. Seligman, "The Four Day Work Week: How Soon?", *Fortune* (July, 1954), 81 ff. A discussion of the probable impact of increased productivity upon hours of work.

J. B. Shallenberger, "Automation," in *Proceedings of the National Conference of Social Work* (New York: Columbia University Press, 1955), 152-159. A general discussion of implications of automation including its effects upon leisure.

G. Soule, *Time for Living* (New York: Viking Press, 1955). An analysis of the problem of how Americans will utilize increased leisure time resulting from advanced technology.

WILLIAM A. FAUNCE (B.A. in Psychology, Michigan State University, 1950, M.A. in Sociology, Wayne State University, 1951, Ph.D. in Sociology, Wayne State University, 1957) is now Assistant Professor of Sociology and Anthropology and Research Associate, Labor and Industrial Relations Center, Michigan State University. He has recently contributed articles to the *Social Forces, American Sociological Review,* and the *Transactions of the Third World Congress of Sociology* (1956).

CHAPTER 26

AUTOMATION AS A MANAGEMENT PROBLEM

JOHN DIÉBOLD, *President, John Diebold & Associates, Inc.,*
Management Consultants, New York

Automation has presented management with a major new problem. As yet management has not faced up to this problem and is hardly even grappling with it in any true sense. This is through no lack of energy or good intentions. On the contrary, the very activity of management in this sphere attests to the progressive spirit and desire for improvement that characterize the modern manager. The trouble lies elsewhere. Automation has turned out to be a much more complex and difficult problem than was originally thought. This being the case, the current disposition to minimize its revolutionary and novel aspects is more hindrance than help in putting automation to work.

Despite a lively readiness to buy the machines of automation, management is doing all too little hard thinking about how to use them. This is the real crux of the dilemma we find ourselves in today and it is a difficult problem.

It is the exception rather than the rule to find genuine personal concern on the part of a top level manager with his automation program. Ordering the machines all too often seems to be a way of appeasing management's conscience that something is really being done about an important but unknown new field. By and large, managements are accepting the generalities they hear about automation. The unfortunate part about it is that too few of the published accounts are based on practical experience. The danger, it has been well said, is not in having push-button machines, but in being content with a "push-button type of thinking."

There is an almost incredible preoccupation with equipment.

Yet the hard truth of the matter is that we have hardly begun to learn how to put these splendid new machines to work.

In dealing with automation most managements have exhibited a degree of credulity unthinkable in other aspects of business. Performance standards are virtually unknown in this new field. Few managements are even yet consciously seeking performance standards with which to measure the real progress of the automation programs they have embarked upon. It may be the laxity of ten years of economic boom, uncertainty in dealing with this new and highly technical field, or the difficulty of coping with such a radical departure from conventional procedures—operation in areas where there are as yet no bench marks to which management can look for guidance. Whatever the reason, businessmen are simply not treating the time, money, and talent they are investing in automation with anything like the hard, rigorous analysis one should expect where the stakes, both in terms of investment and potential, are so high. Admittedly, much must be written off to research and education—to learning the ropes of this new field—but a great deal more is being chalked up to the "education" account than should properly be placed there.

Automation, as a word and as a fact, is only about ten years old. Surprising as it may seem, even in this new field, most management thinking has already become rigid and cluttered with stereotypes that stand in the way of real progress. New insights are needed. It is up to management to do a good deal more original thinking and to learn to question traditional patterns of operation. If we can get beyond these stereotypes and see through to the true nature of automation, we shall make a genuine and lasting contribution to industrial progress.

Virtually everything that is said or written about automation stresses the machines. Certainly these devices are fascinating. Indeed, it is their very complexity and their technological splendor that deceive us into thinking we have made more progress than we really have. But it is dead wrong to equate the bewildering array of hardware shipments of recent years with true progress in the use of automation.

A striking fact about management's attitude toward automation is its failure to realize that there is a wide gulf between the possession of the machines and the profitable use of these powerful new tools. What is needed to cover this gulf is a great deal of work by a management that has foresight into what automation really is and how it can be used. But first must come a realization that "taking the plunge" of acquiring machinery does not automatically open the magic gates to the world of tomorrow.

The sight of magnetic tape computers operating in business offices is becoming almost commonplace in the United States. Many people, including most of the managements of the companies possessing these computers, think they are using automation, or feel that they are at least well on the way to reaping its benefits. The real facts, once unearthed, are in striking contrast. Few of the business organizations that use computers are doing anything more with them than they did perfectly adequately before by less elaborate methods. Many are doing it at considerably greater cost.

It is proper to raise the question of whether or not the long lead time that even the best automation programs require before results can be expected has not often been taken advantage of to postpone the day of reckoning for programs that simply will never pay out. The high turnover rate and excessive job hopping that characterize this field certainly play their part in obscuring the real facts from management.

Of one thing I am certain. Few managers are yet aware of the true status of their own computer program. It will still be another year or more before a good many organizations recognize the discrepancy between anticipation and reality. But the actual facts are there today if we take the trouble of looking for them.

DELUSION THROUGH DEFINITION

It may come as an unpleasant surprise when I say that I do not believe management is facing up to the problem of automation, especially if you have spent time and money trying to apply the benefits of automation to your office or factory. You may have attended some management conferences on automation. If you have, you have undoubtedly spent a good deal of time listening to conflicting definitions of what automation "really" is. There is much truth in the quip that it is as hard for a group of businessmen to define automation as it is for a group of theologians to define sin.

In my opinion, almost all the defining and counter-defining misses the point and misses it in a very serious way. In fact, in a sense, the definitions themselves are to a considerable degree actually responsible for the misconceptions about automation that are causing many managements to miss the point, to put the emphasis on the wrong place and to lose out on most of the benefits that automation could bring.

One's capacity for wonder would be dull indeed if it were not stirred by stories of gigantic oil refineries centrally controlled from a single instrument room with the help of only a few men, or of the giant computers that can count and calculate in fractions of a second, or of the famous transfer machines of Ford, Renault and Austin that automatically machine a cylinder block from a rough casting in less than 15 minutes. Nonetheless, by stressing the achievements of these remarkable devices, the definitions have obscured the real essential automation—that it is more than anything else, a concept or way of approach in solving problems, and marks considerable departure from many accepted practices of management.

It is not that the new technology is unimportant. On the contrary, without these self-correcting machines most of what is considered to be automation by any definition simply could not exist. But the importance of the machines does not lie in their ability to perform mechanical tricks. They are important because, for the first time, they enable us to organize many different kinds of business operations, in the office and in the plant, into systems, and to control these systems far more precisely than ever before. More than this, they enable us to do new things, as well as to perform old tasks better.

Automation is not a particular group of new machines or devices. It is a new concept—the ideal of self-regulating systems— and a new set of principles. Only when management understands this will it gain the full benefit from automation, but this kind of understanding is still so rare as to be almost an isolated phenomenon in the business world.

SOME AUTOMATION MYTHS

I have mentioned certain stereotypes of automation that block effective management thinking. I would now like to give examples of what I mean. Every one of these stereotypes has some truth in it; that is undoubtedly why they persist. Yet so long as they are swallowed whole, just so long will automation fail to mean what it could mean—in the factory, the processing plant, and the

office. These stereotypes are in large part the result of accepting the common view that automation consists of specific automatic machines. Among the most important of them are:
1. That automation is primarily useful as a labor saving device.
2. That the ultimate in automation can be symbolized by an oil refinery or any other highly instrumented process plant.
3. That because automation is highly technical most of the decisions concerning it must be left to the engineers and technicians.
4. That only companies with large dollar resources and exceptionally long runs of product can afford to automate.

Stereotype 1: That automation is primarily useful as a labor-saving device. This stereotype is especially pernicious because, at first glance, it seems so plausible. For office management in particular, both clerical costs and the difficulty of hiring enough clerks to get the work done are serious problems. Consider these facts for a moment. Between 1920 and 1950, the number of U.S. factory workers increased by 53 per cent, but the number of office workers increased 150 per cent. Today there are almost twice as many office workers as there were in 1940. One insurance company that installed a computer to handle some of its office procedures had been plagued with recurrent shortages of clerical labor for 15 years. Small wonder that automatic data processing equipment is often looked on as the answer to a management nightmare.

And the fact is that the machinery of automation in most cases probably will make some dent on this problem. For just one example, Commonwealth Edison in Chicago has used a computer for billing its utility customers. Although programming the machine required 35 man-years, almost double the original estimate, and "debugging" the computer after it was installed took nearly 15 man-years and 1,000 computer hours, the machine is now at work and the company needs many fewer clerks in its financial and accounting section. This kind of saving has led a lot of management thinking to run this way: "If we buy a computer and let it handle, say, payroll, it will soon pay for itself because we can cut way down on staff and punched card machines. Then, after it has the payroll under its belt, we can begin to let it solve some of those big intangible problems that you can't even put a dollar value on."

The trouble with this kind of thinking is that it overlooks the remarkable potentialities of these machines for doing what cannot be done at all without them. Payroll, after all, can be handled perfectly adequately by punched card machines; the pay checks come out on time and they are accurate, even though the job may take the time of a great many clerks. Putting routine operations on a computer may pay for the computer in the end, but it can rarely do a great deal more than that. Commonwealth Edison, itself one of the best run computer installations, has discovered that the real pay-off is in the use of the computer to solve operational problems. Here, the most important savings apparently are not going to be in the clerical area, but in the realm of decision-making.

Executives will never realize the potential of the new equipment if they persist in thinking of automatic data processing in terms of merely eliminating a few, or even a good many, employees by speeding up old procedures and routines. Mechanization of existing procedures may merely result in compounding and perpetuating inefficiency. The real aim is not to speed up an old job but to do a better job. Part of the challenge of automation to management is learning to use new tools to solve new as well as old problems.

The big benefits from automatic data processing will come only when management learns to use automation for its unique ability to provide better, more accurate and more timely information about the operation of a business.

The thought of more information is a frightening one for the executive who knows that his desk and briefcase are already bulging with information he had not even had time to read, let alone digest. When my firm analyzed the operations of a large publishing company, we found that the president received 621 reports during a three-month period, more than one every working hour. And, as W. W. Smith of General Electric has pointed out: "The mass of reports management receives today may be likened to a daily newspaper printed without headlines and without punctuation, without spacing between lines and words, and without capital letters. The mental gymnastics required to determine exactly what is going on is asking too much."

But management, if it uses the new tools correctly, will soon find that what it has now is not really too much information but not enough of the right kind, because the mass of data is too great to be processed by conventional equipment. We will be able to do two kinds of things that we cannot now do at all or can only do in a rudimentary manner.

First, we will be able to plan in advance on the basis not of guesses and hopes but of facts, because automatic data processing can give realistic answers to "what would happen if . . ." questions. A major domestic U.S. air line asked one of these questions when it put maintenance scheduling on a computer. Whenever a plane breaks down, an airline has a complicated rescheduling problem with literally thousands of possible solutions. In effect, the airline asked a computer about each flight on its schedule: "What would happen if this flight should break down tomorrow? What is the best way to use our other planes while that one is being repaired?" The computer was able to run through all possible solutions in a half hour and to pick the best one in each case. As a result, this particular airline has saved one standby Constellation in each of six airports at a saving, for each plane, of one million dollars a year in lost revenue. When jet planes come into use, the comparable saving will be five million dollars a year for each plane.

In many areas of business today we must make decisions based on incomplete information, and so "fly by the seat of our pants," as the saying goes. For example, recalculating complicated budgets or schedules to determine ahead of time the precise effect of a certain decision is too great and too long a task to be practical except when the decision is a major one. Automation gives us tools to practice this type of "what would happen if" management, allowing us to build models of alternative solutions, and so give us quantitative answers about the effects of contemplated policy changes. Thus, in the example above, the airline was able to experiment on paper—or more exactly, on magnetic tape —and to predict the consequences of alternative procedures.

Second, management will be able to get quick, complete answers to problems whose solutions now are incomplete and often arrive too late to be of great use. Recently, a major publishing company was forced to discontinue two of its publications, both widely circulated and historically important magazines. There were many complicated reasons why this action was necessary, but here is a case where the use of automation for decision-making might have prevented the final disaster.

Because of difficult financial conditions, it was known before the discontinuance of the two magazines that the publication of a third company magazine might be terminated and its resources and circulation thrown behind the other two. It was necessary to determine what proportion of the circulation of this magazine duplicated subscription to either one or both of the other two magazines.

Circulation records were kept on punched cards, but they were of such volume that a cross check of the entire three lists was too expensive. Unfortunately, the sample that was taken proved imprecise, and the company underestimated the number of 'cross subscriptions'. The result was that the company experienced a severe and unexpected cash drain in securing enough new subscriptions to meet the circulation guarantee. This set off a series of chain reactions at a critical moment. All three magazines were finally discontinued.

Although such a serious error in decision-making need not be fatal in a strong business situation, it was in this case. And this case is one where advanced automatic data processing could have been used to make the necessary calculations before the final decision, and possibly have averted the collapse of the magazine.

When automatic data processing is used in this way, it can turn the present art of management into a real science of management. It thus becomes clear that management which puts labor-saving first has its eye on the wrong ball. More efficient operation and better control are the real goals. If labor is saved as a by-product, so much the better.

Stereotype 2: That ultimate in automation can be symbolized by an oil refinery or any other highly instrumented process plant. An oil refinery where instruments and controls far outnumber the human workers gives the impression of being very highly automated. Compared with an automobile factory, it is. Yet nothing could be further from the truth than to say, as one business magazine did recently, that "The men have to be there anyway in case of emergency, so the extra costs of more instruments to read other instruments is not considered justified at this time." This is just one more example of a statement that misses the whole point of what automation really means to business.

Refineries have achieved what fabricating is still struggling for—a conversion from batch to continuous processing so that operation can be made automatic. But they are only beginning to feel the full impact of a second, and more significant, stage of automation, when the automatic operations will be automatically controlled.

The intricate controls that run a refinery *almost* by themselves are in a large sense not really running it at all. As things stand, the variables of processing—temperature, pressure, level and rate of flow—can be maintained at desired values without human intervention through the use of feedback control devices. But the values themselves must still be selected and the control instruments adjusted accordingly. In many cases, it is not possible to determine the relationships among these variables that will hold true throughout an entire process. This means that an operator cannot come to work, set his controls and go home. He must reset them every time a test of the product being processed shows that changes are needed.

Thus, it is the operator who does all the decision-making. Since a refinery is an extremely complex operation, it is more a matter of luck than science when he makes the best possible decision at any given moment. The result is that even such a highly automated industry as refining works most of the time on a trial-and-error basis. The best refinery, if we are to judge from some tests that have recently been run using the instrument records of several first-rate refineries, probably does not operate at optimum for more than a few minutes out of the entire 24-hour operating day. The rest of the time output is fluctuating around—and sometimes quite far from—optimum. Genuinely effective control, it is estimated, could increase yields by as much as 30 per cent.

If you are dealing with a process that allows blending to achieve an acceptable final product, as is the case with petroleum refining and the manufacture of many chemicals, control such as this may not be so bad. If you are manufacturing some of the new complex synthetics, such as dacron for example, it is another story. You cannot make up for sloppy controls by blending because the product is either right or it is not. There is no in-between that can be corrected by adding a little something extra at the end. Such processes require that an exquisite level of control be maintained among many rapidly fluctuating variables, and the only way it can be done is by use of a highly complex, overall control system. Because of the lack of such control equipment today, a great deal of the output of such plants goes into waste, and this waste is reflected in the cost of their products.

To achieve genuinely effective process plant control, all the plant's individual controls will have to be integrated into a single coordinated, self-regulating system. Just as a single machine designed on the feedback principle notes and corrects variations in its output, so an integrated self-regulating system will note and correct variations in the end-product of an entire plant, making precise and instantaneous adjustment whenever the product itself shows any variation from optimum quality. Since the control of a number of variables to produce a desired end is essentially a calculating operation, the integrated operation of the process plant of the future will depend upon an electronic computer to analyze, correlate and correct the operations of the individual control devices.

There are two reasons why no processing plant issues a computer in this way. We do not yet possess a computer that is reliable enough to operate suitably in an "on line" capacity for long periods of time, although we are very close to achieving one. More serious, we simply do not know enough about how process variables affect each other and the end product. We do not yet have instruments that can measure reliably, accurately, swiftly, and continuously all the variables of refinery operation, and we do not yet know how to measure, relate and reduce to equations that a computer can handle, all the process conditions that determine the quality of a given end product.

Once these problems are solved, however, automation is going to change the traditional distinction between plant and office, linking them together as a single interconnected system. Up to now, there has been little real consideration of the plant and the office as essential and fundamental parts of the same whole. Of course, it is merely a truism to remark that the activities of the two are inseparably interrelated. The receipt of an order by the order department means that invoice and production require-ment, purchasing, payables, machine loading, and receivables, are all altered. The complexity, size, and time requirements of a system that is part machine, part manual, do not allow for in-stantaneous adjustment of all the relationships, but rather make it necessary to batch them, and to deal with them in groups at specified time intervals, often after much delay.

Even where management has recognized that these activities are interrelated, it has not had the means to link them together mechanically. Automation provides us, for the first time since the growth of complex organizations, with a means for gradually making these organizations function as better systems—sensitive, flexible, and adaptable to change. The first steps have already been taken. Automatic logging devices, specifically in the form of automatic typewriters, can print out in sequence the reading from each instrument. Readings that represent variations greater than normal are typed in red, and a machine can be programmed so that it will index, skip to the bottom of the log sheet, type the off-normal reading in red, and get back to its business on the upper half of the log sheet, putting on quite a show for the engineers who have been alerted by sight and sound alarms.

It was only a short and obvious second step to attach a paper tape punch to the automatic typewriter. Instrument readings are thus automatically made quickly available, in a form that can be processed through a calculating device which will actually perform the engineering calculations of, for example, yield in a refinery. This is exactly what the Tidewater Oil Company is doing in its newest refinery.

Thus far, equipment has not been developed for the completely automatic control which will be achieved when each of the regulating instruments can automatically correct variations in the variable it measures in terms of variations in all other related variables. Nonetheless, it is an important step to be able to derive accounting and production data directly from the process itself, thereby eliminating the costs, the delay, the inevitable error, and the boredom of manual recording, recopying, and eventually keying information into an office machine for subsequent reprocessing. Once it is possible not only automatically to scan and record variables, but to compute the best course of adjusting them, the feedback loop will be closed, and the entire process can be controlled in terms of changes in the characteristics of the end product. When this *end point control* has been brought about, management will have a far higher level of control over the process, and the interweaving of plant and office will be even more complete.

Once these technological hurdles are overcome, as they surely will be within the next few years, management will be in a position to exercise really tight control over processing operations and to produce increased yields of better products, at lower cost.

Stereotype 3: That because automation is highly technical most of the decisions concerning it must be left to the engineers and technicians. The machinery of automation is indeed extremely complex. The science of communication and control, on which it is based, is easily among the half dozen most advanced frontiers of technology, and the computer is the most advanced piece of equipment yet built in this field. But nobody is asking you to build a computer or to repair one if it breaks down.

Those are jobs that only technicians can do. Management, too, has a unique function and one that is, in its way, as complex and difficult. Perhaps more so. Truly fruitful results from automatic data processing systems require a fundamental change in approach, an understanding that the best applications are not the mechanization or streamlining of existing procedures, and a willingness to rethink the problems of an entire business in terms of ultimate goal and final product. These are not technical problems. They are problems of method, organization and attitude and they require managerial imagination, skill and experience rather than technical proficiency.

Bluntly stated, automation is one of the critical areas in which management must *manage*. But instead of realizing this crucial point, management has become so intimidated by the complexity of this new hardware that it has allowed technicians to take over not only the operation of the machines but the actual decisions about how they are to be used. This has led to a whole series of difficulties.

To begin with, there is the electronics committee. Countless times I have heard otherwise responsible executives say, when I asked them what they were doing about automation: "We have formed an electronics committee." And they have said it in a manner that implied that the problem was taken care of. In practice, this often means no more than that the committee spends a year—or in one case I know of, three years—wandering about the country, attending manufacturers' schools and visiting computer installations. This experience is somehow supposed to be sufficient for making critical decisions about the highly detailed process of automating.

The electronics committee has often been a device for creating the impression of doing something about automation while at the same time avoiding any action that could possibly backfire. Many managements genuinely seem to be looking at automation as if they were walking around the edge of an ice-cold swimming pool, realizing that sooner or later they are going to have to jump in but trying to postpone the leap with as much rationalization and fact-gathering as possible. In one case, the electronics committee of a major oil company, after a year's study, actually presented the final equipment choice to the president of the company by asking him to choose between the machines of two different manufacturers.

The installation of automation equipment should not be made unless a thorough understanding of the business itself, and the functions and needs of its operations, underlies it. Most engineers underestimate both the importance of this point and the difficulties of grasping the complexity of modern business operations. While the businessman may regard the specialized knowledge of the engineer with something approaching awe, the engineer frequently regards the unfamiliar processes of business as something that can be mastered in a few months. Glenn White of the Chrysler Corporation has remarked out of the experience of his company that "We are satisfied that the way to put together a team of people to work on electronics is to take somebody who has good knowledge of how to run your business, a good systems and procedures man, if you please. They can be trained in electronics much easier than somebody who knows electronics can be trained in how to run your business."

Stereotype 4: That only companies with large dollar resources and exceptionally long runs of product can afford to automate. This is not wholly true today and before long it will be even less true, but the reason the impression has arisen is easy to understand.

The automotive industry, an automation pioneer, has had great success with very expensive and very specialized industrial equipment, made for the requirements of a particular product. The large transfer machine, complete with loading and unloading devices, is well suited for this industry where literally millions of identical parts pass through a line before new equipment has to be considered.

Such machines are not so well suited to the estimated 80 per cent of American industry that produces in lots of 25 or fewer identical pieces. Nor are they well suited for industries that frequently redesign their products, since any major change in design means costly readjustments at best and may even mean scrapping this expensive machinery.

But the technology of automation that deals with feedback and control systems is producing a new family of machines guided by magnetic or punched paper tape. These machines will make it possible to obtain the benefits of automatic production, yet retain the flexibility of operation essential for job shop production. They are only beginning to appear. Some 40 different prototypes can today be found in the shops of machine tool manu-facturers and during the next few years they will begin to have an enormous impact on the small lot producers of this country. It is my own conviction that these tape-controlled machine tools are today in the same stage of their evolution as computers were when they first came into use in 1946. Within ten years of their appearance, computers had won wide acceptance in business. I believe that during the next three to five years, numerically-controlled machines will gain the same kind of acceptance in factories.

But even today, multi-million dollar transfer machines are not the only kinds of automation equipment suitable for factory use. Management that uses the excuse of small lots may well be guilty of what I call the fallacy of job shop thinking. For example, my firm was recently asked to suggest automation possibilities for one of the largest manufacturers of shovels in the world. This plant produces hundreds of different kinds of special-purpose shovels in small lots. Each kind, since its shape is unique, was being produced separately. The whole factory was nothing more than a collection of job shops under one roof. Yet analysis showed that about 80 per cent of the company's production was limited to not much more than a dozen different models and that this, the great bulk of the plant's total production, could be made automatic without great difficulty.

The most basic thinking about automation has not yet been done in this plant. Here, it is not simply a question of new ma-chines, but of reordering the old machines and the old proce-dures in a more 'systematic,' more productive way. Although reorganizing existing equipment may go no further than intro-ducing good principles of materials flow, it is an essential first step in integrating the fabricating operation, prior to linking the individual machines, just as such integration must precede the introduction of overall automatic control.

STEPS IN THE RIGHT DIRECTION

It is of utmost importance for management to exorcise these stereotypes from its thinking. But getting rid of false ideas is only the first step toward using automation, in office or in factory, as it should be used. The next steps are positive ones: to understand the factors that make for a successful automation program, and then to apply them.

That they have not been properly understood or fruitfully applied is revealed, I think, by a survey my firm recently completed. We polled an extensive cross-section of business organizations currently using automatic data processing systems. The results show a fairly discouraging picture but one that I think might have been anticipated. For while most computer owners showed a vague general satisfaction with their machines, detailed and precise questions indicated that a great deal of this satisfaction can be traced to a lack of understanding of true potential and a corresponding set of lower goals and aims.

My personal suspicion is that in some instances there was also a natural reluctance to admit frankly that "we made a mistake." One indication that this suspicion is not exaggerated comes from a comment made to me not long ago by one of my colleagues, a man who has worked closely with computer users all over our country. "You can pick up that phone on your desk," he said, "and in half an hour you can buy full daytime shifts on any one of 20 large computers that aren't being used full-time." Companies have spent hundreds of thousands of dollars on these machines and, after they have been installed, find that they are using them only a few hours a day.

If management is to avoid this kind of disappointment, there are four essential steps you must take: 1. Define your objectives; 2. Staff your automation program carefully; 3. Estimate your costs realistically; 4. Train your personnel.

DEFINE YOUR OBJECTIVES

What do you want automation to accomplish for you? This is a question that cannot possibly be answered without a thorough preliminary systems study that will enable you to recognize what problems you want to deal with and how your own organization can best go about handling them.

The first thing to realize is that there is no point at all in automating simply for the sake of automating. It is a rare executive who will admit that his company is investigating automation simply because his golf-partner's company already has equipment installed, or because he has been overwhelmed by glamorous advertisements and newspaper stories, or because he has a vague feeling that "perhaps we had better look into this new thing." Nonetheless, I believe that there are at least as many supposedly hard-headed businessmen who have bought expensive equipment on this basis—and lived to be disappointed in it—as there are businessmen who have gone through the practical, painstaking and time-consuming process of a thorough analysis of the entire organization as an integrated system.

Management that has taken this indispensable preliminary step will never make the mistake of confusing the tangible possession of the hardware of automation with the practical use of the concept. Nor will it commit the fallacy of the "step by step" approach. "We are taking a step by step approach to automation" conveys the impression of caution and proper business reserve. But what it often means in fact is that another uncoordinated misstep is being taken. While caution is a desirable thing, what is so misleading about this approach is that the whole concept of systems analysis and design, which is basic, requires a careful and detailed plan for the entire organization if the benefits realized are to be more than marginal. Following a step by step approach often results in throwing out the previous step and redoing a great amount of work to install the procedures and equipment associated with the new steps. This approach is also likely to mean mechanizing old procedures one step at a time, instead of finding out whether some of these procedures are not outmoded altogether.

A thorough analysis will prevent the management of medium-sized concerns from falling into the trap of trying to use the same approach to automation as that of very large companies. A company like Du Pont, for example, can afford to buy a computer and experiment with it on a single application or a series of separate applications.

In many giant organizations, enough people are involved in a single process like payroll or billing so that at least a superficial case can be made for mechanizing this one job, if for no other reason than that the company is gaining experience in computer use. But the data processing activities of smaller firms are typified not by their routineness but by their variety and interconnection. To lift one department arbitrarily from this interconnected whole for the sake of speeding up its routine processes will never lead to the ultimate goal—the design of the information processing system best suited to the particular requirements of the individual organization.

It is often necessary to do what I call *rethinking* in order to define your objectives. Sometimes what we believe to be our problems are in reality symptoms, the true problem lying far deeper. It is important that we do not try to automate in order to tackle symptoms. A thorough systems analysis and a careful thinking through of objectives is the best means of defining the true problem. At the same time an adequate systems study can save management from the opposite but equally serious mistake of trying to make too fundamental a change in one pass.

Similarly, it is often necessary to redesign the product or the process or both in order to make automation feasible. In stating your objectives, you must be careful not to define yourself out of automation altogether. For example, in designing a highly automated plant for the manufacture of telephones, my firm was faced with the problem of automating the manufacture and assembly of some intricate electrical circuitry. If the task had been defined in terms of assembling the product as it had been designed for hand assembly, the equipment investment would have been prohibitive. By defining our objective as creating a network circuit that could function just as the old one did, we were able to redesign the equipment in such a way that automation proved feasible.

STAFF YOUR AUTOMATION PROGRAM CAREFULLY

It is not enough to realize that an automation program is best run by choosing a man who understands your business, rather than by an engineer who understands automation equipment. It is of crucial importance to choose the very best man you can find to head your program.

My experience has been that most top management is appalled to learn that it will not only have to put a key person in charge of automation but that he will have to spend as much as a year simply studying the problems and learning how to solve them. "Why, I can't spare my best man," executives have said to me. "I have to keep the business running, don't I?"

Nonetheless, it is false economy to do anything else. Consider what responsibility for an automation program means. It means a thorough study of the entire business and a well-grounded understanding of its aims and goals. It involves responsibility for making a detailed systems analysis. While it does not necessarily involve intimate knowledge of the technical details of equipment, it involves something much more important—knowledge of what machines are available, their advantages and disadvantages, their potentials and limitations, so that the equipment finally purchased will be the equipment that best fits the needs of the system. It involves responsibility for setting up a training program within the organization for those who will actually be using the equipment. Finally, it involves running the automation program once the equipment is installed.

To give this responsibility to a superannuated vice president, simply because he can easily be spared from present operations, is practically to guarantee at least partial failure for the whole automation program before it starts.

ESTIMATE YOUR COSTS REALISTICALLY

One important reason that hoped-for savings from automation have failed to materialize is that management has not always been realistic about costs.

Businessmen considering automation tend to make two mistakes. They count the cost of the machinery itself, but they do not recognize the associated, often hidden, costs of using it; and they base their plans for the equipment on the assumption that its cost must be justified by immediate and obvious savings. Since a reduction in labor costs is the most immediate and obvious saving, this consideration becomes the justification for buying the equipment, and then dominates its use after it has been installed. In the process, the possibility of capital savings so substantial as to dwarf any conceivable gains from labor savings are overlooked altogether.

Take the case of a paper-making machine. The introduction of an automatic control system on the city-block-long equipment is not likely to remove a single man from the payroll, since many men will be needed to handle breaks in the paper and to run the new equipment. Yet the new system may allow an increase in the speed of the equipment from, say, 2,000 feet per minute to 2,500 feet per minute. The increase in productivity of the equipment which represents a capital investment of $10 million to $15 million, results in a sizable decrease in the capital cost per unit of useful product.

Similarly, the economics of computer installations are badly understood. Management often does not realize that the cost of installing a computer and converting existing procedures will approach the cost of the computer itself. Briefly, this means that if you are planning to spend a million dollars for a computer, you had better count on spending an additional million dollars, for planning, installation, and conversion costs. These may not all appear in the direct cost breakdowns associated with the computer, but they are very likely to be real costs borne by the organization in assimilating the change. If you fail to take this into account you are bound to be disappointed in the length of time that it takes the computer to pay for itself.

It is also important to understand that once the system is in, the largest expense will be the cost of preparing data for the machines. If additional useful information can be extracted as a by-product of something the computer is doing anyway, the economics may be enormously changed. That is why payroll is seldom the best way to use a computer. To put it on the computer in the first place is complex and time consuming, and once it is on there is no easy way for management to get additional useful data from this application, unless it has been planned as part of a system having several objectives beyond payroll preparation alone.

TRAIN YOUR PERSONNEL

Just as important as finding the right man to head your automation program is the realization that personnel to run the program do not necessarily have to be hired from outside your own organization. In fact, there are excellent reasons for training your own personnel rather than attempting to bring automation specialists into your organization.

The training of personnel in analytical procedures, machine operation, and programming is one of the most critical problems of a successful automation installation. It involves good insight into how to select, train and retrain already employed personnel to work with the new equipment and the new procedures. To go far afield to find help only complicates matters further. The mad quest for skilled scientific personnel that is currently taking place in the U.S. has brought into existence a large group of automation "job floaters," who are unstable and often disruptive in an organization when they are introduced over the protests of good personnel people.

When you come right down to it, one cause of the trouble is that the hardware of automation is being shipped faster than competent people have been trained to operate it. But the answer is not to go into the job market for specialists. Good ones are exceptionally hard to find, and money is no longer an incentive for the already highly-paid experienced man. These men are looking for very specialized kinds of opportunities. The solution is to train personnel from your own organization.

Fundamental to the success of an automation program is the attitude by management that automation is its concern. It is clear that management is willing to do the hard work involved in carrying out its responsibility. What is required is a change in attitude toward automation, a change that will dispel the myths and permit management to exercise real control over the automation program — its objectives, its costs, and its results.

JOHN DIEBOLD, president of Diebold & Associates, management consultants, is credited with shortening the word "automatization" into automation, while a student at the Harvard Graduate School of Business, and giving the term its popularity. He holds undergraduate degrees in Economics and Mechanical Engineering. After receiving his M.B.A. from Harvard in 1951, he became a staff consultant for Griffenhagen & Associates, Chicago, which was later merged with his own firm. He became editor and associate publisher of the magazine, *Automatic Control,* in 1953, the year he founded his own company. He left the magazine two years later. In 1952 he authored *Automation: The Advent of the Automatic Factory.*

CHAPTER 27

POLITICAL ASPECTS OF AUTOMATION

CHARLES W. SHULL, *Professor of Political Science*
Wayne State University

Perspective is essential to the study of automation. The process is complicated and complex in character with the ready result that assumptions about it are sweeping and oftentimes intricate in implications. Viewed from the standpoint of the political process, automation as a social and economic phenomenon seems rampant with emotional factors, with challenge to existing societal organization, and with the sound of alarm.

If automation is conceived as altogether new as a social and economic process, then all the political aspects and significances are novel, and unexpected since they are unexplored and thus incapable of prediction. If automation is seen as a further extension of a process as old as the wheel itself, there comes a false sense of security born of that familiarity which breeds contempt. Conceived of as refinement and projection of the process of technological invention through the adaptation of newer mechanical and technical devices, automation acquires a social and chronological matrix, a historical bed, upon which the entire scope of invention rests and is supported. In each of these concepts and contexts automation acquires political and governmental aspects, significances, and problems.

But any examination of the political aspects and significances of automation must be made on the assumption that it is the present phase of automation in current development which matters. Likewise extant political institutions, ideals and ideas, and governmental forms must be taken as the areas where automation has its impact and impels changes which acquire political motivations and significance. The projected treatment will not end at a point in history, but be projected against the open horizon of the future. References will of course be made to specific events, indicated effects of the progress of automation, but there will of necessity occur speculative treatment of some of the political phases of automation. The mood may be ominous in its doubt of human capabilities on occasion; it may be hopeful in its faith for the future. Where speculation is cast into concepts or expressed as possibilities it is to be understood that all is subject to the ultimate and rigorous revision of time and its inexorable passage.

Attention will now be devoted more pertinently and specifically to the obviously political aspects of automation.

One governmental function which immediately comes into focus is the extent to which national, state, or local must enter into the creation, development, improvement or maintenance of recreational facilities. Here comes for example the matter of parks—national, state or municipal, as facilities for use by people with leisure. Not merely is the matter of the provision of such park facilities an issue for government and political concern and action, but the readiness of access and greater convenience for those seeking to enjoy them is also a further element in this complex of the problem of automated leisure. Should the form of recreation that is desired involve fishing or hunting, even camping in the open with no pursuit of fish or game, there will still be points of rapport and contact with government and in turn with political factors. In addition natural beauty spots, lakes, parks, other places suitable for recreational and leisure enjoyment and utilization are futile and valueless unless there is access and convenience of approach. Roads or highways must be planned and improved to make easier the approach and servicing of these recreational areas. Each of these items carry the matter back to its heart—finance—and that definitely in the public sense of taxation.

This has not been intended as a presentation of the greater number of persons specifically desiring to spend a greatly enhanced leisure engaged in activities ranging from intricate, even costly hobbies, to the guzzling of more beer. Should automation create such leisure for the thousands of persons upon the scale and in the magnitude expected by some observers then the pressures for certain types of recreational facilities will be heightened and intensified. Where these facilities fall into the group which are publicly supported and maintained, it is obvious that there would be need for new parks, the improvement of existing ones, the addition of areas to those already devoted to public recreational use, and the development of highways and roads (to say nothing of skyways) giving more direct, speedier, and presumably safer ingress and egress from these selfsame public recreational facilities. Even the degree of choice of recreation in terms of character and form is not exempt from political determination or governmental fixation. If nothing other than motion on the highways in modern motor cars—just travel or joyriding, traffic results from this mechanically created leisure, there would be a large political aspect to the problem of the recreational use of leisure in the impinging world age of automation.

GOVERNMENT AND EDUCATION

The interrelationships of governmental units to public, private and parochial schools within the complex structure of American tax systems marks out another arena of political controversy, stimulated vitally by the advent of automation and needs for education adjusted to this phenomenon of automation. Money, training, certification and standards in each and every aspect lure the genii of the political, pushed into a new salient of action by the sheer existence of automation and the need for a more skilled and educated worker. The surface has not been scratched in this area.

GOVERNMENT FINANCING

Throughout the various sectors of the problem of automation and its political aspects there has been a recurrent mention of the matter of governmental finance. Necessarily this is true,— that ultimately the financial aspect of an economy or a political community has to come into consideration.

It is axiomatic that in order to finance programs and to carry on enterprises governments must rely upon these forms of revenue—taxes, service charges, or borrowings. It is not intended to insert into our treatment of automation and its political aspects anything approximating a comparative study of taxation systems. Nor is it only taxation which enters the stage as a representation of the fiscal problem raised by automation. Before specific consideration of the matter of tax structure and incidence is reached there are certain factors that need explication. Primary is the matter of cost of the machinery of automation and its basic installation. From every point of consideration this is a costly enterprise with one aspect of the decision to institute the process of automation within a given industrial complex or plant being that of saving ultimately and in long range terms by reduction in labor costs. A quick recovery is essentially vital through recouping of investment in large scale automatic installations. In effect then three routes present themselves as available to achieve this purpose. The first consists in the maintenance of a high operating capacity with the resultant production of articles at the same high capacity of turnout for the finished product. A presumption is set up that there will be a maximum sale and consumption of the products of the automatic and electronic devices of industry and that the pricing will be at an optimum to serve both consumer and producer; this would approximate economic utopia in that apparently under these hypotheses everyone should and probably will be happy in the economic sense. It is then the hope of the introducer of automation to make a rapid recovery upon his fiscal investment.

One method of achievement of this objective has been mentioned. As was pointed out this presupposes a high level of economic activity favorably oriented towards a maximum of return through high production and equally high consumption of the output of automation. The normal level of industrial activity or use of automated equipment must simulate what in lesser technological situations might be thought to portray boom times. Governmental operation itself will be a great force in the determination of the possible existence of the kinds of conditions which could exemplify the high rate of demand that would keep the automatic machinery functioning in the way desired. Whether any contemporary economy could function effectively at such high levels of productivity for its automation to be financed directly from revenue received as payments for its output is seriously open to question and should be regarded as moot.

The second method for the relatively rapid recovery of investment in automation would also involve perhaps more directly action by government, and become infested with political viruses and rampant with manipulative aspects. If rapid payment through ready revenue receivable under the basic conditions discussed above cannot be the route for favorable investment in automation and its various component mechanisms, then the road towards more radical, more enticing depreciation treatment may be tried. Establishment of depreciation rates and methods of claiming them has pertinence for tax purposes, particularly income tax practice, and in public utility valuations for rate fixing purposes.

Standard methods of depreciation may not be fully applicable to the technical evidence of automation, i.e., the tools and machinery. Expensive installations of automatic or electronic machines or other forms of automation may readily be obsoleted by a palpably minor development. This is part of the argument here. Technical obsolescence must be evaluated differently from physical obsolescence or erosion of utility and loss of power to function. Newer approaches to depreciation practice collide at this point with what is permissible since tax law impingement takes a variable monetary toll from the investors in automation. Legislation has been proposed, even enacted, providing for what is known as "fast write off" or speaking more precisely the acceptance of a liberal policy of depreciating certain kinds of industrial property. Since this paper considers the political aspects of automation this area of a swift depreciation must be denominated as one of the more controversial phases of the entire process. To illustrate, there will certainly be a great difference in the volume of taxable income accruing to a company heavily investing in automatic processes and instruments if depreciation on this type of investment can be charged off at the rate of 20% a year within a five year period instead of an average of 5% annually over twenty years. Pressure mounting towards such a preferential treatment of businesses, depreciation wise, is always special; it can only bring repercussions of a political character and constitutes a highly charged phase of a problem of automation.

The third method by which an effort is made to recoup investment by rapid returns lies in the assumption that the reduction of the labor staff and its working load resulting from the introduction of automation will be reflected in lowered labor costs, offsetting the higher price of the newer machinery and its installation. To do this, the risk of technological unemployment and recession type economic conditions must be run. Likewise it is assumed that there will be no rise in the rate of remuneration—wage or salary wise, for the technicians and automation operators necessary to manage and direct plants and businesses which symbolize the entire process of automation. The role of government in this situation is diversely conceived and posited.

Partly this has been described in the portion of this chapter briefly devoted to unemployment as a consequence of automation. If there is a relatively great decrease in the actual working force of a nation under the impact of automation, the matter will certainly become fraught with the political in terms of reactions within and about economic conditions. In such circumstances the full range of speculation as to possible remedies—tax cuts, tax increase for social security or for direct relief, as well as controversy as to the responsibilities of various levels of government will be displayed and conceivably fought over.

The discussion immediately above has not had anything to do with the struggle of newer sources of revenue for governments which would result from the pressure for all the range of activities in the areas of recreation, education, or relief or unemployment insurance. Within this particular matter of general financial support for governments in the future there lies a deep and serious problem so far as the possibility of the maintenance of a standard of living is concerned. How the standard of living can be as high as possible, the security of the nation as great and durable as it can be made, and all this attained at the minimum of cost in terms of public finance is a problem of tremendous magnitude and formidably political in character.

Essential to the financially successful operation of an automated industry would seem to be a high degree of peaceful labor relations. Stability of operation, maintenance of the full bracket of so-called union rights, full consideration of the manpower demands of automation and its operational service, are all fundamental. On most of these matters of labor-management relations there exist legal statutes or administrative regulations. The entire gamut of industrial labor relations is politically and governmentally affected by the very existence of automation.

Beyond these facets of the problem of the financial milieu of automation there appear to be two additional ones worthy of some comment. Automation as an industrial order does offer opportunity for some adaptation in the regimen of business and industrial activity. There may be developing a different pattern of operating shifts and specific working times which will present diverse patterns of utilization of the labor force and managerial personnel. Different forms of application of work policies may appear under the drive and pressure of automation having effect not only upon working conditions but upon the accompanying accounting and fiscal practices.

Beyond these there lies the area of national defense and the entire range of military devices which rest heavily upon some feature or other of automation and in developments where research technique and thought leap out towards still newer and more effective push button applications. The political items already discussed as matters of finance and of personnel policy and its administration will be present in these adoptions of automation by governments.

In addition there will be some problems not readily capable of being subsumed under these headings. In each instance the money in question comes from tax revenue and the accumulation of costs created by automation's advance into the processes of government will enlarge the need for funds on the part of government until the entire matter of tax policy, money management, borrowing power and debt limitations become crucial factors in this area.

Stability in the public personnel will be sought as a political desideratum and familiar themes of bureaucracy will be echoed again as strife to maintain status quo relationships breaks out in the face of indicated efficiencies and economies of automation. Recognition of the obligation of governments to exhibit or evince the picture of the proper type of employer-employee relationships enters into consideration and serves to create a sense of security, stability, even to the point of personal frustrations in favor of a general *esprit de corps*. This security sometimes feeds on mediocrity and there are those who argue that automation intensifies this trend bereaving government of public servants of imagination and daring. If this should happen to the operator and manager of automation itself the end result may be a questionable understanding of his role in government and the magnitude of his responsibility.

GOVERNMENT CONTROL

It may be a difficult matter to define precisely who are the owners with respect to the major examples of automated industry or business. That there are such can of course be substantiated although the lists of stockholders and the roll of directors and even of bondholders may be lengthy, even formidably so. Their significance fundamentally becomes threefold with relation to automation as a economic process. First ownership has something to do with the direction of business and corporation policy and thus contains within this group the power to determine whether or not automation is to be introduced. Likewise ownership bears a relation to aspiration for profit and the desire for favorable conditions to create and enlarge that economic gain. Thirdly top management to a considerable extent has a clear relationship and responsibility to ownership. In addition ownership permits management, perhaps in a negative manner, to hire the technicians whose sphere of specialization is that of automation and automatic devices.

The knowledge of the techniques of automation is different from the demands of ownership, of personnel or labor force, and of management in a generalized sense. Subsequent to a decision to install automation within a business or industrial situation, the base capitulation to the expert in electronic or automatic devices has been made. Expensive and extensive investment has been made by the decision of top management and tacitly at least by ownership; at the same time the intermediate managerial corps has neither the time nor the scientific knowledge or skill to install, operate or maintain the costly equipment of automation. Many of these selfsame technical experts on automation have very minor interests or sympathies with or in other phases of industrial or commercial operations. A peculiar kind of semi-autonomy comes in effect to this group of automation experts who in turn are split into those who invent, devise, and install, and those who later operate, service, and maintain.

The fourth group of major character related to the entire problem of automation is government itself, considered both as a body of people and as an institution for social and legal control. In this latter sense government as a positive controller of economic activity has been described in numerous ways as reacting to the impact of automation. As a group of persons, a public personnel, government is also capable of reaction, of aspiration, of advocacy of policy or antagonism thereto, and throughout its function as a group there runs the realistic knowledge that decisions by this governmental complex dictate, delimit, and control determinations capable of assertion by all other groups.

Lines of communication are not fixed or tangible so far as

these groups are concerned; conflict may involve the owners, the workers, and government, leaving the specific designers, engineers, scientists and technicians un-touched politically, eager only to have the machinery of automation continue in operation without let or hindrance. One of these four prime groups is thus cast into a role that is unique. The operators and specialists on automation are divorced from the responsibility possessed by the owners who may be so widely dispersed under the modern concept of stock holding, as the symbolification of ownership, as themselves to appear to avoid responsibility for either the benefits or the ill effects of automation.

Labor seemingly is not responsible for any of the benefits of automation and may indeed have to struggle long and bitterly to secure any of the fruits of its installation. The disadvantages to labor may indeed seem the advantages for ownership and management at its topmost level. The calculation of goods and evils, of pleasures and pains in the Benthamite sense will produce diverse results for labor and owners although the formulas are identical. Government as a group hovers around the entire process, ready to embrace, to encourage, to assure, to arbitrate, adjudicate, and if need be to prohibit—ultimately to take over if necessity decrees. Possibly the owners in chief as well as their top managers may comprehend the range of implications of automation and have at least a glimmer of recognition of their responsibilities, but the significance of their literal escrowal of power in the hands of the automation experts can probably be said to elude these owners and managers in chief. In the last analysis they cannot escape a social responsibility for automation as a basis for economic and political power and prestige without the vacuum occasioned by their abstention or wilful withdrawal and flight being filled by something uncertain and unknown.

Two Latin phrases rise to mind as we set discussion for the concluding segments of this chapter. These are *Quo Vadis*—"Whither goest" and *Quid Custodiet Custodes*—"Who will watch the watchers"? Direction, bearing, latitude, longitude in the sense of social policy development are involved in the ultimate aspect of politics and automation.

With no attempt to prophesy or to prognosticate three lines of possible development would seem worthy of note as the future of the state and its agent, government, is contemplated under the pressure of automation. In an early work of his, written in other connections and with another purpose, Walter Lippmann supplies a title for the concluding question of this chapter—

DRIFT OR MASTERY?

Obviously no one need do much about the political problems stemming from the introduction and advance of automation; the crises which they generate may presumably be met and dealt with as crises, when and as they arise—even if they do. In other words nothing may be consciously done about any of the matters which have been discussed in the chapter. Drift, foundering, sidewise motion in an age of economic and social stagnation will but intensify and make emphatic the realization of the political aspects of automation. This trend is conceivable, but would not appear to be perdurant as drastic actions, mistaken perhaps in scope, would indubitably be taken in these circumstances. The essential thing to be observed is that drift as a result of failure or hesitance to view the problem in perspective is a possible line of future development *vis-a-vis* the problem of automation and its political significance.

Mastery as a result can be posited in terms of two types of triumph—that of the machines or that of man and his culture in deliberate resolution to hold his destiny in view firmly and consciously. Earlier there has been an allusion to the fact that the great search for engineers, scientists, technicians of all types contains an element of panic and a suggestion paradoxically of an anti-intellectualism which would powerfully affect the character and role of the state as an institution. With the emphasis on full employment, a high standard of living, and military supremacy which has marked the thought of the Western World, the role of the expert in the push-button or electronic age will be exalted.

In either the case of drift or succumbing to the Lorelei of technological sirens, bidding honor the defenders of society in the name of scientific interest or military necessity, the character of the state as an institution of social control will be vastly changed.

The great danger is that in either drift as a philosophical fact or surrender to the machine and to automation as a process, appeals will be made for governmental intervention or for legal action so frequently as to cause the complete erosion of any semblance of private or free enterprise. In the face of demands for this governmental intervention to smooth the path of automation, in the attendant presence of large civil installations in terms of computers and other electronic machines useful to governments, in connection with the intimate use of automation as the heart of modern military science and weapons, government as an institution may find it easier to direct and command in the first instance and in effect totally, rather than to mediate and thus preserve a sphere for others.

This will not be a dictatorship in the usual sense—not even in the terminology of totalitarianism. Automation knows neither a left or a right in terms of political ideology. But if neither the owners, or managers, or the scientific technicians can interpret the significance of automation so that laymen who live by it can judge thereof, then surely some governing faction, probably of pseudo-scientists, will attempt to do it and reach for the power that is the state. This will not be a case of bourgeoisie versus proletariat; it will not resemble any previous revolutionary shift in the exercise or possession of raw political power. In this, the ignorance of the technical on the part of the political, and of the political on the part of the technical, will be maximized to maintain a status quo which can only stifle the spirit of mankind.

Dark as is this prospect, and there is little indeed to suggest any alternate solution upon the horizon, the case is not clearly hopeless. Retreat may occur as the price of the maintenance of automation may prove too great and be gauged excessive. Incentives to improve the evidences of automation may be lost or extinguished as time passes—these are the intangibles, the unforeseen eventualities of the future which will be decisive and seminal.

CHARLES W. SHULL (b.1904), received his A.B. degree from Ohio Wesleyan University (1926), and his M.A. (1927) and Ph.D. (1929)· degrees from Ohio State University. After teaching at the University of Kentucky, he joined the Political Science Department of Wayne State University, where he has been Professor since 1950. He is the Book Review Editor of *Social Science* (published by Phi Gamma Mu, the national social science society). He is the author of *An Experiment with Unicameral Legislature,* and *Reapportionment of Legislature in Your Government,* and a contributer to *Introduction to Political Science,* edited by Joseph S. Roucek (1954).

CHAPTER 28

PERSONNEL ADJUSTMENT TO TECHNOLOGICAL CHANGE

KENNETH G. VAN AUKEN, JR., *Assistant Chief, Division of Productivity and Technological Developments, U. S. Department of Labor*

The development and use of machinery and new sources of power and their increasing efficiency and breadth of application, have had a most profound effect on humans and human organizations. Technology has created an enormous rise in material standards of living; over the relatively recent past it has literally transformed mankind's way of life. Some of the results, better food, shelter, and a long life with more leisure, have been greatly welcomed. Others like unemployment, dilution and individual skills, job insecurity, heightened tensions, and a lack of identification with the production process, have made for a growing feeling of uneasiness, and have necessitated man's adjustment.

In this discussion we will concern ourselves with the direct impacts which technological change has upon workers, the adjustments that have been attempted, and why some types of adjustment may be better than others. In doing this we will focus particular attention upon automation. Like mechanization before it, automation is, in part, a series of advanced technological changes which replace many kinds of human efforts. One of its distinguishing features may be that automation is being introduced more widely and at a faster pace than earlier forms of technology. Such a possibility emphasizes the need for much more understanding of the adjustment process, as it has, and can operate. Greater knowledge in this area can help in developing a tradition of meaningful personal adjustments to this new technology, while minimizing the harsh results and fear of the machine which past change has often brought.

EARLY FORMS OF ADJUSTMENT

Over the past nearly 200 years, the growth of technology has brought an enormous expansion in employment opportunities. Whole new industries have sprung from technological discoveries creating jobs and skills which had never existed before. Directly and indirectly during this period, technology has also caused job loss and periods of unemployment for many workers. Attempts by the individuals affected to adjust to the competition of the machine have taken may forms, some have been rational and moderate, while others have been passionate and violent. The latter often resulted in extreme reactions which slowed, to some extent, the pace of technological progress. For example, in England, during the mid 1700's, a sullen mob of worried spinners smashed into the mill of James Hargreaves, and ruthlessly destroyed the first workable multi-spindle frames. The threat to their job security they saw in this early machine drove the spinners to a violent form of adjustment. Similar extremes of adjustment, based on the fear of unemployment, are chronicled in the rise and growth of the factory system in most major nations.

In the United States, in the midst of almost continuous economic growth and chronic labor shortages, there has been sufficient displacement to keep alive a certain suspicion of and opposition to technological change. The largest single group of workers displaced by the machine undoubtedly were the unskilled common laborers. Between 1910 and 1950 about 5 million of their number were replaced by some form of technology in factory and farm.[1] What happened to the individuals over these 40 years is, of course, not known in any detail. A majority probably improved their lot through better jobs or self-employment, a few no doubt suffered intense hardships, while others retired from the work force.

Skilled journeymen were sometimes as much affected by new technology as were the unskilled. The Owens automatic glass bottle blowing machine, for example, largely destroyed the bottle

---

1. U. S. Department of Labor, Bureau of Labor Statistics, *Economic Forces in the U. S. A. in Facts and Figures* (Washington, 1954), 28.

blowers craft. In contrast to the violence of the spinners, the glass craftsmen sought to adjust by accepting wage cuts to bring their craft methods into competition with the machine, but to no avail. Improved machine could produce faster and cheaper than human labor. Between 1910 and 1924 more than 8,000 skilled bottle blowers had left their trade.[2] In addition, hand cigar makers, cutlers, woodworkers, and many other craftsmen who had invested much of their lives in learning an exacting and fulfilling trade bowed before more efficient machinery. A tiny fraction of these skilled workers were able to carry on in the dwindling specialty shops; most watched younger, less skilled machine tenders take their place. The accumulated frustrations and bitterness of displaced men welled up into a fixed anti-technological attitude, which in some trades has only recently begun to soften.

The fear of and resistance to technological change which grew in many worker groups was in part a function of the earlier social attitudes. In the late 1800's and the early part of this century, individuals were conditioned to live within a strictly defined code of "swim or sink." Little or no help was available to the displaced man. Those with outmoded skills often were forced into jobs, when they could be found, which were far below their former status, leaving them in a state of active dissatisfaction. Society then was not only largely inactive in helping the technologically displaced, but it felt, in accordance with the mores of the time, that any efforts to help would be quite wrong. Each man must fight his own battle! The loss suffered by those technology displaced was overshadowed by the help the machine gave to the growing number of semi-skilled workers, and to the rising standard of living.

Thoughts of technological unemployment, arising part from reality and part from accumulated fears, have nonetheless been slow in dying. In 1928, Secretary of Labor J. J. Davis wrote a press article analyzing unemployment of that day primarily in terms of machines replacing men. In the early 1930's, the U. S. Congress held hearings to determine whether or not labor saving machines could be causing the Great Depression. Recently, in connection with automation, there have been implied predictions of widespread labor displacement. For example, at the

---

2. Millis and Montgomery, *Organized Labor* (New York: McGraw Hill, 1945), 432.

automation hearings before the Congress in the fall of 1955, five trade union witnesses testified that "short-run" dislocation problems could have very serious results for workers.[3] But except for employment changes resulting from business recessions, there has, over the past 20 years, been little evidence of outright technological unemployment. Apparently, this is not a major problem of automation. Social attitudes toward unemployment in general have undergone considerable change, especially after the depression experience. The idea of 'every man for himself' in respect to joblessness has been tampered by a number of important national and state laws. Today our society through its government structure gives people some income while out of work, helps find them new jobs, and when necessary, generates conditions in the economy which create additional jobs. What well may become a problem of first importance under automation are the adjustments, not to job losses, but to the changeover of jobs and the very jobs themselves which the new technology requires.

IMPACT OF TECHNOLOGY ON JOBS

When new technology, in the form of automation or other forms, is introduced into an establishment or one of its sections, some kind of labor displacement will most likely result. Whether the displacement will be *external* in character, layoffs and unemployment, or of an *internal* character, in which people are absorbed in other spots within the establishment, depends on a whole complex of circumstances within and without the immediate area of the establishment itself. Among the circumstances are the level of business activity in the firm, general economic conditions, the firm's future prospects for expansion, the number of unfilled jobs at the time of the introducing the change, and the skill of management in arranging for the change. Over recent years, the growth of our economy has greatly reduced the likelihood of external technological displacement. In many cases, sheer economic necessity like labor shortages or

---

3. Hearings on Automation and Technological Change before the Subcommittee on Economic Stabilization, Joint Committee on the Economic Report, Congress of the United States, October 1955. (Reuther, 101; Beirne, 339; Kennedy, 461-462; Coughlin, 213; Carey, 220.)

plans for future expansion have dictated against layoffs. In addition, the attitude of our society toward unemployment has changed to a point where management, if possible, would prefer to absorb displaced workers rather than face the bad public relations which might redound to their firm. As a consequence, mangement has been often stimulated to do the kind of advance planning for technological change which has successfully minimized or even ruled out entirely the possibility of external displacement.

The Department of Labor's Bureau of Labor Statistics has published at this writing four case studies of the introduction of automatic technology. The studies analyze the adjustment process with special reference to the matter of job reassignment which arises in the case of internal displacement. They were made by examining company records and alternately interviewing officials of local management and of the local labor union, where such existed, and then reciting all of the facts from the earliest planning for the change through to its installation and full operation. The kinds of establishments covered in the published studies were:

a. An electronics plant using printed circuitry and automatic inserting machines in the manufacture of television receivers.
b. An insurance company introducing an electronic computer in its statistical sections.
c. A bakery using highly mechanized equipment.
d. A petroleum refinery introducing a new process making use of controlling instruments.

INITIAL PHASES OF ADJUSTMENT

Within these case studies one can recognize certain discrete steps in the adjustment process: planning, consultation, communication and finally reassignment. We will examine here the three initial phases and consider reassignment later on. The fact that there was practically no external displacement in these BLS cases resulted in part from the planning phase. Where management-union relations were reasonably good the planning was aided by joint consultation between management and worker representatives. The forms of communication, in which the workers were notified of the changes, varied considerably between the plants. Each of these phases deserves discussion in greater detail.

Planning in connection with technological change has traditionally been concerned with the financial and technical aspects of the change, and to some limited extent with the people involved in the change, but only insofar as the people bore directly upon the financial or technical aspects. That is to say, there would be some advance notions as to where workers might sit or stand depending on the new machinery, and how many workers might face job loss, but no great concern with the people themselves. More recently, there has been a growing realization that advance planning for new technology must be oriented to a greater degree toward the individuals who will be primarily affected. For example, in regard to the effect upon employment, the case study of the television plant showed that management paid considerable attention to the timing of the change. They introduced their new equipment at the beginning of their peak production season and thus avoided any layoffs. In the insurance company planning started two years in advance of the actual change for both technical and personnel matters. Unemployment posed no problem here because of a severe shortage of clerical help, however, since 133 persons were to have their jobs replaced by an electronic computer, serious consideration was given to their reassignment. Starting almost a year before the actual change, an official of the section in which the computer was to be installed met with the personnel director and his staff on a regular basis. Here they discussed and planned for the personnel transfers and other matters they knew would arise. The personnel planning in the insurance company study covered each phase of the adjustment process, and showed considerable superiority over less complete plans in that it resulted in a smooth and almost painless transition into automatic data processing.

Consultation with employees or their representatives has been a very useful and important part of advance planning. Among the BLS automation case studies, two, the bakery and the petroleum refinery, report instances of management and union representatives jointly working out personnel plans, including compensation changes, for a proposed technological change. However, advance consultation of the type mentioned here is not widespread and depends largely for its success upon a mature relationship and a mutual respect between the parties. It has, perhaps, been used most frequently in the ladies garment industry where management and union over the years have achieved a highly workable relationship. The consultation involved in this industry covers both worker and payment adjustments which arise from technological change.[4]

In the bakery and petroleum refinery, union-management relations were advanced to a point which permitted a more or less objective review of planned changes. In these cases, the changes might have resulted in unemployment (at the bakery) and in discontent among the workers affected (at the refinery). However, the agreements reached between the parties in advance of the automation installation greatly minimized both possibilities, and in both cases was instrumental in: (a) a more willing acceptance of the changes by the workers themselves, and (b) in easing the transition between the old and new with a minimum of displacement problems, either external or internal. This essentially democratic adjunct to the planning process, although infrequently used thus far, can well be an area of considerable future growth.

Communication of what has been planned to the workers who will be affected is a most necessary step in the methodology of adjusting to change. In actual practice, both the timing and the manner in which notification has been given, has varied, producing a variety of worker reactions. For many years, an issue between trade unions and management has been whether or not unions should, as a matter of course, be informed in advance by employers of coming technical changes affecting their members. A union viewpoint on this matter has been expressed by the research director of one of the large textile unions, as follows: "The first basic union demand has been to be notified and given adequate information about impending changes. The usual proposal is that the union receive the data concerning specific, proposed innovations. The contracts in the textile industry usually require information on the approximate date of the installation, its nature, the proposed duties and job assignment and expected earnings and the provision made for the affected employees."[5]

---

4. For examples see K. G. Van Auken, Jr., "Plant Level Adjustments to Technical Change," (*Monthly Labor Review*, April 1953), 387-391.

5. Solomon Barkin, "Trade Union Attitudes and Their Effect Upon Productivity," *Industrial Productivity* (Industrial Relations Research Association, 1952), 116.

In the BLS case studies, we see some of the possible variations in communicating or giving of notification. As cited earlier, in two of the cases, notification was automatically a part of the union-management joint planning for change. At the insurance company reported on, where the employees were not represented by a union, advance notice was given in two ways. First, a month prior to the physical installation of the computer, the vice president in charge of the division affected held a meeting with the personnel and told them in straight forward terms of the coming change, and of the fact that some would be displaced although no one would lose his job or suffer a loss in pay. Additionally, all employees of the company's home and field offices were informed by the company's employee publications. Here the computer was described, including the possible effects it would have on personnel. In neither method of notification was any attempt made to gloss over any of the innovation's implications.

In the TV plant, the vice president in charge of production informed the production foremen about 2 weeks in advance of the change. At about the same time, officers of the local union were similarly informed. The workers who were to be affected by this change, however, got their information via the "grapevine." The union here did not appear to have played an active role in this conversion to automation. It believed that automation would benefit workers as a group and put its emphasis on obtaining for the workers a share in production gains.

Another way of handling the matter of advance communications before technological changes occur was described by Frederick K. Leisch.[6] Apparently, no notification was given directly to employees in his company concerning the installation of a proposed computer, unless they asked questions. Three months before the machine was to be installed, executives, department heads, and supervisors were informed of the change to come, followed by a series of more detailed explanatory meetings with this same group. Mr. Leisch said of these meetings:

By this means, we simultaneously prepared our supervisors both to cooperate in whatever operational changes might affect their specific sections and, also, to tell their rank and file

---

6. Executive Vice President, A. C. Nielsen Company, at the Symposium on Electronics and Automatic Production, jointly sponsored by the National Industrial Conference Board, Inc., and Stanford Research Institute.

people anything the latter *might wish to know.* [Emphasis supplied]

Here were five ways in which communications to employees were handled. Three the direct approach, the two others less direct.

REASSIGNMENT, A LATER PHASE OF ADJUSTMENT

With the introduction of technological change like automation, the planning and communication and other first steps are then followed by a shifting of people. Some are moved to new jobs which have been established directly in connection with the innovation. Others, who are internally displaced, are transferred within the establishment to different but not necessarily new jobs. These moves require a selection process in order to pick those who will remain with the new technology, and in some cases special training or retraining to fit workers to both the newly created jobs or the jobs to which transferred. At this point, it must be said that not all of the new jobs arising from the innovation will be filled by persons already within an establishment. Some would obviously be recruited from outside, especially where skills or training hitherto not used are required.

In the case of the people who are retained to work on new jobs created by automation, or to remain in their old jobs which have been coordinated in the new process, the problems of change-over are relatively simple. In most cases the change is accompanied by increased wages, a higher status, and the challenge of working directly with a new process, one which is a focal point of attention within an establishment. As will be discussed later, occasionally people moving to automated jobs meet with disturbances which must be dealt with. For those displaced workers who are absorbed in other spots within the establishment, the transition can be more difficult. Among the BLS case studies, one of the most effective ways of carrying out the transition was found in the insurance company. The 133 persons whose jobs were taken over by the computer were given considerable latitude in the selection of their new jobs in other areas of the company. Each was interviewed concerning his job preference, and each was permitted one or more interviews with the supervisors of other departments. Based on a mutually satisfactory choice by both the employee and the supervisor, each person was placed in another job with at least equal rates of pay compared to their former job. At the bakery studied, the procedure involved in transferring those persons not retained in the automated jobs was somewhat different. The collective bargaining contract already provided that a worker would not suffer a pay loss because of a transfer to a lower skilled job. Because of heavy displacement by automatic baking equipment, it also became necessary to create a number of new jobs into which some of the displaced workers could be absorbed. Consequently, a number of cleaning jobs, euphemistically titled "sanitors," were established. Many displaced workers were transferred to the sanitor jobs and others of similar inferior status. Concerning this, the study states, "Although, in most cases, transfers represented a downgrading to a lower rated job, there apparently was little discontent over the shift since workers were guaranteed their rate of pay at their old job. The employee was consulted and if an alternative proposal was practicable, he was given an opportunity to consider it. Any new worker coming into the lower rated job received the established rate for that job."

In the report of the automation changeover in the petroleum refinery, the reassignment of workers affected by the introduction of advanced technology was governed largely by the collective bargaining contract and the interpretations of it reached jointly by worker and management representatives. The major change, in this case, directly affected 164 persons. Based on seniority within their individual skills, 62 per cent were transferred to jobs without change in pay while 38 percent could be absorbed only in lower paying jobs. According to the report, some choice was available to those displaced; it states: "Insofar as possible, the workers were given a chance to choose from the available jobs."

In all of the reassignments reported by the BLS, even those where there was loss of pay and status, the element of some participation in the transfer procedure by the workers themselves stands out. Although this democratic process was more limited in some cases than in others, it does illustrate a principle of good job adjustment practice.

ADJUSTMENT RESULTS: PEOPLE AND JOB STRUCTURE

In line with the foregoing discussion of adjustment, we naturally would like to know something about the selection process—why some employees are chosen and some are rejected for work with automatic technology. Following this there should be some attention given to training and retraining. Both of these topics, however, are bound up in a broader consideration of technology's effect upon (1) the job structure, and (2) the individuals involved. Individuals may be defined as the workers on the spot using an older system before automation was introduced. Job structure, on the other hand, refers to the hierarchical listing of jobs within a company or a section of a company. With regard to the people involved in the BLS case studies there occurred, as we have seen, considerable internal displacement, or parallel transfers in the same job level, including some downgrading in both job status and wage level. Upon the job structure, these changes resulted in an increase in the relative proportion of skilled or higher paid jobs.

The same results upon both the job structure and people affected occurs in an important automation case study made in Europe.[7]

Much has been written recently about the fact that automation will bring about a shift of workers from the semi-skilled job levels—the machine tenders—to job levels of higher skills. The higher skills meaning maintenance people for complex machinery, draftsmen, programmers, and the like. This is consistent with the Labor Department's findings to a degree. In their case studies, however, the shift was not immediate and generally did *not* involve an upgrading of the less skilled people on the spot.

In the accompanying table there are listed some of the data findings from the BLS case studies of automation with specific reference to both people and job structure, before and after. Also included are the data from the French study of the Renault automobile plant in Paris. In this table the reader will note that for each case the workers are divided between what we might call approximately higher skilled and lower skilled jobs. The distinction between higher and lower skills was made on the best evidence available which meant the use of some subjective judgment and also accepting rates of pay as indicators of skill levels. The differences in skills are, of course, relative to each case separately.

The table lists all of the case studies which quantify in sufficient detail this aspect of automation impact. It is possible that

---

7. A. Lucas, "Automation at Renault," *Case Studies of Automation*, No. 12 (European Productivity Agency, Paris, April 1957).

other studies with sufficient data have been missed, but most of those made in Europe and the U. S. have been reviewed and found quantitatively incomplete.

The Immediate Effects of Change on Individuals in these illustrated cases was to reduce, almost immediately, the number of people required, especially those in lower skilled or lower paid jobs. As stated before, this did not mean unemployment for them, except in one slight instance, but rather an absorption in other spots within the establishment. The ease of the transition was greatly aided by advance planning, and by the fact that the companies introducing automation were expanding and dynamic and actually needed most of those internally displaced to staff their expansion.

To summarize, some of those displaced were transferred to jobs of similar skill level as in the insurance company, other transfers meant a drop in job status as in the bakery. For the most part, there was no change in pay rate even where the job status dropped. This was due to company policy in some cases and union insistence in others. In the oil refinery, however, 62 of the people who were transferred ended in jobs with lower pay rates (although about half of these were covered by a 6 months maintenance of pay provision in the union contract). Of the lower skilled people who were not transferred, only 20 or about 12 percent were retained in jobs of higher skills and/or pay. Among the higher skilled/paid employees there were also a few transfers where the individuals were unable to meet requirements for the new jobs.

The net effect of the change on the job structure shown in each of the illustrated cases was to increase the proportion of higher paid and/or skilled jobs in relation to the total group. As can be seen from the table this resulted even where there was an absolute decrease in the higher skilled paid group. The structure changes here were influenced far more by the decrease in the number of lower skilled/paid jobs, than by any increase in the better jobs except in the case of the Renault automobile plant. Here, the form of presentation used in the table does not completely describe the changes. The automated engine block line required a significant increase in higher skilled personnel, but the semi-skilled jobs, although almost the same in number, were on average, of a lower order than those needed before. This was caused by a rise in the number of material handlers needed

Number of persons on spot of change before and after introduction of automation, selected case studies

Spot of change	Skill or wage level[1]	Number of Persons				Net change
		Before		After		
		Absolute	Ratio	Absolute	Ratio	
BLS Study, The Introduction of an Electronic Computer in a Large Insurance Company, 1955 STATISTICAL SECTION	Over $5,000/yr.	17	9	13	16	−4
	Under $5,000/yr.	181	91	67	84	−114
BLS Study, A Case Study of a Large Mechanized Bakery, 1956 BREAD BAKING LINE	Skilled operators	12	24	6	29	−6
	Semi-skilled operators	39	76	15	71	−24
BAKERY, BULK MATERIALS HANDLING	Skilled	0	—	2	65	+2
	Laborer	24	100	5	36	−19
BLS Study, A Case Study of a Modernized Petroleum Refinery, 1957 PRODUCTION GROUP	Highly skilled {Stillmen, Operators}	25} 67} 92	46	29} 71} 100	62	+8
	Semi-skilled {Helper, Coke, Cleanout}	32} 76} 108	54	55} 15} 68	38	−39
European Productivity Agency Study, Automation at Renault, April 1957 ENGINE BLOCK LINE	Highly skilled	4	4	26	21	+22
	Semi-skilled	107	96	100	79	+7

[1] The higher and lower skilled/paid jobs indicated relate to the individual establishments or group and are not necessarily equally high or low as related to the other cases shown.

to insure a smooth flow between the automated lines. There had been no equivalent jobs previously.

Selection of Personnel was a necessary step because the changes resulted in the creation of a number of new skilled jobs. The bulk of these jobs were filled by retaining the skilled personnel who had been attached to the former process or method. The remainder were filled by new hires. The selection of persons to fill the jobs of considerable skill increase which automation had necessitated was primarily based on prior training and education. The requirements for filling jobs like engineers, electro-machinists, and programmers precluded the lower skilled people entirely.

The small proportion of the lower skilled people on the spot who were able to better their lot were selected on the basis of seniority and in some cases, higher aptitudes. In these particular cases, no elaborate selection system was needed, since the higher job skills to which the few persons moved were higher only in a very relative and marginal sense.

Job Content Changes were dealt with in these cases to some degree. The upgrading of the marginally higher jobs was based as much on increased responsibility as on increased skill. To illustrate from one of the case studies not included in this table: A group of women workers in a TV plant had been assembling electronic parts in to TV chassis using wire, soldering irons and other small hand tools. They had developed considerable manual dexterity at this work. With the introduction of machines which automatically inserted the electronic components into boards on which the wiring had been preprinted, the women then had only to monitor the machine line, to attend to possible machine jamming, and to perform a simple dip solder operation as the automatically assembled boards reached the end of a conveyor belt. Nevertheless, the new jobs brought pay increases approaching 15 percent. It is likely that these pay increases were primarily compensation for the constant attention required by, and the responsibility for the damage which could result from inattention to, the expensive machinery.

The superior skilled jobs like engineers, electronic technicians and maintenance specialists which were brought into the job structures were a requisite part of using automatd techniques. The question of changing any job content was not at issue in connection with most of the higher skilled jobs. It was simply necessary to establish such jobs in order to operate the automated process.

The Retraining required by automation applied more to the semi-skilled people than to the higher skilled. For the small number of affected persons who were moderately upgraded, retraining amounted to a short on-the-job break-in. For example, in the automatic bread line of the bakery studied, a few weeks was all that was necessary to train the top operators. The study states on this point: "The period of training was principally necessary for adjustment to the pace of the new system rather than for adaptation to new skills." In the case of five persons who were upgraded for work on the insurance company's electronic computer, the on-the-job training lasted somewhat longer. At the oil refinery, operators and helpers received primarily on-the-job training for the new equipment they were to use. On the other hand, the stillmen, who already qualified as highly skilled, attended an intensive 6 months instruction course.

For most of the highly skilled jobs in the companies studied, however, retraining was not applicable. These jobs required a number of years of formal professional or technical education. The individuals selected to fill them had already all or most of this education.

It is important to realize that the shift to higher skills which automation will bring will be gradual, not affecting all people in the same manner. The need for higher skills will probably be at least as great among the producers of automation equipment as within the factory of the user of this equipment. However, a greater growth in automation's development is required before evidence of a significantly higher demand for skills will become apparent.

The majority of people affected by automation in the BLS case studies were not eligible for, nor did they move to higher skilled jobs. Certainly retraining, as reported in the studies, should be fully exploited to help as great a number of individuals as possible make the shift. In many cases, however, the new skills will require much more than retraining. An increase in skills will depend primarily upon improvements in vocational guidance, formal training and education, and other matters bearing upon how people choose and prepare for their careers.

INDICATIONS FOR FURTHER STUDY

Mass production techniques of the past 40 years have widely introduced what Adam Abruzzi refers to as the "trivialization of work." Machine paced operations, finely subdivided tasks, careful analysis of the time and exact motions necessary to do each small operation, have placed many workers in the role of a quite simple mechanism. This highly rationalized and efficient industrial procedure has achieved constantly increasing output per man-hour, but in many cases has removed from working men and women the important non-material stimuli of responsibility, pride of workmanship, status, and sense of social usefulness.[8] In the design of equipment and production systems which do not take into account the whole man, including both his psychological as well as his economic role, there will result dissastisfactions which, in the end, can mitigate against the very efficiency of operation which the designs seek to achieve. Cumulatively, such dissatisfactions among workers can equal the active distastes for new technology that arose from the threat of displacement. Thus, adjustment to an advanced technology like automation will require not just the mechanics of reassigning the affected personnel, important as this phase is, but also the adjustment of the equipment and the systems of production to more fully accommodate the total role of working man. Returning to the BLS case studies on this point, in the insurance company two of the former supervisors indicated a preference not to remain in the field of automatic data processing. Although the reasons for this preference were not made entirely clear, it can be assumed from their selection of nonautomated jobs that working with an electronic computer would not bring them satisfaction. Quoting from the bakery study: "A dividing machine operator professed to a 'feeling of confusion' as he worked amid the more automatized equipment about him. He was transferred to another department where, employed as a general handyman, he continued to receive [his former rate of pay]." In *A Case History of a Steel Mill* by Charles R. Walker, the changeover to a highly automatic method of producing steel pipe is described. Concerning their reactions to the new system, two of the workers are quoted in the following way:

---

8. J. A. C. Brown, *The Social Psychology of Industry*, (Harmondsworth, Middlesex: Pelican Books Ltd., 1956), 38.

'On the old mill I went home after work, rested, and my muscles were no longer tired. I also had nothing to worry about. On this new mill your muscles don't get tired but you keep on thinking even when you go home.'

Another man said:

'I'd rather have to work hard for eight hours, than have to be tense for eight hours, doing nothing with my muscles the way I do now.'

In other words, for muscular fatigue technology has substituted tension and mental effort.

In theory, a pure form of automation will develop systems of production which will remove the tension producing repetitive activities from the work place. In its present state of development, automation may well be in a transition period. As shown from case studies, under today's automatic technology jobs of a "trivial" nature remain, and workers are asked to perform tasks which for many do result in greater tensions. Research into these vestigial anomalies, leading to a better resolution of the problems they create, is needed. Additional study must be given to the effects which technological innovations have upon informal social systems which occur within a plant or establishment. Many authorities have put forth the hypothesis that the destruction, by new processes, of a social order among workers can only lead to great human difficulty in the utilization of the new techniques. This requires empirical testing. The problems of older workers adjusting to new technology also need greater emphasis as our work force continues to age.[9]

Although some work has been started by capable researchers in each of these areas of the adjustment problem, much remains to be done. We cannot fully understand or clarify the goals of our technological society until we have explored the major facets of man's adjustment to it. The focus for future research must increasingly bear on what can be done to adjust technology to mankind working with it so that the needs of man, in the sense of the social group, and in the sense of the individual's psychology, can be more nearly met.

---

9. A start is being made in this direction in a Department of Labor study of adjustment, by age group, of white collar workers to electronic computers.

Selected Bibliography

Adam Abruzzi, *Work, Workers, and Work Measurement* (New York: Columbia University Press, 1956) . A critique of current work rationalization methods.

S. Barkin, "Trade Union Attitudes and Their Effect Opon Productivity," *Industrial Productivity* (Industrial Relations Research Association, 1952) , pp. 110-129. A union official analyzes union-management relationships and their attempts to achieve greater productivity.

J. A. C. Brown, *The Social Psychology of Industry* (Harmondsworth, Middlesex: Pelican Books Ltd., 1956) . A discussion of social and emotional problems of modern industry.

H. A. Millis and R. E. Montgomery, *Organized Labor* (New York: McGraw Hill, 1945) . A detailed analysis of trade union growth.

U.S. Congress, *Automation and Technological Change*. Hearings before the Subcommittee on Economic Stabilization of the Joint Committee on the Economic Report, October 1955.

U.S. Department of Labor, Bureau of Labor Statistics, "Studies of Automatic Technology:" No. 1. *A Case Study of a Company Manufacturing Electronic Equipment,* October 1955; No. 2. *The Introduction of an Electronic Computer in a Large Insurance Company,* October 1955; No. 3. *A Case Study of a Large Mechanized Bakery,* September 1956; No. 4. *A Case Study of a Modernized Petroleum Refinery,* September 1957.

C. R. Walker, "Case History of a Steel Mill," in *Man and Automation* (The Technology Project, Yale University, 1956) . The introduction of automation in steel pipe making. Effects on individuals.

W. F. Whyte, *Money and Motivation* (New York: Harper, 1955) . An analysis of worker reactions to incentive plans.

KENNETH G. VAN AUKEN, JR., Assistant Chief, Division of Productivity and Technological Developments, U.S. Department of Labor, received his B.S. degree from Tufts College (1941) and a certificate (1944) from the Harvard Business School. He is the author of numerous articles, published in the *Monthly Labor Review* and other periodicals.

CHAPTER 29

AUTOMATION AND PUBLIC ADMINISTRATION

HERMAN LIMBERG, *Senior Management Consultant,*
Division of Administration,
Office of the Mayor,
City of New York

In a paper I prepared two years ago[1] I reported that no one had as yet come forward with a plan, prediction, or slogan for "push-button government." A short time later, however, I read a press release, issued by one of the major equipment manufacturers, in which several references were made to "automated government" and "push-button administration." To uphold the accepted concepts of government and dispel the vision of mechanical monsters planning and directing the activities of American public agencies, I revised my earlier findings. My revised report indicated that no one in public administration had conceived or was disseminating such insidious propaganda.

Administration, whether business or public, is basically and primarily dependent upon the human mind and the interrelationships of human beings. Luther Gulick has defined the science of administration as "the system of knowledge whereby men may understand relationships, predict results, and influence outcomes in any situation where men are organized at work together for

---

1. Limberg, Herman, "Automation—Implications for Public Administration in America," presented at the tenth annual meeting of the National Conference of Professors of Educational Administration, University of Arkansas, August, 1956. Published in the NCPEA Report, "Automation—Its Meaning for Educational Administration," 1957.
Most of the material in this chapter is based on the paper here cited.

a common purpose."[2] The key word in this definition is "men." This does not negate the utility of machines for achieving administrative objectives, but places machines in their perspective as tools, rather than masters, of human beings.

Automation is a symbol of the present era of human progress. It is the "new technology" which has often been proclaimed as the herald of the "second industrial revolution." Its achievements and potential are tributes to the human mind.

During the past few years, the new technology has been widely applied to governmental operations. Its introduction and use have not, however, been subjected to the clamor and fanfare which have marked the advent of automation in private industry. While several levels and branches of government have expressed their interest in the industrial and commercial implications of automation, the implications of automation for public administration have not as yet received intense consideration.

Governmental study and appraisal of the ramifications of automation have been channeled into two major areas of concern to public officials. Such studies have been necessitated by pressures emanating from outside and from within governmental enterprise. External pressures are directed toward the prevention and solution of economic and social problems created by industrial and commercial application of automation. Internal pressures stem from problems arising from the application of automation to the administration and operations of governmental agencies. In presenting the total pattern of the public administration approach to automation, this chapter is divided into the two primary classifications of the subject matter: (1) External Implications and (2) Internal Implications. The various aspects of the subject are considered in their relationship to each of these classifications.

To provide a frame of reference within which the subject of this chapter may be viewed in its proper perspective, there is appended to this chapter a digest of the definitions and concepts of automation which are currently in circulation. Basically, there are three types of automation: (1) "Detroit automation" or "continuous automatic production," in which automatic machines are combined with transfer or handling equipment, (2) "feed-back

---

2. Gulick, Luther, "Science, Values and Public Administration," in Luther Gulick and L. Urwick (Eds.), *Papers on the Science of Administration* (Institute of Public Administration, New York, 1937), p. 191.

automation," which incorporates a self-correcting control system, and (3) integrated data processing, in which extensive mathematical computations are directed by programs on punched cards or tapes. Any one type, or combinations of two or all three types, may be used in actual applications.

Although standardization of the definitions and terminology of automation would do much to facilitate more general understanding, the basic concepts serve to point up the implications for public administration. Perhaps the thinking about automation has not yet been "stabilized," but the experience already gained from its use imparts special meaning to Thorstein Veblen's observation that "invention is the mother of necessity." The invention of automation has mothered the necessity to explore its immediate and long-range effects and to plan and prepare to meet and deal with them effectively.

EXTERNAL IMPLICATIONS

Governmental interest in the social and economic problems raised by automation and in the means to anticipate and solve them has already been shown in a number of recent instances.

During the fall of 1955, a Congressional inquiry into the impact of automation was made by a subcommittee of the Joint Economic Committee. At a press conference in March 1955, President Eisenhower stated that he expected the proper agencies of government to continue earnestly their investigations of automation and its developments, and that if it appeared necessary he might appoint a special commission to study the subject. The Bureau of Labor Statistics of the United States Department of Labor has been conducting a series of case studies of industrial and commercial applications of automation covering various labor aspects. This department's Bureau of Apprenticeship is seeking to expand the number of joint labor-management programs to assure an adequate supply of skilled workers to meet the needs of automation. In May 1954, the New York State Department of Commerce sponsored a conference on Automation and Industrial Development to consider "automation in industry with primary reference to its implications for industrial management and for the economy of the State of New York."

These governmental inquiries and surveys, as well as the studies made by labor and industrial organizations and the conferences and reports of trade and management associations, have served to identify and project many of the problem areas of automation. Periodicals, such as Business Week and Fortune, have also contributed to the published materials currently in circulation.

Most important and complex are the problems which pertain to the worker. Among these are the possibility and extent of displacement and unemployment, the need for training for upgraded skills and newly created jobs, shift of employment opportunities, reduction in the work week, and the effects on the health and welfare of the worker.

There will be a displacement of workers in both factories and offices, but the actual size of displacement on the whole appears to be relatively low. Many displaced workers will be assigned to other jobs requiring higher skills, some may be transferred at the same wages to assembling, shipping or maintenance work, others may be downgraded. Many displaced workers will be absorbed by normal turnover during the transition to automation. Displacement will also have a serious effect on future job seekers, since a number of job opportunities now available will disappear with the spread of the new technology.

To a large extent the effects of displacement will be offset by the creation of new jobs, existing personnel shortages in certain technical fields, and increased demands for maintenance workers, engineers, electronics experts, electricians, mechanics, pipe fitters, tool makers, analysts, programmers, supervisors, and management personnel. More salesmen will be needed to dispose of the increased output. The atomic energy, guided missiles, chemicals, plastics, and electronics industries have been searching for more workers, despite gains made in automation. Nevertheless, displacement, even in generally favorable economic circumstances, may result in readjustments and hardships for large numbers of workers.

Warner Bloomberg, Jr. has suggested "displacement insurance" to provide "some additional incentive to manufacturers and business executives to introduce the new technology at such a rate as to minimize employee displacement." Such a program, says Bloomberg, "would help cover the costs of retraining; the expense of changing to another company, if this were necessary, plus the minimum expenses of living until the new work commenced."[3]

---

3. Bloomberg, Jr., Warner, *The Age of Automation*, League for Industrial Democracy (New York, 1955).

Training programs will be required to prepare workers to fulfill new job responsibilities. With so many changes taking place, narrow, specialized training may prove to be a handicap for future job requirements. As Professor Peter F. Drucker has stated, "under automation a school could do a student no greater disservice than to prepare him for his first job."[4] Hence, existing patterns of education and vocational preparation will probably require revision. The shortage of engineers and scientists, which is becoming more acute as automation advances, presents not only a problem of training at the college level, but raises questions as to the adequacy of high school training where fundamental skills in science and mathematics are begun. The recruiting and training of teachers in these subjects also requires attention.

Organized labor is seeking a shorter work week, longer vacations, earlier retirement, and longer schooling as ways of sharing in the rising productivity and maximizing employment opportunities.

Consideration must be given workers who face special difficulties in obtaining jobs. It may become necessary to find new job opportunities for the physically handicapped. Workers who may be too old to adjust to new jobs may be forced into retirement before they reach age 65.

Bloomberg foresees a possible reversal of the tendency towards earlier retirement. He believes that "the next generation of older workers, who are likely to be better adapted to the new modes of production and more flexible in their skills, will not face sudden obsolescence; and the qualities most needed among workers in the automated plant—responsibility, experience, know-how—diminish least, if at all, as the employee grows older. Such workers could stay at their jobs far past the present age of retirement without loss of status or efficiency. They may prefer not to retire. Instead of giving the younger workers a chance by getting off the top of the ladder at an earlier age, the older workers—a substantial power in the unions and in political organizations—might well demand more educational opportunities for their children—opportunities which would delay their sons' entrance into the labor market."[5]

---

4. Peter F. Drucker, "The Promise of Automation," *Harper's Magazine*, CCX, (April, 1955), 41-47.

5. Bloomberg, *op. cit.*

An optimistic view of automation is contained in an editorial of the New York Times of December 7, 1956. Entitled "Labor Boosts Automation," the paper reported that Local 1 of the Amalgamated Lithographers of America, AFL-CIO and leading lithographing firms jointly sponsored a show at which new techniques and machines for increasing productivity were demonstrated. "This dramatic episode," said the editorial, "culminates years of ALA experience which has convinced it that 'laborsaving' devices, in the long-run, create more labor than they 'save'—and can bring higher wages and shorter hours of work." Automation may open up new vistas of industrial medicine. While a reduction in accidents and chronic disease is foreseen, an increase is anticipated in new and intangible problems that may affect the health of people on and off the job. A shorter work day or work week will necessitate greater emphasis on health maintenance programs, including home safety, health education and counseling. Automated industry, with its trend toward more and more skilled work, should be concerned with the effects of mental and emotional stress on its workers.

From the foregoing observations of the potential and reported effects of automation on the worker several implications for public administration may be deduced. The need for review of social security and unemployment insurance systems to ascertain their adequacy becomes obvious. Free governmental placement services may require expansion. The sufficiency of government subsidized and operated training and educational facilities should be studied. The Social Security Act may need amendment to provide earlier social security payments to workers forced to retire before age 65. Governmental mediation and conciliation services may face reorientation.

Increase in leisure time will create demands for expansion of recreational facilities such as parks, museums, and public libraries. Public health, welfare and educational programs may need enlargement and supplementation to meet pressures induced by the consequences of automation.

Although the implications which emerge from an analysis of the effects of automation on the workers are extensive and serious, they do not call for charting of unexplored areas of public administration. Both the philosophies and machinery of government involved in the effective handling of the problems are well established and in full operation. Federal, state and local governments have long been committed to finding and providing solutions. Under the Employment Act of 1946, both the legislative and executive branches of the federal government are charged with the responsibility to adopt and maintain a high level of employment. It is also important to note that the subcommittee of the Joint Congressional Committee on the Economic Report stated in October 1955 that its "best and by far the most important single recommendation . . . is that the private and public sectors of the nation do everything possible to assure the maintenance of a good, healthy, dynamic and prospering economy, so that those who lose out in one place as a consequence of progressive technology will have no difficulty in finding a demand for their services elsewhere in the economy."[6]

Other problems which will necessitate continuing governmental attention will include provision of adequate public services to accommodate the conversion, movement and relocation of factories and offices, protection of the consumer interest, and assistance to small business.

The conversion, movement, and relocation of automated factories and offices will affect the planning and regulatory functions of municipal and local governments. Layout changes required by automation may necessitate building structures of a pattern different from those now in existence. Zoning, building codes, and licensing requirements may need revision. Typical of governmental approach to the problems raised are the studies of efficient land use, zoning requirements and desirable transit patterns and facilities conducted by the Department of City Planning of The City of New York. In its April 1956 Bulletin, this agency published the results of recent surveys and projections "in keeping with the new technological developments and economic currents." The implications of automation for municipal government are embodied in the Bulletin's statement that "a knowledge of changing employment volume, composition, and location, and a realistic appraisal of the level of employment in 1960 and 1970, can be exceedingly helpful in suggesting solutions to vital city problems concerned with future commercial and industrial needs."

Protection of the consumer interest involves continuing governmental studies of purchasing power and price trends, expan-

---

6. From hearings of Subcommittee on Economic Stabilization, a subcommittee of the Joint Committee on the Economic Report, Congress of the United States.

sion or contraction of markets, shifts in the scale of consumer preferences and changes in the consumer-price index. Many adjustments will undoubtedly result from the introduction of new products and services not available or possible before automation.

For small business there will be gains as well as problems. The comparative adaptability and flexibility of small business, and the demand for components and parts for the new machines, will benefit many small firms. Drucker believes that automation should strengthen the competitive position of the small company, "if only because mechanical machine-setting enables the small organization to offer a more complete and diversified range of products at a competitive cost." He believes, too, that automation "will create opportunities for countless businesses to specialize in servicing equipment."[7] On the other hand, many small firms will find it difficult, if not impossible, to meet the challenges of automated competitors. Governmental programs for assisting small business may require augmentation to finance and counsel new and expanding enterprises, and existing companies facing bankruptcy or failure.

Continuing research on the progress of automation and its implications, and provision of additional services and facilities may necessitate increased governmental expenditures. This may well be the most serious and vexing of all the external implications for public administration. The solution may lie in the internal implications of automation.

INTERNAL IMPLICATIONS

Interest in automation is but another manifestation of the drive for better management in government which has received tremendous impetus in recent years. The Hoover Commission studies of the federal government, the comprehensive review of the City of New York by the Mayor's Committee on Management Survey, and similar studies of state and local governments throughout the country, give ample testimony of the intensive search for more efficient and economical public administration. Motivation for this search is found not only in the growing complexity of government, increasing demands for new and expanded services, and the concomitant rise in operational costs, but also

---

7. Drucker, *op. cit.*

in the prevailing philosophy that "good government is good politics." Automation appears to offer many of the solutions sought.

In submitting the executive budget for 1958-1959, Mayor Robert F. Wagner of The City of New York stated in his message to the Board of Estimate and the City Council: "In the rapidly changing field of automation New York City is constantly on the alert. All of the latest developments in the electronic field are under continuous study and evaluation so that we may avail ourselves of the most modern facilities. The magnitude of the activity of The City of New York in the use of modern electronic machines may be measured by the fact that the annual rental paid for equipment now in use totals more than $1,500,000. The new installations on order or in process will increase the figure beyond the $2,000,000 mark and will represent the use of machines valued at $10,000,000. What we have done in the field of automation to keep our city in the forefront of modern communities is unmatched elsewhere in municipal government and compares most favorably with what has been done in private industry."

The key to successful application of automation is adequate planning which entails thorough study and analysis of all of its aspects. Realistic, rather than dramatic thinking is called for. Published findings of studies and experiences of both private companies and governmental agencies reveal a multiplicity of areas which should be probed. For the purposes of this chapter, six of the principal areas will be considered: (1) applications, (2) systems and procedures, (3) organizations, (4) personnel, (5) costs vs. savings, and (6) approach.

APPLICATIONS

Continuing study of potential and actual applications of automation is an essential phase of planning to achieve the benefits which can be derived from the new technology.

Although the "Detroit" and "feed-back" types of automation may ultimately be used in the operation of such governmental facilities as electric power plants, transit systems, and repair stations, integrated data processing offers the most extensive potential because of the nature of most governmental activities. An example of a "feed-back" application is the installation, completed in April 1958, which operates the lighthouse on Orient, Long Island.

The electronic device automatically turns on the light one hour before sunset and turns it off one hour after sunrise. On April 17, 1958 the New York Times reported that "automation has come to Orient Point Light."

From experience and surveys thus far recorded, applications of electronic computers to data processing may be classified generally into accounting, procurement, inventory control, production scheduling and control, document handling, scientific and engineering analyses, and management planning and reporting.

Accounting applications include payroll, billing and accounts receivable, disbursements and accounts payable. Recent studies by General Electric and other companies point to the feasibility of preparing budgets on the computers. A number of government agencies are now considering these applications as well as the possibilities in tax assessing and collection, and appropriation accounting.

New York City is contemplating the installation of electronic machines to replace punched card equipment for processing of payrolls, withholding tax, pension and social security deductions. The City's Bureau of the Budget is installing electronic equipment to expedite the preparation of budgetary changes and to provide more effective budgetary controls. In Los Angeles, the City Housing Authority prepares monthly rent bills on electronic installation for the preparation of property tax rolls.

Procurement, inventory, production scheduling and control applications are in the planning stage or in operation in a number of governmental agencies. The Army Ordnance Tank-Automotive Command recently installed a computer to control stocks of tank and auto parts all over the world. The Air Materiel Command uses a computer in the procurement, supply, storage and issue of materials. The Planning Section of the Air Force prepares assignment schedules of all aircraft and determines supply requirements with a similar system. The New York City Transit Authority uses a number of electronic machines to control its inventory now valued at $15,000,000.

Electronic document handling, involving the processing, sorting and delivering of records and authorizations is being studied by several groups with a view toward reducing substantially the costs of such clerical operations. Initial attention has been directed toward the processing of checks. In California, the State Treasurer's office uses electronic accounting machines to process all checks returned to the State. Applications in this category offer a vast potential for government operations.

Typical of scientific and engineering applications are the computations for map making by the Army Map Service, and the analyses made by the Atomic Energy Commission in the design and construction of power and research reactors. The Bureau of Public Roads of the U. S. Department of Commerce is planning to use computers for computations for earthwork and bridge and hydraulic design. The State of California is using a data processing machine for engineering calculations for highway construction. The Weather Bureau uses computers in weather forecasting.

Management planning and reporting applications involve the processing of various types of interrelated data to aid management in developing forecasts and making basic decisions. Such processing is generally characterized as operations analysis. In this category may be placed the Federal Bureau of the Census activities which use electronic computers to project population data, economic and industrial statistics in a variety of classifications and indices. Such applications will make possible the compilation and issuance by governmental agencies of more adequate and more timely reports of data of major significance and value in constructing business barometers.

Integrated data processing techniques offer many possibilities for facilitating and improving governmental planning. The various factors affecting budgeting, such as income to be realized from various sources, objectives to be achieved, and expenditure distributions might be integrated to simplify and expedite budget preparation and control. Integrated processing of population statistics, employment trends, industry migration patterns, available land areas, zoning restrictions, transit schemes, tax rates, and other pertinent data would not only cut years from planning time, but would also greatly enhance the end results. Similarly, planning for schools, health and welfare centers, hospitals, sanitation districts, and police and fire stations, would be simplified and made more effective and economical by coordination through the medium of electronic computers.

Another fertile area for automation lies in the field of intergovernmental relationships. Coordination of federal, state and local governmental activities in the administration of such functions as health, welfare and education could be made more effective through integrated processing of data of common interest and significance. Automation can also make a major contribution to the solution of regional problems of city and local governments in metropolitan areas through integration and coordination of interrelated data. An equitable sharing of the costs entailed would make possible the utilization of automation techniques by the smaller governmental entities which could not otherwise afford them.

While the foregoing are only a few of the examples of actual and potential applications of integrated data processing in public administration, they should serve to point up the essentiality of thorough and intensive study before such applications are effectuated.

SYSTEMS AND PROCEDURES

Complete analysis of systems and procedures must precede determinations of the advisability of a changeover to integrated data processing and development of the programs to be fed into the computer. Every procedure will have to be examined closely for its essentiality and logical sequence, because the elimination of only one step may save hours of complex programming and expensive machine time. It is also conceivable that the addition of one line of data in an integrated system may obviate the need for setting up or maintaining a separate bank of information in one or more departments.

Although systems and procedures analyses will entail the expenditure of thousands of manhours, there are many benefits to be realized from simplification and improvement even if it is found that electronic processing is not feasible or advisable.

ORGANIZATION

Existing organizational patterns will require extensive study to ascertain the need for revisions as data processing is integrated. Since the new technique will, in many instances, cut across and obscure existing departmental lines, it may become necessary to revise the division of functional assignments, and to eliminate some departments and create others.

The high speed of automated operations will necessitate the creation of an organization capable of moving fast, primarily because of the substantial costs of lost production time. Centralization of data processing facilities, which in most cases will be a basis for their justification, will not only make possible, but will require decentralization to the lowest possible levels of decision-making authority. Top management planning will, therefore, have to anticipate deviations from basic policies and provide reasonable guides to direct action when deviations are necessary. The inflexibility of data processing will also call for more and more decentralization of decision making.

An insurance company recently reported that automation of its operations resulted in "decentralized-centralization." Centralization was characterized by home office control of practically all operational functions. Decentralization was effected through the rapid development of branch offices and the allocation to them of greater administrative responsibilities. As a result, field offices were able to offer faster and better service to policyowners.

Decentralized-centralization in government may be illustrated by applying the concept to a public school system. Public school principals would be authorized to make decisions and take required action affecting their respective schools in conformance with centrally established policies and standards. Expeditious decision-making and action by the principals would be made possible and would be based on centrally produced data. Such data would include school population, health conditions, attendance levels, utilization ratios, and other pertinent factors for each unit in the school system. Analysis of comparable data of all the schools would be an important guide to decisions at each school. An electronic computer at school headquarters would be the core of such an organizational pattern for which rapid production of up-to-the-minute information would be the prime essential.

As yet no general pattern has been evolved as to where the computer installation should be placed in the organization, although various arrangements have been tried. It seems obvious, however, that automation will lead to many changes in organizational concepts of public administration.

PERSONNEL

The most challenging aspects of automation will be the human rather than the mechanical. The effects on government personnel will be as extensive and ramified as those previously considered in connection with the external implications.

Problems of displacement, upgrading, transfer and training must be met and solved. The content of jobs affected by automation, and job evaluation and classification plans will require close scrutiny. Recruitment and training for new jobs created by integrated processing will present problems.

An important guide in approaching the personnel situation is contained in the report of the Subcommittee on Economic Stabilization of the Joint Congressional Committee on the Economic Report. This report, which may be considered a mandate to the federal government, states that "when in the interests of economy and efficiency, the federal government finds it necessary to displace faithful employees, the Subcommittee feels it must be a model employer in handling personnel problems such as retraining, reassignments, and severance allowances." The principle here enunciated may serve as a guide for state and local governments, too.

Early in 1956, the United States Civil Service Commission undertook a survey of the effects of automation on federal employees and federal personnel practices. Referring to this survey, John F. Powers, president of the New York State Civil Service Employees Association, in an article in the Civil Service Leader of May 8, 1956 proposed that "it might also be well if the (New York) State Commission, like its federal counterpart, could conduct some preliminary surveys as to the probable effects (of automation) upon civil service personnel."

In a special release published in the New York World-Telegram and Sun on May 23, 1956 James A. Campbell, president of the AFL-CIO American Federation of Government Employees, called upon federal agencies to adopt a "more human approach to the problem of automation." He proposed a policy statement pledging maximum job protection, a general policy of training present employees to operate automation devices, and advance planning for retraining or reassigning affected employees.

In the central office of the Veterans Administration, classification and qualification standard technicians have been working with departmental and Civil Service officials to "visualize the type of positions which would be involved, the definition of an appropriate series and methods whereby individuals to be engaged in electronic data processing and allied fields may be idenfied, trained, developed, appropriately assigned and adequately paid."

The growing awareness of the personnel problems which automation will pose emphasizes the seriousness and extent of this major area of the implications for public administration.

COSTS VS. SAVINGS

Careful and detailed cost studies and projections of possible savings should be made before decisions are reached to change over to integrated data processing.

In estimating costs, consideration should be given not only to the expenditures for equipment, but also to the costs involved in the analysis, planning and development which must precede the installation of the new systems. Recruitment and training of personnel, and, where necessary, new locations to accommodate the new equipment should be included in the estimates. Operating costs, including more exacting maintenance requirements, of the new systems should be scrutinized. Costs will vary with the extent of automation, the time required for planning and installation, the types of machines to be used, and decisions to rent or buy the machines.

Indications of the cost potential for government may be gleaned from recent studies in the oil and insurance industries. A recent survey of twenty-five oil companies which have installed electronic computers revealed that preparatory costs ran between $150,000 and $1,500,000, in addition to equipment costs which ranged from $32,500 to $4,500,000. At the Life Office Management Association's Annual Conference in 1955, it was reported that costs, even for a modest installation may range from $1,000,-000 upwards, and that the eventual cost of a large computer installation will be calculated by the volume of work it processes.

Projections of savings should encompass intangible as well as tangible results. Although reductions in labor costs will be substantial, experience has shown that such savings will be neither the only kind nor the most important. In many instances it has been found that labor savings were insignificant compared to savings resulting from improved quality of product, greater productivity, and faster and more effective planning. Major economic benefits are also to be derived from the ability to perform more and better services at practically no increase in cost.

For many of the questions which appraisals of costs and savings will raise, there may not be any answers for several years, since many of these answers can be discovered only when the new system is in full operation.

In the study of costs and savings very serious attention should be given to the impact of automation on government budgets.

Problems will be encountered in providing funds for planning and installing new systems. Justification for the necessary expenditures may also present problems, because resulting savings will probably not be reflected in subsequent budget reductions. Mitigation of the pressures which may be expected in opposition to the expenditures will depend on the extent to which present budgets will be stretched to include new and expanded services and improved performance.

Provision of additional and better services without spending more than present budgets allow may prove to be the measure of success in meeting the external and internal implications of automation for public administration.

APPROACH

Since automation of governmental operations will, in the main, mean integrated data processing through electronic computers, integration and coordination should be the guiding principle of the approach to the new technology.

The first step should be directed toward integration and coordination of thinking and planning. A top level committee, representative of the entire governmental structure, should be designated by the chief executive to study over-all aspects of automation, and develop the basic philosophy, principles, objectives and yardsticks upon which change-over decisions should be made. Department committees should be set up to study the implications of automation within individual departments and to report, through their respective department heads, to the central committee. These committees should be composed of specialists trained in the fundamentals and applications of automation. Top level coordination and interdepartmental cooperation will greatly facilitate the exploration of interrelated systems and the promotion of effective utilization of available manpower.

Typical of this approach in government is the action of the Governor of Puerto Rico, who appointed a committee to study the implications of automation and report its conclusions to him. This committee, in turn, organized a survey team of government employees trained in data processing techniques. Private industry, in general, has followed this approach.

In the conduct of planning surveys it is important that the approach be cautious as well as progressive. Government, like business, should be on guard against embracing automation because it is in the current mode or because its miracles are wonderful to behold. In addition to the costs entailed, there are limitations and risks which are inherent in automated systems and which should be carefully explored. For example, electronic systems impose a heavy dependence on equipment and a very small group of key personnel. Equipment failure or loss of key personnel may result in serious operational breakdowns. Provisions must therefore be made for coping with such contingencies.

The ultimate product of the planning committees should be a master plan in which all feasible and justifiable applications would be blended into an over-all, integrated management control system. This should also include a time schedule for preparatory actions and for effectuating installations.

During the planning and transition stages, which may require a minimum of two to three years, the committees should issue reports of progress for the benefit of government officials and personnel, as well as the civic associations and employee organizations, who will be following the developments with intense interest.

The possibility of borrowing experts from private industry to aid the planning committees should be explored.

Establishment of a central clearing house for the compilation, dissemination and interchange of the vast amount of pertinent literature currently emanating from a variety of sources would greatly facilitate the research which is entailed in planning. Invaluable assistance toward this objective could be rendered by such groups as the American Association of Planning Officials and the American Society for Public Administration. Already active in the field is the Municipal Finance Officers Association of the United States and Canada.

For public administration automation poses problems which may be classified as (1) External and (2) Internal. In each classification, thorough and careful analysis and planning by governmental agencies are essential.

The economic and social problems created by industrial and commercial applications of automation compromise the external group. Effects on the worker appear to be of primary significance. Some of the factors which bear close study are displacement, training, shifts of employment opportunities, reduction in the work week, and more leisure time. Other areas of government interest include effects on the consumer and small business, and the need for converting and relocating factories and offices. To deal effectively with such pressures, government may find it necessary to: modify social security and unemployment insurance systems; broaden educational, health and welfare programs; provide more recreational facilities; revise master plans of cities and industrial areas; and expand programs for aiding small business and protecting the interests of consumers. New and expanded services may require increased governmental expenditures and, as a concomitant, a more intensive search for the means of financing such expenditures.

Problems classified as internal are those which emerge from the application of automation to the operations of government departments and agencies. Automation in government will, in the main, mean integrated data processing. Adequate planning will be the key to successful utilization of the new technology. Before decisions are made, actual and potential installations should be studied, and detailed surveys should be made of organization, systems and procedures, and effects on personnel. Realistic projections of costs and savings should be made. Estimates of outlays should include costs of preparation and installation as well as operation. Decisions favoring automation should not be predicated solely on anticipated reductions in government budgets. It is most probable that the benefits of automation will be reflected in more and better services, possible only through automation, without burdensome increases in budgets. Government's approach to automation should be guided by the principles of integration, coordination, and caution. Plans and decisions should be formulated by top-level coordinating committees working with departmental specialists. For public administration, automation offers a potential as vast as outer space, which it will undoubtedly help conquer; but decisions to "automate" should be based on feasibility, operational limitations, costs and savings, and demonstrable benefits.

Selected Bibliography

American Management Association, *Keeping Pace with Automation* (New York: The Association, 1956).

Warner Bloomberg, Jr., *The Age of Automation* (New York: The League for Industrial Democracy).

John W. Carr, "Qualifications for Supervision of an Electronic Office," *The Office XLII* (April, 1956), 14-23.

Robert T. Collins, "Automation Advances in Automatic Production," *Advanced Management* (May, 1955). (See also other important articles on automation in this issue.)

Peter F. Drucker, *The Practice of Management* (New York: Harper and Brothers, 1954).

———. "The Promise of Automation," *Harpers Magazine,* CCX (April, 1955), 41-47.

Educational Policies Commission, *Manpower and Education* (Washington: National Education Association, 1956).

Haskins and Sells. *Data Processing by Electronics* (New York: Haskins and Sells, 1955).

———. *Introduction to Data Processing* (New York: Haskins and Sells, 1957).

George Kozmetsky, and Paul Kircher, *Electronic Computers and Management Control* (New York: McGraw-Hill, 1956).

H. S. Levin, *Office Work and Automation* (New York: John Wiley & Sons, Inc., 1956).

Ted F. Silvey, "The Technology of Automation," *Congressional Record,* A-4262 (June 15, 1955).

H. Solow, "Automation: News behind the Noise." *Fortune,* LIII (April, 1956) 150 pp.

U. S. Department of Labor, Bureau of Labor Statistics, Report #2, "The Introduction of an Electronic Computer in a Large Insurance Company." (October, 1955).

David O. Woodbury, *Let Erma Do It* (New York: Harcourt, Brace & Company, Inc., 1956).

HERMAN LIMBERG is Senior Management Consultant in the Office of the City Administrator of The City of New York. His duties include the development and installation of major programs of economy and efficiency. He has also been coordinator of the records management and forms control programs and consultant on training. Prior to his present assignment he was management coordinator and chief of the training and education section in the Division of Analysis of the City's Bureau of the Budget. He received A.B. and Doctor of Jurisprudence degrees from New York University, and is a member of the New York State Bar. During World War II, he served as Industrial Consultant with the War Production Board and the Smaller War Plants Corporation. A specialist in organization, human relations, systems and procedures, management controls, and training, he is a Lecturer in management at the Baruch School of Business and Public Administration of The City College. He has taught at Brooklyn College, N. Y. U. Management Institute, Pace College, and the New York State School of Industrial and Labor Relations at Cornell. He has served as chairman and discussion leader of American Management Association Seminars, and in 1956 presented a paper on the "Implications of Automation for Public Administration" before the National Conference of Professors of Educational Administration at the University of Arkansas. He recently received from Mayor Wagner of The City of New York a citation for distinguished and exceptional public service. Dr. Limberg is a member of the Society for the Advancement of Management and the author of numerous articles on management which have appeared in The Office, Office Management, Purchasing, Systems, Advanced Management, The School Executive.

CHAPTER 30

WORK AND AUTOMATION*

BERNARD KARSH, *Associate Professor of Sociology,*
University of Illinois

It would be presumptuous to assert that we now have sufficient
evidence to point conclusively to the changes in the concept of
work which automation will generate. Indeed, experts are not
at all agreed on what automation is—whether it is simply an
extension of the normal trend of technological change or
whether it is something new. Nonetheless, it might be fruitful to
speculate on the impact of automation on the concept "work,"
and one way this can be done is to examine some aspects of the
"meaning of work in the age of the machine." Clearly, automated
processes are related to machine processes, and the impact of the
machine upon work should yield some clues by which a tentative
assessment can be made of automation's effects.

The relationship of work to automation can be developed
along these lines: (1) Work is one of the principal sources of
status in our society. The kind of work that we do, to the extent
that it can be readily identified and given a meaningful label by
our associates, gives us, in great measure, our place in life; (2)
The subdivision of labor which came with the extension of the
machine has to a considerable extent removed that label so that
now, unless one follows one of the professions or still retains
one of the traditional crafts or skills, status must come from
some other source. For the unskilled or semiskilled worker—the
machine tender—it comes largely either from the organization

---

*This article makes use of material presented by the author in *Current
Economic Comment,* August, 1957.

for which the work is performed or from the physical accouter-
ments of our culture, the things that money can buy. The work
itself no longer serves the essential function it did in an earlier
time; (3) Work satisfaction is bound up with the degree to
which the worker can exercise his judgment on his job and thus
have some control over how he spends his time and effort.
Workers will go to great lengths to increase the "judgment-con-
tent" of their work, much to the annoyance and frustration of
the industrial engineer and even the top corporation executive;
(4) Just as the machine substituted mechanical power for muscular power, automation promises to substitute the judgment of
electronic sensing devices for human judgment in a large category
of manufacturing and data-processing operations.

SOURCES OF STATUS

Work is the inescapable fate of the overwhelming majority
of men and women. Despite a recent cultural tendency to de-
emphasize hard work and to accentuate "having a good time"
(presumably, the antithesis of labor), work still remains the most
important segment of adult life. The impact of work routines is
found in almost every aspect of living and even in the world of
dreams and unconscious fantasies. It is not an overstatement to
suggest that work is not *part* of life, it is literally life itself.

A man's work is one of the most important things by which
he is judged and certainly one of the more significant things by
which he judges himself. The work that he does is what Everett
Hughes calls a combination calling card and price tag.[1] The im-
port of this is emphasized by the fact that one of the first things
we want to know about a stranger is what kind of work he does.
Our judgment of him is based, to a considerable degree, on an
identification of his work. And his judgment of himself is also
given in similar terms. The press agent is likely to describe himself as a "public relations counselor," the plumber may become
a "sanitary engineer," and the farmer becomes a "food producer."
Recently a Chicago restaurateur announced that his waitresses
are now called "hostesses," the hostesses "food service directors,"
the busboys "table servicemen," and the dishwashers "utensil

---

1. Everett C. Hughes, "Work and Self," in John H. Rohrer and Muzafer
Sherif, *Social Psychology at the Crossroads* (New York: Harper and Brothers,
1951), 313-14.

maintenance men."[2] Even social scientists emphasize the "science" end of their names. These hedging statements in which people pick the most favorable of several possible names for their work imply an audience. "And one of the most important things about any man is his audience, or his choice of the several available audiences to which he may address his claims to be someone of worth."[3]

These comments suggest that man's work is one of the most important parts of his social identity, of his self, indeed of his fate, in life.

It requires little more than passing familiarity with the organization of factory production to note the dissimilarity between the role of the machine operator or tender and that of the craftsman producing finished articles for the trade. Two differences seem most notable: mechanization and standardizat'on give the worker little but negative control over work methods or quality of product, and the minute subdivision of labor blurs out the part of any single workman in the total process. Ask a worker who is operating a production lathe in Department 22 to tell you exactly how the work he does fits into the work that is done in Department 23 or Department 21, or in the other side of the building, or in Plant 2, or into the work done in the home office in New York. The typical modern factory or industrial establishment is host to so many specialized functions that it is unlikely that any single individual can describe the processes necessary to manufacture and market the product or even the processes that take place in a single department to make only a small part of the product. It is even difficult for a worker to describe his own activities at work in relation to the activities of his neighbor at the next machine.

The specialization of function found in the modern factory removes any semblance of a "property" relationship between the producer and the product of his toil. Indeed, one can argue, as does Peter Drucker, that it is no longer the worker who is responsible for the final product. It is now the factory. The factory has become the unit of production and the worker does work— he does not make a product. There was a time in our society, albeit many years ago, when a shoemaker actually made shoes. But in the modern shoe factory, the workman cannot say that

2. *Time,* April 15, 1957.
3. Hughes, *op. cit.,* 314.

he is making shoes; he is operating a machine which makes
parts of shoes. And what kind of satisfaction is the worker supposed to get from eight hours a day of buffing an already finished shoe or running a machine which stitches welts, two of the
more than 2,000 operations involved in modern shoemaking.
Examples such as these lend support to the usual dreary picture
of routinized, standardized, time-sequence-paced, elementally described, machine-controlled work.

The impact of the machine upon work and its meaning was
stated by Adam Smith. He wrote:

> The understandings of the greater part of men are neces-
> sarily formed by their ordinary employments. The man
> whose life is spent in performing a few simple operations
> . . . has no occasion to exert his understanding . . . He
> generally becomes as stupid and ignorant as it is possible
> for a human creature to become.[4]

Almost a century and a quarter later this notion was rephrased
by F. W. Taylor, whose efforts to rationalize man's work brought
him world-wide fame. Taylor sought to reduce work to its simplest components and then describe them for the operator. A
latent purpose was to take from the worker his opportunity to
exercise judgment over his work. Taylor taught a Pennsylvania
Dutchman, whom he called Schmidt, to handle 47 tons of pig
iron a day instead of the previous 12.5 tons "and make him glad
to do it."[5] Every detail of the man's job was specified and a
"scientific manager" was assigned to supervise every moment of
his work—when to pick up a pig, how to carry it, how far to
walk with it, when to put it down, when to rest, how to load
it onto a car. Taylor got the optimum amount of work out of
Schmidt by systematically prescribing each component of the
task. By exact precalculation, even to three decimal places, he
got the desired response.

Taylor recognized that this kind of mechanical regime would
have some impact upon the man who followed it, and prescribed
the kind of worker who could best carry out his directions:

---

4. Adam Smith, *The Wealth of Nations* (New York: Modern Library,
1937), 734.

5. Frederick W. Taylor, *The Principles of Scientific Management* (New
York: Harper and Brothers, 1947), 40-48.

"One of the very first requirements for a man who is to handle pig iron as a regular occupation," he wrote, "is that he shall be stupid and so phlegmatic that he more nearly resembles an ox than any other type."[6]

In most respects, these dire predictions have not been realized as their predictors stated. But one wonders if the predictions have failed largely because workers refused to let the machines (and the logic of efficiency in which the machine is central) take from them what judgmental faculties they still possessed, and if this refusal has not been expressed in the ingenious ways workers find to exercise judgment designed to "beat" this logic. If only a small fraction of the ingenuity which the worker applies to beating the industrial engineer could be tapped *by* the engineer and put to work *for* the company, perhaps the productivity of the American industrial system—ignoring for the moment problems of distribution and consumption—would make the promise of the "full life" almost a reality and would do it without automation.

It is often said that modern man is alienated from his society—that he suffers from a loss of roots, that he finds it difficult to attach lasting sentiments to any of the values of his society. In substance, it is said, man finds little satisfaction from the world in which he lives and that his "depersonalization" has come as a result of the industrial system which he has created. If it is true that one of man's principal sources of satisfaction and identification derives from his work, then we can hardly escape the conclusion that the machine has indeed alienated man so far as it has made it difficult for him to achieve a readily identifiable status as in an earlier time. When a man describes his work as a carpenter, a blacksmith, a bookkeeper, a dentist, a bricklayer, a plumber, or even a janitor, his audience can fairly well assign a status to him in terms of his work and relate that status to their own. But when a man describes his work as, for example, a "stud torquer," a "gin pole operator," a "back winder," or a "facing tacker," what status does this give him? What identification—what price tag or name card—goes along with these kinds of work? Mostly either a question—"What's that?"—or perhaps even laughter. Thus, the stud torquer, when asked "What do you do?" is more likely to say, "I work for the Ford Motor Company," and the "gin pole operator" is likely to respond with, "I

6. *Ibid.*, 59.

work for the ABC Construction Company." That is, such a worker gets his status, his identification not from the work that he *does*, but from the organization that employs him. One of the current best sellers is W. H. Whyte's *The Organization Man*, and one wonders just how much this new creature is really a product of a status-requiring and machine-denying technology.

The idea of alienation also involves the satisfaction one gets from one's work as a result of the amount of control one exercises over it. There is good research evidence to show that for many people work satisfaction is bound up with the possibility that the worker can maintain some control over decisions of what work to do and over the disposition of his time and routine. Indeed, if we examine those occupations which carry high prestige, the skilled crafts and the professions, we find, as we go up the ladder of skill, greater and greater degrees of judgment exercised by the worker and wider and wider areas of decision control brought to bear by the worker on his work. Certainly, not all workers covet control-retaining occupations. Many probably prefer rudimentary and directed activity rather than judgment-making work. But it does suggest that our society puts a prestige premium on the occupations which require higher degrees of individual control and the use of individual judgment.

This is not to argue that a specialization of function—a minute division of labor—removes the value of skill in the organization of production. What is most often removed is the comprehension of the community—one's audience—with respect to the work that we do, and therefore there is introduced a difficulty in assigning some appropriate status to the worker. A semi-automatic machine accomplishes a number of distinct but continuous processing operations, to the open-mouthed awe of a visiting dignitary. It may depend for its continued operation on a workman who checks dials and gauges, makes decisions, and bears the responsibility. But the machine is likely to be far more spectacular than the workman and the machine gets the bulk of the credit. The worker's pride is very nearly pointless unless it is upheld by the esteem of the community. Thus, specialized work that is insulated from recognition by the worker's significant community—those whose opinions of him mean the most to him—violates the sentiments of "craftsmanship."[7]

---

7. Wilbert E. Moore, *Industrial Relations and the Social Order* (New York: Macmillan Company, 1951), 232.

There is another characteristic of the dependence of the worker upon the machine that has significance for the loss of workmanship: the tendency to desocialize the worker in favor of strengthening his orientation toward the machine. This tendency toward social isolation in the shop may be the product of painfully close supervision in an attempt to keep the worker devoted to duty, it may be the result of the "stretch-out" (the compounding of duties on a single worker), or it may be the deliberate or unconscious spatial arranging of job assignments to make it more difficult for the workers to develop informal relationships among themselves.

Subservience to the machine is never so apparent (whether "real" to the observer or not) as when the machine or the conveyor belt forms the sole aspect of the environment taken into account by those responsible for the assignment of tasks.[8]

The monotony of routinized factory work often produces a kind of crazy racing against the clock, the organization of slowdowns as a tactic in the silent war of the worker against the time-study man and the industrial engineer, ordinary horseplay to break the monotony, and the spectacular and violent eruption of wildcat strikes, ostensibly against "speed-ups" but often generated, in part, by the utter boredom of routinized and trivialized labor. Dan Bell has said that if "conspicuous consumption" was the badge of a rising middle class, "conspicuous loafing" is the hostile gesture of a tired working class.[9] Sociologist Don Roy has described how, in many machine plants, workers play the "make-out" game.[10] A worker will "put-out" at a breakneck pace in order to fulfill a piecework quota so that he can be free for the rest of the day. Piecework is often preferred to "day work" or a flat payment of an hourly rate. The pieceworker designs an hour-by-hour schedule of work completions by watching the clock, breaking up the continuous flow of time into specific short intervals. He achieves a victory over the despised

---

8. *Ibid.,* 233.
9. Daniel Bell, *Work and Its Discontents* (Boston: Beacon Press, 1956), 15.
10. Donald Roy, "Work Satisfaction and Social Rewards in Quota Achievements," *American Sociological Review,* XVIII, No. 5 (October, 1953), 507-14.

time-study man when he "makes out" early. And he also gets even with the foreman. "Since worker inactivity, even after the completion of a fair day's work, seemed to violate a traditional supervisory precept of keeping the appearance of being busy even if there is nothing to do," writes Roy, "making out in four or five hours could be used as a way of getting even with the foremen for the pressure that they applied when the quota was unattainable."[11]

This kind of worker behavior has its counterpart even on an assembly line—a machine-paced activity. Though far more difficult to control, it still permits the assembly-line worker to exercise some individual judgment, some decision-making role, with respect to his work.

One of the very interesting findings made by Walker and Guest in a study called *The Man on the Assembly Line* was the way in which the men sought to "buck the line" by introducing variety in their own mechanical work rhythms. One way was to accumulate a number of subassembly items which they called "banks" and then take a short rest. Another was to free themselves, if only for a few moments at a time, from the mechanical harness to which they were hitched by "working up the line" very fast and then catching a breather. The most popular jobs in the plant were those of utility men, foremen, and repairmen —those least resembling assembly-line work. The utility men, who act as substitutes for the line men at various times, spoke of getting an idea of the whole line, of meeting and talking with different workers, and of knowing all the job. Wrote the authors,

> To one unfamiliar with assembly line work experiences the difference between a job taking two minutes to perform and a job taking four might §eem far too trivial . . . [but] for the worker one of the most striking findings of this study is the psychological importance of even minute changes in his immediate job experience.[12]

---

11. *Ibid.*, 512.
12. Charles R. Walker and Robert H. Guest, *The Man on the Assembly Line* (Cambridge: Harvard University Press, 1952), 146.

THE AGE OF AUTOMATION

There is practically no research data which is addressed to the impact of automation on the meaning of work. But we do have some clues, and these will be discussed in the following pages.

"Automation" is defined in various ways, but for my purposes, it may be summarily defined as the accomplishment of a work task by an integrated power-driven mechanism entirely without the *direct* application of human energy, skill, intelligence, or control. Automation is more than simple mechanization, and it is advancing on at least two levels of technology. One level is the so-called *continuous flow automation* which is typified by machines that replace men's muscles, eliminate physical effort, and substitute electricity for manpower and that are coupled to devices which eliminate human judgment in the administration or direction of the control of the machine. This is what Ted Silvey calls machines to run machines. It is the technique of the automatic mechanism which can watch what the machine is doing, make sure it follows the instructions, and automatically correct mistakes. Another phase of automation is sometimes called *business automation*. This is the use of computing and decision-making machines, so-called "electronic brains," which hand out administrative, statistical, and clerical policies. Conceivably, an entire plant can be operated under the administration of these types of machines. Doing so would involve analyzing sales reports, ordering and checking the flow of materials, scheduling and controlling the operations, making out invoices, recording payments, and so on—all by machine. Experiments are now going forward at an increasing rate to combine these two phases into a single computer-run system. Whether this technology is old or new is not relevant for our purpose here. Its most important consequences will be felt and are now being felt as a result of the tremendous pace with which it is being developed and put to use.

These technical changes may have several effects. With automated processes doing the work of large numbers of human workers, managers of enterprise may no longer have to worry about a large labor force. This means that new plants can be located away from major cities and closer to markets or sources of raw materials and fuels. It is often cheaper to build a completely automated factory from the ground up than to automate piecemeal an existing factory. Thus, there is an impetus toward constructing new plant facilities either on the peripheries of the large cities or away from them altogether. This, in time, may leave the giant, sprawling industrial metropolis as a dwelling place for human beings instead of a location for unsightly factories. The radio manufacturing, chemical, and automobile industries are already developing along these lines. And since automation involves continuous flow of materials and information in a highly integrated and coordinated fashion, space utilization tends to be more efficient and factories of smaller size.

When the myriad of special purpose production machines, each tended by a single worker, is replaced by a single, huge, multi-purpose, punch card or tape controlled machine, communication between workers and the formation of work groups becomes more difficult. At least one British union, for example, has already asked for "lonesome pay" for workers overseeing automated processes. Trade union solidarity is fostered, at least in part, by the intimate relationships between workers thrown closely together in the workplace and by the development among them of common perceptions of their divorcement from management. With fewer workers in a given plant, and with these workers spatially isolated from each other, the supervisor, presumably, can know more of the workers in a more personal way. He is in a position to treat them as individual human beings rather than as clock numbers. The result may be a greater social and psychological identification of the worker with the supervision and the values of management, rather than with the union steward and the values of the labor movement.

Labor is cheap in relation to the tremendous cost of an automated machine. Workers can be laid off when they become unproductive, but it is enormously expensive to permit the machine to be idle. To write off the big capital investment, more and more of the automated plants may expand shift operations in order to keep the plant running 24 hours a day. And as Bell suggests, more and more workers may find themselves working odd hours. When this occurs, the cycles of sleeping, eating, and working, and social life become distorted. A man on the regular eight-to-four shift follows a cycle of work, recreation, and sleep, whereas during the same day the fellow on the four-to-twelve shift is on a cycle of recreation, work, and sleep, and the night man goes to bed around mid-morning, gets his recrea-

tion in the late afternoon and early evening, and goes to work
at midnight. Where workers alternate these routines, friendship
patterns change sharply. When the wife and children follow
a "normal" routine while the man sleeps through the day,
home life becomes disjointed.

It is often said that the era of automation represents a
"second industrial revolution" in the sense that a computer or
a feedback-controlled transfer machine takes from man the necessity that he use his brain, or a good part of it, in his work, just
as the first industrial revolution, ushered in by the power-driven
machine, represented an extension of man's muscle power. However, the drill press or lathe operator, and even the assembly-line
worker, still maintains some control over his work. But as a
button pusher on an automated machine, a man now stands
outside his work and whatever control existed is finally shat-
tered. Restricting output, for example, becomes a very difficult,
if not impossible, thing to do. An oilworker simply cannot slow
down a cracking tower in order to get some satisfaction from
the "boss." There is very little, if anything, the man who tends
the huge broach in the Cleveland engine plant of Ford can do
to affect the operation of that machine. A modern continuous
tin mill operates almost wholly independently of the worker
who watches lights, dials, gauges, perhaps a television picture
tube or a spectroscope. In operations of this kind, muscular
fatigue is replaced by mental tension, interminable watching, and
endless concentration.

A single instance may give some clue to this problem. The
New York Post reported the experiences of a Ford worker:

> Then there are workers who can't keep up with automa-
> tion. Such as Stanley Tylack. Tylack, 61 and for 27 years
> a job setter at Ford, was shifted from the River Rouge
> Foundry machine shop to the new automated engine plant.
> He was given a chance to work at a big new automatic
> machine. Simply, straightforwardly, he told his story: "The
> machine had about 80 drills and 22 blocks going through.
> You had to watch all the time. Every few minutes you
> had to watch to see everything was all right. And the
> machines had so many lights and switches—about 90 lights.
> It sure is hard on your mind. If there's one break in the
> machine the whole line breaks down. But sometimes you
> make a little mistake, and it's no good for you, no good

for the foreman, no good for the company, no good for
the union."

And so Stanley Tylack, baffled by the machine he
couldn't keep up with, had to take another job—at lower
pay.[13]

Mr. Tylack's experience suggests, of course, that the tedium
of routine tasks may be preferred by some workers. It also
suggests that we may not yet know what kind of skills, experience, personality, and similar worker attributes are required to
operate the new machines. It seems clear that a skilled tool and
die maker who, by virtue of his high skills and seniority, is
entitled to take a dial watcher's job may not possess the psychological requirements of that job at all. But this need not be a
problem in an age of automation since the machine tender and
dial and light watcher can be replaced by other machines, and
this is happening very quickly.

Here is where there may be a real social gain for these new
processes. Diebold and Drucker have pointed out that automation requires workers who can think of the plant as a whole. It
requires engineers and designers who have an over-view of the
entire process which the factory is intended to perform. The
locus of attention is shifted from the machine to the whole
plant and the individual "cut-and-fit" method of production gives
way to a continuous-flow process which eliminates the contribution of the machine tender or batch mixer. There is left, on the
one hand, the unskilled worker, the broom pusher, whose job
may be too menial to automate, and, on the other, the highly
skilled worker who designs, constructs, repairs, and programs
the machine. There is less trivialization, less minute specialization of function, and a need to know more about more than
one job. There is a need for highly trained technical personnel
who are able to see the process as a whole and who can conceive
of the factory—not the machine or the individual worker—as
the unit of production.

American industry now employs almost 18 million non-super-
visory white-collar workers of which the largest single group,

---

13. Cited in *Automation and Technological Change*, Hearings before the
Subcommittee on Economic Stabilization of the Joint Committee on the
Economic Report, 84th Congress, 1st Session (Washington: U.S. Government
Printing Office, 1955), 103.

the clerical workers, comprises almost half, 8½ million. These are the stenographers, secretaries, cashiers, bookkeepers, typists, receiving and shipping clerks, telephone operators and office machine operators. The next largest group is composed of the professional and technical workers—the engineers, accountants, auditors and the like.

The importance of white-collar workers in manufacturing industries has grown dramatically. While the number of production workers increased by 5 per cent between 1947 and 1956, the number of non-production workers, most of them white collar, increased by 52 per cent. In the chemical industry, an industry particularly subject to automation, production workers increased 6 per cent while non-production increased 67 per cent; in steel the production workers increased 3 per cent while the non-production went up 37 per cent. In the automobile industry, the number of production workers declined by 3.5 per cent while the number of workers engaged in the processing of information —the non-production workers—increased 24 per cent.[14] Changes of similar magnitude have occurred in food, textiles, fabricated metals, aircraft and oil refining.

The increased professionalization required by this technology suggests that some of the loss of status attending earlier technological changes may be regained. And though the union organizer may argue with the engineer or white-collar worker that his problems are no different from those of the plant workers, so long as the community thinks that the engineer or the programmer occupies a higher status level, for all practical purposes he has a higher status. Thus, there may be in the making a reversal of the historic trend toward status dilution for a large category of industrial workers. The problems which this may raise for the labor movement may be just beginning to be felt.

White-collar and professional workers remain the largest and most important untapped source of further trade union growth. But the labor movement, thus far, has hardly begun to touch this potential. Indeed, there is good evidence to conclude that it may not even know how to tap it. A union organizer will not recruit a status-conscious computer controller by trying to convince him that his problems are the same as the worker whose job was too menial to automate. And where unions have auto-

---

14. Jack Stieber, "Automation and the White Collar Worker," *Personnel*, (November-December, 1957).

matically included in existing bargaining units the new high status occupations, a shift in the social composition of the union may develop. The auto workers union has already begun to face this problem and has been forced to reconsider its traditional ways of doing things in order to accommodate it.

Also very important may be the elimination of what Taylor began about a half century ago. Wage incentive plans are based on the premise that human worth at the workplace can and should be measured by the number of units the worker produces. But with automation the human worker will no longer be concerned with nonsystematic cut-and-fit methods, and he no longer produces the units. The machines automatically do this according to instruction fed them by other machines. Highly skilled workers will be more concerned with designing and operating whole production systems in which banks of integrated machines are central in the measurement of worth. And in this kind of production system the human worker stands outside direct production activity. His new concern will be with adapting the production system to special situations, and this is an activity whose value can be measured neither by production units nor by time units. Hence, incentive plans, with their involved measurement techniques, may vanish. Adam Abruzzi suggests that a new work morality may arise in place of a morality based on a unit-worth concept.[15] The "one best way" definition of worker worth, Abruzzi asserts, will give way, the stop watch and the slide rule will disappear as instruments for the measurement of worth, and fractionalized time or production units will no longer be useful measures. Worth will be judged on the basis of organization and planning and the continuously smooth functioning of the whole system. The individual worker loses his importance and is replaced by the team, and greater value is placed on the operating unit as a whole. Whyte's *The Organization Man* may indeed replace the traditional individualism which has characterized American society for so long.

---

15. Adam Abruzzi, *Work, Workers, and Work Measurement* (New York: Columbia University Press, 1956), 297.

SELECTED BIBLIOGRAPHY

Adam Abruzzi, *Work, Workers and Work Measurement* (New York: Columbia University Press, 1956).

Daniel Bell, *Work and Its Discontents* (Boston: Beacon Press, 1956).

Everett C. Hughes, "Work and Self," in John H. Rohrer and Muzafer Sherif, *Social Psychology at the Crossroads* (New York; Harper and Brothers, 1951).

Wilbert E. Moore, *Industrial Relations and the Social Order* (New York: Macmillan Company, 1951).

Donald Roy, "Work Satisfaction and Social Rewards in Quota Achievements," *American Sociological Review*, XVIII, No. 5 (October, 1953).

Adam Smith, *The Wealth of Nations* (New York: Modern Library, 1937).

Jack Stieber, "Automation and the White Collar Worker," *Personnel*, (November-December, 1957).

Frederick W. Taylor, *The Principles of Scientific Management* (New York: Harper and Brothers, 1947).

Charles R. Walker and Robert H. Guest, *The Man on the Assembly Line* (Cambridge: Harvard University Press, 1952).

BERNARD KARSH is Associate Professor of Sociology in the Department of Sociology and the Institute of Labor and Industrial Relations at the University of Illinois. He received his M.A. and Ph.D. degrees from the University of Chicago. He has contributed many articles in various scholarly journals and is co-author of *The Worker Views His Union* (University of Chicago Press, 1958). His book *The Union Organizer* will be published by the University of Illinois Press in the Fall, 1958. He was contributing author to the recently published volume, *Labor and the New Deal*. He has also contributed to the *Nation* and other non-academic magazines.

Dr. Karsh has for many years conducted research among rank and file union members in a wide variety of settings. He has also been active in adult education teaching classes of union members at the University of Chicago, Roosevelt University, the University of Kansas and Indiana University as well as at the University of Illinois.

Chapter 31

AUTOMATION AND SOCIAL STRATIFICATION

JIRI NEHNEVAJSA, *Department of Sociology,*
Columbia University, and ALBERT FRANCES,
Bureau of Applied Social Research, Columbia University

Specifically, we think of automation as an independent variable which somehow (functionally) accounts for changes in the dependent variable—stratification. Let these major variables be defined before their relationship can be plausibly considered.

Automation may be defined as a *principle* and as a *factor,* a distinction which we find advisable because it may avoid some of the pitfalls one notices in scrutinizing the literature. As a *principle,* automation pertains to pure and applied *theories* relating to the possibility of sensitizing machines to external and internal stimuli (feedback) so that, independently of their operators, they react to their environment and to their own parts (automatic control and monitoring) according to a given rationale (program). Automation as a *factor* refers to the application of automation principles to *specific operating conditions.* This application takes the form of utilizing self-setting and self-correcting machines, and the employment of programmed processes and procedures.

This chapter will not be concerned with the possibilities pertaining to automation as a principle. As a factor, however, it promises to be highly relevant to our present aims. At this time we are not interested in the question whether automation represents an advanced stage of industrialization, or whether it is best considered a new kind of phenomenon (different from previous forms of industrialization in kind, not in degree), leading into a *new* industrial revolution. This question is inherently uninteresting here because we are not trying to find historical analogues but to underline plausible relations between automation and social stratification.

DEGREES OF AUTOMATION

We have said that automation is a variable. As such, it can presumably take any one set of *values* over some specifiable range. We shall be concerned only with ordered relationships and not with numerical values. Thus it suffices to say that automation can vary from zero to some maximum.

But this is satisfactory only on one level of abstraction, and certainly does not meet the conditions required for a thorough analysis of the issue. Above all, automation is not unidimensional if we consider it as a factor. From the standpoint of the total *social system*, by degree of automation may be meant some measure of *relative density* of automated units as the ratio of A-units[1] to other production units in a given society. Automation may also pertain to an *industrial complex*—such as the automobile industry, the steel industry, etc. On this level, its degree is reflected in some appropriate index of density of A-units within the complex. Next, it may correspond to a given *plant* or *factory* or *organization,* that is, the A-unit.

Even this differentiation is not quite enough. Within the A-unit there can be various degrees of automation—which have been the subject of many technical and non-technical classifications. We shall tentatively use the terms *level, span* and *penetration.* Level of automation pertains to the degree to which the machines employed are self-setting, self corrective and programmed. Span refers to the degree to which production processes are automated, and penetration stands for the degree to which the secondary needs of a factory are attended to by automated devices as well.[2]

It is easy to imagine that the combination of these variables within the A-unit, in an industrial combination of these variables within the A-unit, in an industrial sub-system and in the total

---

1. Throughout this paper we shall use the abbreviation *A*— as symbol for automation. Thus, when we refer to automated units, or to the members of an automated unit, or to the functions of automation, we shall write A-units, A-members and A-functions.

2. James R. Bright, *Automation and Management,* (Boston: Harvard University, 1958) 41; John Diebold, *Automation. The Advent of the Automatic Factory* (Princeton, N. J.: D. Van Nostrand, 1952).

society, shall result in numerous characteristic types of automation. This may perhaps explain the diversity of opinions on *effects* of automation.[3] Our main objection, at this time, is that some specific situations of fact taking place in factories undergoing automation or having achieved automation should not necessarily be the irremediable effects of automation for, in the total situation, automation is only one factor.

STRATIFICATION

As a dependent variable, the concept of automation implies that, somewhat, the values of appropriate indices may vary over some specifiable range. Like automation, stratification is not a unidimensional concept. Numerous hierarchies result from differential evaluations and these, in turn, are defined over inequitable distributions of rights, duties, privileges and obligations. Furthermore, the empirical characteristics of these stratification hierarchies depend greatly on the *evaluator*—whether some objective indicators are employed, or evaluations of actors are made by others, or whether actors evaluate themselves.

For the sake of this discussion we shall be speaking mainly of the occupational hierarchy. The reader is somewhat free to decide which of the above constructs he prefers. For one, occupational stratification is relevant in its own right.[4] Secondly, occupational standing may be considered an indicator of a more complex fabric of evaluations, of social position or social class, and thus merely a parsimonious measure of some generalized stratification hierarchy.

To say that social stratification is a dependent variable furthermore calls for specification of the frame of reference. Here we shall deal mainly with *the continuum of discreteness—fluidity,* which means that under some structural conditions systems are generated in which relatively discrete strata exist, while other conditions are conducive to fluidity, so that a continuum rather than strata is the appropriate referent.

---

3. *Automation and Technological Change*, Report of the Sub-Committee on Economic Stabilization to the Joint Committee on the Economic Report, 84th Congress of the United States (U.S. Printing Office, Washington, D.C., 1955); Arthur Turner, "A Researcher's Views on Human Adjustment to Automation," *Advanced Management*, (May 1956), 21-25.

4. Bernard Barber, *Social Stratification, A Comparative Analysis of Structure and Process,* (New York: Harcourt, Brace and Co., 1957), 39.

The specific question is then whether automation, as defined in a limited way, tends to increase or decrease the discreteness of the stratification system, or whether a social system becomes more or less rigid in the presence of varying amounts of automation. And then, more rigid than what? We shall assume time zero, the present, to be our point of departure. With respect to this anchorage, changes in the stratification system may be tentatively assessed.

PART II: FUNCTIONS OF AUTOMATION

NO COMMON BOUNDARIES

In studying the relationship between the A-unit and stratification we find that the two referent systems *Do Not Have Common Boundaries*. To speak of stratification we must take as relevant subject matter the actors and the space which may be affected by the A-unit.

First, we shall have to consider the A-members. Even so, the significance of these members transcends the unit itself. They live in its neighborhood, they form or integrate in families and associations and are actors in a network of social relations taking place in their temporal and spatial environment. Second, their quality of A-unit-members characterizes other individuals within the relevant environment (or collective) as non-members. The area thus becomes directly and indirectly qualified by the A-unit. It can be thought of as an *automated island*[5] which includes the A-unit (plant, factory etc.) and its functionally related field around it. For the sake of brevity we shall call it an *aut-land*, which involves a notion conceptually similar to the idea of "metropolitan area" containing the metropolis as the focal point and its functionally relevant environment.

The functions of the A-units upon the stratification system are primarily performed through the aut-lands—considered as collectives—distinctly characterized by automation and consequently apt to be studied in their analytical, structural and global properties.[6] They will also manifest themselves on the *social cate-*

---

5. This phrase has also been used to express the concentration of automated machines in a given manufacturing process within a plant, or the areas in which automatic machines are in operation.

6. Paul F. Lazarsfeld and Herbert Menzel, *On the Relation Between Individual and Collective Properties* (Bureau of Applied Social Research, Columbia University, 1958).

gories[7] more susceptible of being directly affected by their social proximity to the A-functions, such as the category of the industrial workers. Theoretically, A-functions will generate chain reactions gradually extending throughout the social system[8] but at this time we need not be concerned with the extreme implications of circular latent functions of this kind.

CONTINGENT AND PERMANENT A-FUNCTIONS

A-units affect their aut-land and proximal social categories in two different ways:

A. By the specific A-functions resulting from special circumstances exclusively related to a given degree of automation in the A-unit. These functions can be defined as transitory because they are altered when the A-degree is modified. Further modification may entail even their total disappearance. We shall refer to them as *contingent*-A-functions.

B. By the A-functions inherent pertaining to automation, whatever these may be, so that the higher the A-degree, the more these functions will be operating and effective. We shall call them *permanent*-A-functions. It is obvious that the differentiation between *contingent* and *permanent* functions is of the utmost importance. Otherwise we may mistakingly ascribe to automation a great number of effects on industrial organizations or on society on the empirically cogent grounds that a sample of automated industries produced these effects while in fact it is possible that they are the product of a set of circumstances not directly related to automation and that changes in these circumstances will cancel those effects. Present difficulties in isolating permanent-A-function stem in part from the fact that the principles of automation have been very irregularly adapted to the countless variety of industrial situations. Automated processes have been imposed upon complicated networks of economic and social relations originally prepared to meet the contingencies of the mechanized factory but rather inadequate to face the issues brought forth by automation. In many plants today relatively advanced automated machinery operates together with dissimilar mechan-

---

7. Robert K. Merton, *Social Theory and Social Structure* (Glencoe, Illinois: The Free Press, 1957), 299-300.
8. W. F. Ogburn, "Technology as Environment," *Sociology* and *Social Research*, Vol. XLI, No. 1 (September-October 1956), 3-10.

ical devices. For example, the Lorain (Ohio) works of the National Tube Division of the United States Steel Corporation adopted automation in its Number 4 Seamless Mill. However, the mechanical plugging process within the mill was an anachronism of the kind which, in Walker's words, "are characteristic features of the transition to an age of automation."[9] In this and similar cases automation intervenes as a factor among many others. This is because the principles of automation have not advanced enough to serve some specific industrial requirements or because its full application would not serve the utility goal of the organization.[10] In general, every industry and every plant faces the problem of automation in terms of internal and external circumstances confronting its production at the time when automation is considered, as well as over some relevant future planning horizon. Accordingly, *automation principles are adjusted to those circumstances rather than the reverse*. Or, as Bright stated, "we have not realized how automation's impact will vary widely with different products, manufacturing techniques, industries, degrees of mechanization, and sales environments."[11] This is why today there are as many kinds of "automation" as industries, and thus the functions of automation are still highly elusive. Not only does every A-unit exhibit its own contingent functions but also many functions of automation— either contingent or permanent—*are barely distinguishable from those of advanced mechanization*. Perhaps by means of nation wide research we could obtain some sort of typology, but even so the information obtained would not suffice to draw highly reliable and valid conclusions such as to infer the effects of automation on social stratification. The survey would reflect many opposite contingent functions pertaining to various kinds of A-units, many of them actually neutralizing each other or becoming neutralized by other unit-functions, owing to their minimal weight. In both cases they need not ultimately affect the evaluation underlying the stratification system or its basic structure. Moreover, continued changes in the A-units modify the contingent functions almost from day to day.

As it is, however, there is no doubt that the progressive ap-

9. Charles L. Walker, *Toward the Automatic Factory, A Case Study of Men and Machines,* (New Haven: Yale University Press 1957), 17.

10. James R. Bright, *Automation and Management,* (Boston: University, 1958), 79-87.

11. James R. Bright, *op. cit.,* 87.

plication of automation principles in industry and administration will result in stratificational changes. Whereas mutually counteracting contingent-A-functions appear and disappear at every point of the spectrum (as for example the typical dissatisfaction of workers during the trial-and-error installation period which almost every survey reflects) the permanent-A-functions increase proportionally to increments in automation. This must be so by the very force of the nominal definition of the permanent functions, that is, if there are such mechanisms as permanent-A-functions, they must increase when automation increases, either as manifest or as latent functions. Moreover, contingent functions are somewhat random in character, intensity, direction and time, while permanent functions are constant. Thus, only permanent functions can exert the kind of continued pressure upon the social system which is likely to generate basic changes in stratification. It may be that, at present, these functions do not exceed the level of inertia imbedded in the structure they tend to modify. In this case, to be sure, no substantial modification takes place. Minor changes can be either absorbed by or somehow compensated for, in the social structure. There is no reason to doubt, however, that as automation continues to increase, its permanent functions will also increase to the point where its impact on the social structure will be patent. Interesting as it may then be to study them, it is more important to have some way of predicting such changes before they actually take place. Consequently, preliminary theoretical steps toward the isolation of these permanent-A-functions must be undertaken to give at least some indications to researchers in the field. This paper will be mainly concerned with the possibility of applying existing social theory to an eventual delimitation of some presumed permanent functions.

AUTOMATION AND ROLE PERFORMANCE

To start with, we should like to inquire into the effects of some (tentatively assumed) permanent-A-functions upon the role performance of A-members. In order to establish a few points of reference we shall use some of the categories which have been customarily used as indices of job evaluation:[12]

---

12. Richard C. Smith and Matthew J. Murphy, *Job Evaluation and Employee Rating* (New York: McGraw-Hill Co., 1946), 42-52.

a) skill requirements (manual and technical)
b) effort requirements (physical and mental)
c) responsibility requirements (production, equipment, time, safety, etc.)
d) working conditions (endurance to)

These four indices, here considered as variables, are also qualifiers of role performance. The question now is how these variables are affected by the A-functions in general, and by the permanent-A-functions in particular.

Some writers on automation have emphasized that the automated unit would require in aggregate *less* physical effort but more *skill*.[13] This means that the A-unit-member would be upgraded according to an evaluation whereby physical effort lies at the bottom and mental effort at the top, or by which the hierarchical evaluation of a role is positively related to the amount of skill required for its performance. By implication, it has been argued that an important portion of the labor force would be displaced by automation because of lack of adequate technological knowledge. Along this line, the A-unit would supposedly need only superskilled labor of the kind not frequently available among average industrial workers.[14]

This theory has been effectively contested in the Bright report.[15] According to Bright, generally and quantitatively speaking, *the more automated a unit is, the sharper the overall decreases in skill, effort and responsibility requirements.* Whereas set-up and maintenance men *increase* in skill and responsibility requirements in low and medium automation types, further automation increases tend to reduce these requirements save for a *few* positions. Engineers and technicians rather than increase their skill are concentrated upon narrowing areas of specialization, with the exception of those engaged in research and high level planning operations. As for today's largest group in the industrial category—the production workers—the more auto-

---

13. John Diebold, *Automation: the Advent of the Automatic Factory,* (Princeton, N. J.: D. Van Nostrand, 1952).
14. Frederick Pollock, *The Economic and Social Consequences of Automation,* (Basil Blackwell, Oxford, 1957). Norbert Wiener, *Cybernetics, Or Control and Communications in the Animal and the Machine,* (New York: John Wiley and Sons, Inc., 1948) 187-9. P.D. Many of Dr. Wiener's views were less pessimistic in his subsequent publications.
15. James R. Bright, *op. cit.,* 170-211.

mated a unit is, the more skill, effort and responsibility requirements *decrease.*

This has been explained in the following way: *First,* automated jobs seem to require less knowledge of art, less dexterity, less knowledge of theory, less experience, less physical effort, less judgment, less exposure to hazards,[16] less responsibility concerning safety of equipment, of the product and of other people, and less decision making. William A. Faunce has observed that some old timers complained about increases in noises and other unpleasant conditions in the automated factory where they worked. These may be objective circumstances pertaining to a given type of automated factory (contingent functions). Yet their discontentment might well be owing, as Faunce suggests, to resistance to change and to the dissolution of team groups, relative isolation of individuals and diminishing of physical activities etc. which may count for "subjective" discomforts having little to do with the physical environment.[17] In general, a layman can get a rough idea of the production worker in the highly automated factory by comparing him to a kind of switchboard operator or to a watchman[18] whose job is to check a series of controls and state their condition at periodical intervals. This is not so, however, when the amount of automation is medium or low.[19]

Second, the production worker controls the work pace even less than under mechanized assembly-line processes. In the A-unit, planning requirements—necessary to avoid downtime losses, bottlenecks etc.—are so strict that processes have to be timed at the exact required measure, independent of the abilities or willingness of the operator and generally at speeds far exceeding any the human operator could develop.[20]

At present, the above is, to be sure, a somewhat risky generalization. Perhaps a nation-wide survey of A-units would not

---

16. James R. Bright, *op. cit.,* 186-7.
17. William A. Faunce, "Automation in the Automobile Industry," *American Sociological Review,* Vol. XXIII, No. 4 (August 1958).
18. James B. Bright, *Ibid.,* Appendix X; Charles A. Walker, *Ibid.,* 29-39.
19. Low automation may be such as not to affect or to affect inversely the internal structure of the A-unit. This is more so in the intermediate stages between automation and mechanization. See for example W. Lloyd Warner and J. O. Low, *The Social System of the Modern Factory* (New Haven: Yale University Press, 1947), 66-89.
20. Charles R. Walker, *Ibid.,* 130, 29-30, 33-4, 104; James R. Bright, *op. cit.,* 186, in more automated units.

quite justify it, but, here again, we are referring to permanent-A-functions. If these characteristics could be confirmed as permanent-A-functions their principal effect would be the reduction in role performance requirements. Furthermore, if our assumption were correct, it would coincide with other opinions whereby a maximum of automation implies a minimum of human participation in production processes, except for a fraction of the labor force, assuming that the level of production is held to its pre-automated level.

Displacement can result from *changing performance requirements, psychological inadaptability* and *functional displacement:*

a) *Changing performance requirements.* This issue is concerned with the problem of adapting the role-actors to automated tasks, and to the socialization of new recruits. The retraining of old workers to their new jobs has been a matter of company policy. Walker mentions that at the Lorain Works "thousands of individual conversations were held with members of the crews to give them technical information and to prepare them for their new duties"[21] but it should be remembered that the plant exhibited a medium-low type of automation. The crews were trained in both automatic and manual controls. Bright writes that "contrary to much speculation, the training of the operating working force (judging by these 13 plants) is not a problem."[22] Whereas automation has originated some specializations which ordinarily could not be filled by former mechanical workers (such as programmers and some maintenance roles) for the rest, automated tasks are easier to perform than their mechanical counterparts. This is perhaps one of the reasons why efforts of labor unions to preserve seniority rights have been generally successful, except for the implications resulting from psychological inadaptability.

b) *Psychological inadaptability.* In general, adjustment has been considerably more difficult for old workers with long training in mechanized processes. Although effort requirements seem to have decreased in an *absolute* sense, in a *relative* sense they may appear increased. We underline relative because we want to stress the subjective character of this increase: in most cases it may not be as the worker perceives it to be. In fact, once accustomed to the new work, many former mechanized workers have

---

21. Charles R. Walker, *op. cit.,* 156.
22. James R. Bright, *op. cit.,* 125.

come to realize that it was not as difficult as it had seemed at the start. It can also be assumed that the effective downgrading of skills, with its corresponding loss of independence, self-esteem and self-confidence, has affected the morale of many oldtimers. Changing conditions of work, which in many cases entail the dissolution of teams, may also count for psychological strain. These circumstances are not likely to hinder new recruits, however, unless they have been pre-socialized toward internalizing values applicable to mechanical types of performance. Difficulties in psychological adjustment can develop either into internal reactions against automation or into loss of A-membership. Published research findings lead us to conclude that the former usually occurs during the initial period, when even outright sabotage can be expected and the latter during the A-unit consolidation period.[23]

If discontented old members leave the A-unit, management probably would not be the least embarrassed unless massive labor union opposition took place. This would be so if most of the old actors were forcibly separated or compelled to resign. It seems, however, that most of the mechanized workers achieve a satisfactory level of adjustment.[24]

c) *Functional displacement.* All in all, the major factor of displacement seems to be the reduced manpower requirements of the A-unit. At present the effects are not obvious because of the slow pace of automation and perhaps the constant pressure of the labor unions. In order to diminish resistance, management has planned manpower reduction policies so that vested interests would be the least hurt. This involves first the gradual elimination of temporary help, and, as a second step, the limitation of new hiring. Thus, after a time, the total payroll shrinks to the desired proportions. These methods, moreover, optimally adjust to the technical requirements of automation.

The argument that automation invariably determines production increases does not alter the fact that automation reduces manpower needs per the same amount of production. Even if the man/hour rate of production augments, more or less the same number of workers will be needed. Whether the present configuration of our economy will tolerate constant increases in production that could neutralize reduced manpower needs

23. James R. Bright, *op. cit.,* 210-1.
24. Charles R. Walker, *op. cit.,* 192; Robert R. Bright, *op. cit.,* 198-204.

is a question far beyond those issues raised here. There are grounds to believe, however, that the social system in its present form will tolerate only a given margin of production increases, and that beyond that margin substantial changes will occur either by design or by default in the social structure.[25]

LEVEL OF REWARDS

In spite of the fact that role performance requirements decrease with automation, and thereby recruitment requirements do too, evidence indicates that the rewards have tended to increase.[26] The question is raised now: if less skill, effort and responsibility and less hardship are involved so that actors are generally more substitutable than skilled workers in mechanized factories (since even training takes less time), how is it that members of A-units should be receiving higher compensations?

For one, we may argue that this is but an arithmetic by-product of increased production rates per man, but assuming the persistence of some form of profit motive this is a lame explanation. If the change in reward level is not attributable to the more exacting performance and recruitment characteristics of the roles on the whole (for, as we have pointed out, the expectations of a *few* roles—maintenance and programming—become somewhat more demanding, at least initially), we shall say that the increased rewards mirror *increments* in favorable evaluation of the roles themselves, that they reflect status up-gradings of the roles.

This, in itself, is a rather revolutionary function. What we are asserting is that roles become less demanding but of higher prestige! Whereas the worker under conditions of mechanization tends to be evaluated, for the sake of distribution of rewards, according to his production, the A-unit worker is so removed from production itself—due to the disparity in the man/output ratio that he must be rated in terms of his functional part in the process.

It may be said that the A-worker's role increases in functional significance for the A-system although it requires less effort and less skill. As a result of enormous increases in produc-

25. Ernest Nagel, "Self Regulation," *Automatic Control* (New York: Scientific American and Simon and Schuster, 1955), 8-9.

26. James R. Bright, *op. cit.*, 204-9.

tion speeds and capacities, worker *turnover* or instability would constitute a much more severe problem than it does under mechanized conditions.

However, jobs have rarely been rated according to their functional significance. External communication is of the utmost importance to an organization but this has not made better paid positions for local telephone switchboard operators. It follows, we think, that in the A-unit evaluations are not made in terms of individual roles but in terms of the whole role-system constituting the A-unit. In other words, rewards do not necessarily correspond to either performance requirements or functional importance, but become a function of the A-unit which gives increased status to its members.

From another point of view, it has been argued that compensational increases in A-units are due to both labor union pressures and managerial policies. The fact, however, that such circumstances should occur precisely in A-units and not *elsewhere* in the industrial system can be taken as an indicator of status-gains among the actors in A-units. If status gains were a permanent function of automation they would be comparatively weak in minimal automation stages and increase with it. However, there is the both theoretical and empirical possibility that certain contingent-A-functions work toward status limitation or compensational discreteness. It can also be that other functions unrelated to automation and pertaining to dynamic characteristics of the unit or the complex itself (such as contractual wage negotiation between union and management) arrive to neutralize the permanent-A-function of status and compensational gain. But a neutralized force is still a fully operant force. Increases therein require increases in neutralization. If not the equilibrium will end and the force, or in this case, the function, becomes significant.

STABILITY, CONTINUITY AND PERMANENCY

In order to effect further differentiation between A-units and mechanized units, we should also mention the *factors* of employment continuity, compensational stability and employment permanency, which seem to be reasonable candidates for the title of permanent-A-functions.

The A-unit tends to secure more continuity in employment than the mechanized unit. Several reasons could be mentioned in support of this hypothesis. Long term planning requirements, downtime avoidance and high cost operation and maintenance ask for a permanent force of reliable employees. Our idea of the increased damage attributable to turnover could be reiterated. Moreover, since employees are related to machines and not to production, eventual changes in rate of production will not affect the employees. *Whether production is high or low, care and maintenance of the highly expensive machinery must continue. In this sense payroll costs are minimal if compared with machinery costs.*

By the same token, the effective distance between operators and production will affect changes toward fixed annual wages. This will lead toward a higher compensational stability at a higher compensational level.

Where employment permanency is concerned, mechanized factories often force the earlier or unexpected retirement of the worker who can no longer hold up, owing to age or illness, under the physical hardships involved in his performance. It is beyond doubt that a vast majority of automated jobs allow later retirement. Workers who in mechanized factories would have been forced to give up their activities through loss of dexterity or endurance—thus becoming economically and socially downgraded to lesser positions or to idleness—may in the A-unit continue their active life for longer periods.

RECAPITULATION

For the sake of clarity, we can now summarize our previous assertions as follows:

If the circumstances referred to above are permanent-A-functions, a higher type of automation in the A-unit would imply:

1. Higher financial rewards
2. Increased compensational stability
3. Increased employment continuity
4. Greater employment permanency
5. Better working conditions
6. Lower manpower requirements for given production levels
7. Lower performance requirements

For a high proportion of A-members, increasing automation

will result in less demanding role performance but higher role-status.

The question is now to determine how these assumed permanent-A-functions may affect the stratification system of the aut-land and the proximal social categories.

PART III.   AUTOMATION AND STRATIFICATION

MOTIVATION TOWARD A-MEMBERSHIP

In the first place, the labor force in its entirety will have the sub-category of *A-members* as a new sector. This means that every member of the labor force will either belong or not belong to an A-unit. Membership will thus produce objective, subjective and reputational criteria of stratification. Whereas the objective criterion enables the researcher to delimit A-membership sub-categories for the purposes of description and analysis, the subjective and reputational criteria affect in their own right the referent stratification system.

Secondly, A-membership can be subjectively evaluated in terms of a positive-negative continuum of responses. We may conclude that, in general, non-members will be positively motivated toward A-membership if A-functions contribute to "the occurrence of favored uniform evaluations of the actor and/or his actions in his action system."[27] Subjectively, industrial workers have arrived at a categorization of "ideal employment conditions" which approximately express their own descriptions, evaluations and expectations[28] as well as their desiderata. If we compare our above-mentioned 8 propositions with some empirically obtained lists of workers' preferences, such as Wight Bakke's,[29] we shall see that they match at many counts.

Collated with mechanized units, A-units seems generally nearer the "ideal employment conditions." We must take into consideration, however, that many contingent-A-functions may occasionally take forms highly unattractive to the worker. On the other hand, we assume that the higher the degree of automa-

---

27. Hans L. Zetterberg, "Compliant Actions," *Acta Sociologica* (Vol. 2—fasc. 4, Copenhagen, 1957), 188.

28. Hans L. Zetterberg, *On Theory and Verification in Sociology* (New York: The Tressler Press, 1954).

29. E. Wight Bakke, *The Unemployed Worker* (Yale Univ. Press, 1940), 12.

tion, the more pronounced the permanent-A-functions will be. Therefore, if lower performance requirements and greater status-gains are permanent-A-functions, A-membership will finally be the object of positive evaluation. This, of course, with the exception—operating as a special case—of psychological inadaptability. Furthermore, we have argued on the impact of the A-unit as a whole on role statuses of the workers. In an experimentally oriented culture which places high evaluations on science and technology (which it does precisely for being experimentally oriented: it attributes high standing to roles which function to discover new or more efficient ways of reaching social goals), the A-unit as a global entity must be evaluated chiefly by *generalization*. Concretely, this means that the high standing of science and technology "rubs off" on actors performing in roles associated with sub-systems utilizing the newest products of science and technology. And finally, the A-unit seems to be somewhat awe-instilling since the *automation* principles are as unknown to the layman as nuclear energy.

This, in fact, has the effect that although internally (within the A-unit) there may be an overall degrading of skills, this is not *socially visible* in the aut-land because the whole A-unit is rather incomprehensible.

AUTOMATION AND BUREAUCRACY

If more non-members are motivated to secure membership, and, at the same time, the A-unit implies lower manpower requirements, *competition for A-membership will accordingly increase as automation itself increases.*

Since as already stated, performance requirements in general are lower, the basis for role recruitment will have to be raised proportionate to the competition. The disparity between low performance requirments and high recruitment requirements will have to be solved by the introduction of *some prerequisites* not directly relevant to the role performance but somehow relevant for the A-system. What shall these be? The performance of the A-member is not evaluated in terms of *what he actually does* but in connection with the automated processes he supplements. Included in this consideration are the costly disturbances which could take place if the A-member does not faithfully comply with his oversimplified performance requirements. This means that *loyalty* will have to be rated over and beyond *skill*. Accordingly, recruitment requirements will tend to be based, among other items, on the candidate's presumed *reliability, strict devotion to regulations, continuity, discipline, methodical performance and conformity with a prescribed pattern of action.* This is the extended meaning of what automation writers and personnel specification writers understand by "flexibility" versus the assumed "rigidity" of the mechanized worker. In fact, a content analysis of literature about automation would find the word "flexibility" and the analogues "adaptability" and "adjustability" extended ubiquitously so far as employees are concerned. These A-member attributes are rewarded by management with *life long tenure in the absence of disturbing factors which may decrease the size of the organization* and with a *working life planned for him in terms of a graded career throughout the organization of promotions, seniority, pensions, incremented salaries etc. . . . . all of which are designed to provide incentives for disciplined action and conformity to the regulations.* We have underlined the above paragraphs because they happen to be literal quotations from Merton's analysis of bureaucracy.[30] *Thus we arrive at the proposition that one of the major permanent functions of automation is the gradual bureaucratization of the industrial worker.*

The favorable disposition in the value system about this trend, evident in polls assigning high desirability to public service positions, and the quantitative move from individualistic independence to corporate security[31] indicates little or no resistance to be expected.

AUTOMATION AND RIGIDITY

At this point we can operate with the following tentative hypothesis:

A. A-unit-membership as object of positive motivation
B. A-unit-membership as status-giving factor
C. A-unit-membership is a factor accounting for transformation of blue collar into white collar (bureaucratization).

---

30. Robert K. Merton, *Social Theory and Social Structure* (Glencoe, Ill.: The Free Press, 1957), 197-200.

31. National Opinion Research Center, "Jobs and Occupations: a Popular Evaluation," *Class, Status and Power*, Eds.: R. Bendix and S. M. Lipset (Glencoe, Ill.: The Free Press, 1953), 411 and fol.

Moreover, if A, B, and C stand for permanent-A-functions, they will increase when automation increases.

How can we relate them to the stratification system of the aut-land? We have conceived the aut-land as to consist of the A-unit and its functionally related field. The A-unit may be physically located in an ecological dense or fluid area. In the first case, *social visibility* may be low, thus the differentials between members and non-members will be less conspicuous. Since the A-unit is not the only organization exhibiting status-giving and bureaucratic characteristics (official and corporate bureaucracies show similar trends) A-members may tend to identify with the social categories most in harmony with their own subjective evaluations. In low visibility areas this transfer may be less noticeable than in ecologically fluid areas of high visibility. Furthermore, the aut-land is *homogeneous,* that is, primarily composed of members of one social category—for instance, industrial workers—the withdrawal of the sub-category of A-members therefrom may produce striking social consequences.

In general terms, we could tentatively assume that the higher the type of automation in the A-unit, the greater social visibility and the greater homogeneity in the aut-land will result in greater differentiation of the A-member's sub-category and higher bureaucratization trends in it. In turn, increased differentiation can be considered as *one of the factors* contributing to stratification rigidity.

Other factors toward rigidity are the restrictive recruitment requirements of the A-unit, the special conditions created by changed performance requirements and the A-member's status-gains. At the same time, restrictions on social mobility within the A-unit, whose members are unable to ascend to levels beyond their technological knowledge, internally increases the rigidity of the system.

But for the aut-land to become more rigid as result of increased automation it is necessary that the assumed permanent-A-functions prime over other factors the functions which may neutralize those of the A-unit. For instance, labor union policies —as mentioned before—may neutralize permanent-A-functions.

At several instances throughout this chapter we have mentioned that both contingent-A-functions and non-A-functions of every kind can counteract the weight of the proper permanent-A-function. It is not only theoretically significant but indispensable for prediction and practical purposes to arrive at a clear distinction of these "counteracting" types of functions. And this, among other reasons, because the theory of permanent and contingent functions can be made essentially operational in practice. At the same time, and for purposes of practical efficiency, the theory aims at connecting with the Mertonian proposition of *functional alternatives*[32] with all its yet unrealized implications. This is not the place to inscribe the lines of the connecting bridge (the theory of functional analogues) whose consequences may prove so relevant to industrial sociology. We may advance, however, that whereas contingent functions of all kinds can be functionally substituted by its analogues—for instance, one of higher functional significance—permanent functions cannot be substituted unless the causal item itself is entirely removed, which in the case of technological change and automation is utterly impossible. This works as a special case to the postulate of *Non-Indispensability of Cultural Items,* to which Merton refers.[34] Conversely, contingent functions could be artificially produced so that highly dysfunctional permanent functions become neutralized by them. How to arrive at the scientific introduction of compensatory functions (contingent in kind) lies within the working goals but not the present possibilities of the sociologist. However, since the beginning of organized society, this kind of readjustment has been attempted, although the analytical premises upon which these readjustments could be successfully made were hardly known.

SOCIAL SYSTEM STRATIFICATION

So far we have been referring to the A-unit and the aut-land. Now the question is whether the rigidity trend signaled above will increase with the extent of automation in a given branch of industry following an index of automation over the entire social system.

We should be inclined to think so, although we must fear in mind that automation is only one factor among a constellation of factors the ultimate of which is the whole social structure. Now again, and even more so, the functions of one single factor can be compensated, neutralized, or minimized by others. The permanent A-function, however, will be increasingly relevant in

---

32. Robert K. Merton, *op. cit.,* 34.
34. *Ibid.,* 34-35.

the social system as the automation index increases. Obviously we cannot be concerned here with all potential changes in the social system, but simply with underlining trends produced by the permanent A-functions. At this time we may at the very best formulate some problems.

Considering the category of industrial workers, we can legitimately inquire into the consequences of a possible transfer of the A-segment of this category from blue collar, class solidarity and semi-dependent characteristics to white collar, organizationally loyal and bureaucratized characteristics. Is this functional or dysfunctional to the social system? If the latter, what kind of contingent functions could neutralize or minimize these trends?

Along these lines we could further ask, first, whether the blue collar, solidarity oriented and unionized segment of the industrial labor force will remain the same *if* the A-members leave its ranks, and, conversely, how will the white collar, loyalty oriented and bureaucratic category be affected by the new membership. Can this produce the crystallization of this white collar segment into some social aggregate, group or action-collective?[35] Second, which would be the manifest and latent functions of this move on labor union composition and policies, on industrial and ecomic relations, on the political structure, on the social system and finally on the stratification system? Perhaps the most important, and the final, question is how the social system is going to respond to the changes introduced by some contingent and all the permanent A-functions. Yet, with increasing automation throughout the social system, visibility of automated roles obviously grows. Furthermore, comprehension of *automation principles* will increase—if for no other reason than that the educational system will bring into production actors who will have internalized values of the automation era. If status gains are to occur for all A-roles to start with, further stages of automation should be marked by more precise differentiation *within* the collection of automated roles and actors in such roles. Specifically, a high level of automation in the total system would seem to produce sharp differentiation between highly specialized maintenance and programming roles, and all other roles in the A-system. But even sharper might be the boundary between the total labor force (now in automated processes) and the remain-

---

35. Robert K. Merton, *Ibid.*, 299-300.

ing semi-skilled and unskilled labor force in processes which do not lend themselves to automation.

Thus, the occupational hierarchy might well become *more rigid* in that upward mobility would decline under these conditions. This should in effect signify a trend away from *democratization*. Should this be one of our conclusions? The answer is yes and no.

In a social system presently characterized by discrete strata— as most underdeveloped societies are—automation will likely reinforce these strata-distinctions. It is entirely plausible, and even likely, that recruitment requirements for automated roles may be made compatible with evaluations accounting for strata-discreteness in the first place (such as caste lines). This assumes that underdeveloped countries *could* move toward automated processes without the intervening mechanized stages, a possibility which cannot be disregarded.

In a social system characterized by stratification continua rather than by fairly discrete divisions between status groups (wherein upward mobility has been operative), we think that automation will tend to decrease mobility and increase discreteness of strata, at least in *occupational* terms. But this need not mean that the stratification system as a whole need to become more rigid. Compensatory effects are possible as well as probable: if rigidity increases in the occupational hierarchy but correspondingly *decreases* in other important sub-systems, for instance, in politics, or in patterns of informal social intercourse, the *overall* impact of automation on system stratification may be *anything whatsoever*.

Our final conclusion: Assuming present and constant values of all other indices characterizing the social structure (for instance, present likelihood of skilled workers becoming congressmen, governors, or of being guests at dinner parties of industrial managers, and so on), increasing automation leads to increasing rigidity in the stratification system as a whole. But assuming adaptive processes in the whole social system—and therefore that other indices change their values—*increasing automation may have any effect whatsoever*—it may lead toward increasing equality or decreasing equality. This is not a negative conclusion for, translated into positive terms, it holds that increasing permanent functions can be at any given time controlled, neutralized or minimized either by contingent functions of the same item, or by other functions. As an oversimplification of this, remains the principle whereby one single factor (in our case the *factor auto-*

mation) cannot change the social structure without interference from all the other social factors, and unless it primes over all of them.

If it holds that automation can bring about just *any overall* change, the conclusion is inescapable that *appropriate manipulation of the relevant indices* may produce changes most compatible with the goals of the social system and the purposes of the individuals in it.

This is, however, something that social planners may want to ponder—a problem far beyond that of analysis although meaningfully attacked *only* through analysis.

JIRI NEHNEVAJSA was born in Czechoslovakia where he attended the University of Masaryk at Brno. He completed his higher education in Switzerland in the Universities of Lausanne and Zurich. In 1951 he joined the research staff of the Conservation of Human Resources Project (Columbia University). The same year he was appointed Instructor in Sociology at the University of Colorado, where he became Assistant Professor in 1953. Since 1956 he has been Assistant Professor of Sociology in Columbia University where at present he is in charge of the School of General Studies Department of Sociology. Professor Nehnevajsa is a consultant to the United States Air Force Office of Scientific Research, the System Development Corporation and the Rand Corporation. He is also associate editor of the "International Journal of Sociometry," of the journal "Group Psychotherapy" and the editor of the "Lesenbuch in Soziometrie," Opladen, Germany. He has contributed numerous articles to journals and books in sociology, social phychologie and social research.

ALBERT FRANCES, Lecturer in Sociology at Columbia University, has been working on a number of projects at the Bureau of Applied Social Research. A research associate of Prof. J. Nehnevajsa in social stratification and political sociology, he has conducted studies in the Department of Correction of the City of New York. From 1954 to 1956 he was the Director of Public Relations of Brown-Raymond-Walsh, U.S. Government prime contractor for the construction of the American Air and Naval base program in Spain. Since 1957 he has been associated with the U.S. Department of State under a service contract. Albert Frances also has the L. en D. (Law) degree from the University of Madrid, and is the author of several publications in the fields of sociology and law.

CHAPTER 32

THE TECHNOLOGICAL AND ECONOMIC PROBLEMS OF AUTOMATION IN THE U.S.S.R.

K. KLIMENKO, *Academy of Sciences of the U.S.S.R.*, and
M. RAKOVSKY, *Deputy Minister, Automation Tools
and Equipment Industries, U.S.S.R.*

At the present time, Soviet industry is having to face a number of problems of vital importance to technological progress. Thus, the directives issued by the Twentieth Congress of the Communist Party of the U.S.S.R. on the sixth five-year plan (1956-60) point out that: 'in order to ensure future technological progress, increase productivity and ease in the strain on workers, the rate of mechanization must be speeded up considerably and industrial production methods must be automated on a large scale. . . . It is essential to press on from the automation of separate units and operations to automation of the technological processes in the workshops and the creation of completely automated plants.'

The increased interest now being shown in the automation of manufacturing processes is the outcome of this policy, which the Soviet Government is steadily pursuing in the field of technology. The progressive mechanization of the various technological processes which has been going on for over twenty-five years in all branches of industry has paved the way for the transition to automated production.

As a general rule, side by side with mechanization, there is still a certain amount of manual work involved in connexion with the operation and supervision of production processes. When

---

* Reprinted from: "Social Consequences of Automation," *International Social Science Bulletin* (UNESCO), X, 1, 1958.

a whole series of operations is mechanized, manual work is replaced by work done by machines and mechanical devices, so that it becomes possible to use, for basic and auxiliary operations alike, a set of machines for the manufacture of a given volume of goods; but here too, regulation and supervision of the manufacturing process require constant manual intervention. Automation, however, means the introduction of machines, which not only carry out technological operations but also ensure the supervision of such operations, and check the quality and accuracy of the machine work, the handling and even the packing of the finished article. According to Lenin, the replacement of manual work by machines ' . . . is the essence of the progressive development of human techniques. The higher the degree of technological devolopment, the more manual labour disappears and is replaced by a series of increasingly complicated machines. . . .' Automated production makes it possible, with the help of machinery, to ensure a continuous manufacturing process, untouched by human hands, but under human supervision. This is a revolutionary change, not only in techniques, but also in technology and the organization of manufacturing processes.

Automation is no longer 'a technique of the future'—it is already a reality in Soviet industry. It provides the ideal means of perfecting manufacturing processes, raising the productivity of collective work, turning out products in plenty and, hence, of solving the most important economic and political problems involved in the building up of communism in the U.S.S.R. When account is taken of the most recent scientific achievements—particularly the use of atomic energy for industrial purposes, the employment of semi-conductors, the development of radio-electronics, and so on—and in view of the fact that the latter will make it possible, in the very near future, to effect considerable improvements in the various automatic plants, it becomes evident that the prospects of automating production are really unlimited. Under the socialist economic system, social barriers are entirely abolished, as are certain anomalies which prevent the automation of all forms of production. In the U.S.S.R., automation is the outcome of the problems inherent in building up a communist society. Since systematic technological progress meets the workers' basic needs, they are wholeheartedly behind it. In a socialist system, automation not only reduces appreciably the amount of work required, but also makes working conditions far less arduous; in fact it does away with what is often very hard physical work by exercising control over manufacturing processes; hence the need to improve the technical qualifications of those employed in automated production. For such persons, the result is greater material well-being and removal of the barriers between physical and intellectual work, which is known to be one of the most important factors facilitating transition from socialism to communism.

One very interesting result of automation is that it makes it possible to meet national manpower needs more fully and to raise the workers' standard of living. Taking basic technological operations alone, it has been estimated that, under the present five-year plan, automation will make it possible to release some two million workers. The mechanization of accounting and computation will enable from 350,000 to 400,000 persons to be released and these will be completely absorbed by other firms or branches of industry. Moreover, since there is no unemployment in the U.S.S.R., automation promises to shorten the working day.

By increasing production capacity and output, automation greatly reduces production costs and thus ensures that the capital outlay involved can be recovered quickly. Such expenditure, sometimes very high in certain branches, is recoverable within two or three years, or even in a few months, as a result of savings made in production costs. Provisional general estimates show that at the end of the sixth five-year plan, it will be possible to effect direct savings amounting to some 25 to 30 thousand million roubles on raw materials, fuels, electric power and wages alone, as the result of automation plans now in hand. In the iron and steel industry, for example, the development of automation will enable the foundries, within five years, to increase production by 2.5 million tons of pig-iron and 3 million tons of steel. In the mechanical engineering industry, automatic equipment will sharply reduce metál waste. With the same volume of manpower and raw materials, the automation of certain branches of light industry and food production will make it possible to increase the tempo and volume of production, to reduce costs substantially and to raise the quality, while ensuring that the factory workers enjoy improved and healthier conditions of work.

As things stand at present, a number of basic factors contribute to the development of automation in the U.S.S.R. Outstandingly important in this connexion is the large increase in the volume of production planned for the sixth five-year plan, as well as the fact that firms in many branches of industry have become specialized; this facilitates the creation of entirely automated mass production sectors, workshops and firms. The rapid tempo of electrification also plays an important part. The need to modernize or completely replace outdated equipment is also a telling factor in the extension of automation operations. The amount of equipment to be replaced is large, and its replacement and modernization must of necessity go hand in hand with maximum automation. All this will enable the output obtained with the equipment in service to be increased considerably and the objectives of the sixth five-year plan to be not only reached, but even exceeded.

The trend towards the automation of production in the U.S.S.R. is due to the Soviet Government's policy of releasing the maximum number of workers from unhealthy and arduous work. Working conditions are particularly hard and dangerous in the chemical industry, the petroleum industry and metallurgy. It is not surprising that these should be the branches of industry in which automation has developed most rapidly and, in some cases, reached a very advanced stage. For instance, when one of the workshops of the Efremov Synthetic Rubber Factory, where the use of mercury was found to be harmful to the workers, had been completely automated, it was possible for the whole staff to be withdrawn from the premises in question.

Striking achievements are also to be noted in the automation of some other branches of industry. For instance, all the hydraulic plants coming under the Ministry of Electric Power Plants of the U.S.S.R. have been fully automated and telemechanized to an extent corresponding to 65 per cent of their total capacity. In this ministry's large electric plants, the regulation of combustion in boiler units is automated up to 77 per cent of total steam-production capacity and 96 per cent of the stoking of the boiler units is automatically controlled. Large-scale operations are under way with a view to automating manufacturing processes in several branches of light industry and food production. Automation of new technological processes is most effective where the equipment is adapted to automatic control. Thus, in the first Soviet atomic power plant, the employment of absolutely reliable automatic controls has made it possible to regulate rapid thermo-nuclear processes. In mechanical engineering—the branch of industry in which it is most difficult to introduce automation—over 100 automatic production lines are now in operation, in addition to a great many automatic and combined machine-tools. Outstanding success has also been achieved in the automation of a number of other manufacturing branches.

It is interesting to hear what certain business representatives of capitalist countries who have visited the U.S.S.R. think of the automatic equipment being used in this country. For example, Neville L. Bean, one of the American experts who visited the industrial firms of the Soviet Union, told an editor of the review *Machinery* that the experts who had inspected the automatic production line at the State First Ball Bearing Plant had been amazed at what they had seen. 'Any technological organization in the world,' said Mr. Bean, 'would admire such a perfectly designed production line.'

Technological advancement calls for automatic control of manufacturing processes and necessitates the creation and operation of various automatic instruments and devices, machines, sets of machines, automated sections and fully automated firms. Knowledge of the process of automation has become a *sine qua non* of technical efficiency. The invention of automatic remote control devices and means of protection against accidents has opened up entirely new vistas and has made it possible to resort to a number of technological processes formerly considered out of the question. It is, in fact, impossible to do certain things by hand at very high or low pressure, at high temperatures or at very great speeds. Modern aviation—and particularly jet-propulsion—the manufacture of synthetic products, the catalytic processes of the chemical industry, steam production at high parameters, research in the field of nuclear physics and, lastly, work on the fuller use of atomic energy in industry have become possible only through the introduction of various automatic devices. Continuous processes in the iron and steel industry, chemistry, and petroleum industry and power plants require automatic control, because they are based on very strict adherence to a technological system which can be controlled only by intricate automatic machines.

That is why efforts in the U.S.S.R. to apply a new technique and to make technological progress in industry are primarily efforts to achieve complete automation of production. The solution of problems raised by such automation is bound up with the introduction of a wide range of standardized automatic measuring instruments and control devices. A big industrial firm needs a single, central control board which should be at some distance (anything up to several hundred yards) from the workshops and units; this means that there must be a considerable number of telemechanized devices; in the final count, it is remote control, combined with the use of regulatory machines and devices that determine the success of automation.

In the recent years, Soviet industry has succeeded in producing a number of new types of specialized equipment. For 1956, the production of the factories of the Ministry for the Precision Instruments and Automation Devices Industry alone is estimated at over 28,000 electronic potentiometers, nearly 300,000 pyrometers, nearly 2,500,000 apparatuses for the regulation, recording and measurement of gas pressure and rarefaction, and nearly 60,000 different models of consumption meters and output regulators. Work is in progress on the production of radioactivity level meters and densimeters, vacuum gauges and radioactive and ionization pressure gauges, as well as other apparatuses of the very latest type. Devices are being made which use very high supersonic vibrations, radioactive rays, thermo-electric effects, the ultra-violet zone of the spectrum, and the photocolorimetric principle. Particularly difficult problems are raised by the construction of computing machines and devices which henceforth enable various technological processes or production units to be controlled automatically. Computers, for instance, will be used in mechanical engineering and particularly in the programme setting of machine-tools. In the most up-to-date examples of these machines, the original information is recorded on magnetic tape (or some other 'memorizing' system), from which the mathematically computed order is transmitted to the operating machines. Many metal presses and machine-tools, working to a set programme, will doubtless be widely used in the very near future in Soviet industry. The further technical improvement of the design of electronic machines will largely depend on the introduction of high-quality semi-conductors, the utilization of magnetic effects and radio engineering and also on research in physics and mathematics.

Owing to the introduction of automatic devices, machines can now replace, not only manual labour, but also some types of intellectual work. This branch of science—known as 'cybernetics'—is connected with the use of various electronic computers. The Soviet BESM machine performs 8,000 arithmetical operations per second, and the new machine at present being designed in the U.S.S.R. will be able to perform 20,000 operations.

Apart from the various computation problems which are particularly important in scientific research, electronic machines will open up enormous possibilities for the direct control of industrial operations on the basis of programmes set in advance, and also for the translation of foreign languages, bibliographical work, traffic control, and so on.

Integrated automatic systems are planned, from the outset, on the lines most conducive to subsequent development and maximum efficiency. Modern technology provides many opportunities in this respect. Thus, all technological problems involving closely allied principles can, even though they may arise in the most widely differing branches of industry, be solved by similar methods. Through such standardization of the simplest automatic devices, the latter can be combined to form the most diverse and intricate automatic units. Industrial materials intended for a great variety of purposes and differing widely from the technological point of view have many common static and dynamic properties. Thus, it is possible for open-hearth furnaces, diffusion apparatuses or saturators in the sugar industry, tunnel furnaces in the silicate industry, coke oven and bakers' ovens to be classified in the same group. By studying the different production materials and classifying them scientifically, it will be possible to place many of them in a single group, which will make it far easier to develop efficient and reliable designs for integrated automation. Many other technological problems will have to be solved in the near future. Theoretical research has been undertaken in order to provide a scientific basis for the principles and methods of rational designing of automatic systems and, among other subjects, the theory of automatic regulation and control is also being studied. It cannot escape notice that often, in inventing effective and interesting automatic devices, certain questions connected with technology and the organization of production are completely overlooked. However, Soviet experience in operating many automatic production lines, and particularly that of a well-known automobile piston factory, has shown that it is not enough to establish a system of automatic machines; it is necessary to examine thoroughly and change radically the whole technological process and the order of work on the production line, to train qualified staff beforehand, to provide for the control of production operations and also for their complete servicing when required. In other words, the automation of production, as it is understood in the U.S.S.R., does not consist only in constructing various types of automatic machinery, but also includes the preparation of projects at a level suitable to the new equipment and the production system. At the same time, evidence of the economic soundness of the new automatic devices must be supplied and the most rational variant singled out.

Owing to its technological nature, automation in capitalist countries has certain peculiar features, which are an illustration of the competition between individual firms and the special ways in which the various branches of industry develop. Moreover, automation was first introduced in capitalist countries some twenty-five to thirty years ago and, at the present time, much of the automation equipment can be considered as out of date. Things are quite different in the U.S.S.R. Starting with the latest equipment, the Soviet Union can bypass whole stages in the development of automation. It would be a mistake for it to take account only of foreign trends and forms of automation. In particular, as has already been said, the U.S.S.R. should aim primarily at introducing standardized automatic units, combinations of which would provide all branches of the national economy with reliable, inexpensive and relatively simple automated systems. This, of course, is feasible only where there is general planning of industrial development.

The automation of industry in the U.S.S.R. calls for extremely well regulated co-operation between scientific research institutes, designing offices, bodies responsible for studying projects and all ministries. Specialists working in universities and institutes have a very important part to play; their contribution to the development of automation will make it possible to establish industry on a firmer scientific basis and to raise the standard of training for specialists in the automation of production.

Naturally, in such collective work, the activities of the various people taking part must be closely co-ordinated. The task is entrusted to a special government agency—Gostekhnika, the State Technical Organization—which is responsible for directing all work connected with automation, adopting a single policy in the field of technology and preventing any wasteful duplication of work.

In view of the outstanding importance of automation to the national economy, all workers in Soviet industry and science are faced with a number of major economic problems. In the early stages, the main attention was centered on technical problems, but now that a certain amount of success has crowned the efforts made to develop different types of automation and the stage of its general application to the most diverse branches of industry has been reached, economic considerations are coming to the fore.

When once the most arduous, dangerous and unhealthy production sectors in the U.S.S.R. had been partially automated, the question of in what order the automation process was to be further applied naturally arose. The question cannot be answered except on the basis of a detailed economic analysis. It is impossible, for instance, to disregard the fact that there are in the U.S.S.R. several million persons engaged on loading, unloading and transport operations. Mechanization and automation would mean releasing 75-80 per cent of these workers. In the mechanical engineering sector, nearly two million workers are engaged on manual operations. The light industries and the food industries at present employ nearly four million persons, a figure which could easily be reduced by 25-30 per cent through partial automation, and by 60-70 per cent integrated automation. When weighing up the possibilities of automating any particular branch of industry, account must of course be taken to the technical level and degree of mechanization of the branch concerned, as well as of the extent to which the technological processes can be adapted to the requirements of automation.

The preliminary data obtained by extrapolation of the results of partial automation of certain technological processes shows that the period required for recovery of the capital outlay varies from eight months in some branches of industry to two years in others. The planning services of the Ministry for the Precision Instruments and Automation Devices Industry which have for several years past, been working out plans for partial automation, have reached the conclusion, that the period required for recovery of the capital outlay varies between ten and fifteen months for most of the manufacturing processes in the chemical industry; between eighteen and twenty-two months in metallurgy; between ten and twenty-five months in the non-ferrous metals industry; between nine and sixteen months in the power production units (mainly in small electric power plants and thermal plants) ; and from fifteen to twenty months in the food industry.

For instance, the integrated automation of the Urals Aluminum Plant would cost close to 27 million roubles, while the annual savings thus effected would be about 20 million roubles. The integrated automation of the Dniepropetrovsk Electrode Factory would cost about 4.5 million roubles, and the resulting annual savings would be about 3 million roubles. The outlay for the installation of automation in a number of electric plants—the Bryansk, Shakhty, Vladimir and Kiev plants, and No. 11 thermal plant of Mossenergo (Moscow Power System)—was recovered in periods ranging from six months to three years. Automation costs of synthetic rubber and ammonium nitrates production are recovered within eighteen months to two and a half years and those of synthetic fibre production in as little as seven to eight months.

Another interesting point is that the annual economic yield deriving from the automative equipment is equivalent to about double its initial cost—which is a clear proof of the advantages of increasing the production of automatic equipment.

It is very important to distinguish between capital investment in scientific research work on automation and capital expenses that enter directly into the operating costs of the automatic plant and to keep the two separate. It is sometimes said that certain automatic production lines are expensive, but in fact much of the cost is due to the expenses incurred by the scientific and experimental work involved. For instance, the cost of equipment used in atomic energy production and more especially in sectors using computing machines is very high and is not immediately recoverable. It is more expensive to produce electricity by atomic power plant than by ordinary power plant; but this is due mainly to the outlay on scientific and experimental work, which is recovered gradually, as production expands. It is for this reason that Soviet economists consider that the cost of research and experimental work should be reckoned separately, and not be included in the production cost of the first automatic plant. When the cost of any new automatic production line is calculated, all the expenses involved in the project should not be included since, once the first tests have been made, many similar units can be set up on the same pattern.

Capital investments connected with the running costs of the automatic equipment are quite a different matter; these are naturally recoverable very quickly as a result of the rapid fall in production costs. Soviet economists propose that approximate dates should be set for the recovery of capital outlay for every different branch of industry and type of equipment; and that, over that period, the said equipment should be considered to be paying its way (5 to 6 years, for instance, for machine tools, and a longer period for power plant, etc.). These dates should allow for the economic aging of the equipment ('psychological' depreciation) as well as for technical depreciation.

The time-limits for the recovery of capital outlay affect the assessment of the economic yield of the automatic machinery. Experience shows that the automation of production in any form gives conclusive results. The best Soviet open-hearth furnaces, for instance, with partial automation and intensified technological operations, produce 10.8 tons of steel per square metre of hearth a day, whereas the corresponding figure for the U.S.A. is only 7.8 tons. Automation of the open-hearth furnaces in the Magnitogorsk metallurgical combine has led to lowering the fuel consumption by 5.8 per cent and the consumption of refractory materials by 13 per cent, with a 3 per cent increase in furnace productivity. In the case of blast furnaces, the use of humidified blast with automatic control and regulation of the parameters has made it possible to raise the furnace productivity by 15.8 per cent, and cut coke consumption by 5 to 8 per cent. The amount saved in the synthetic rubber industry by automating the contact furnaces is about 10 million roubles a year; while the automation of the synthetic alcohol plant at the Sumgait synthetic rubber factory has increased the daily output of alcohol by more than 70 per cent, and halved the number of workers.

Frequently, the introduction of automation, besides cutting the cost of labour, materials, fuel and consequently, also production costs, increases output and reduces capital outlay as well. Thus, the automation of a blast furnace costs 500,000 roubles, and puts the output up by 8 per cent. Hence, the result of automating twelve blast furnaces, at a cost of 6 million roubles, is equivalent to that of building a new blast furnace at a cost of about 100 million roubles.

In the production of artificial fibre, automation has made it possible to reduce raw material costs by 7 per cent, labour requirements by 26 per cent and production costs by 10 per cent, while at the same time improving quality and variety considerably. In the chemical industry, automation besides increasing productivity and cutting production costs, stabilized the technological processes, improved the quality of production, reduced breakdowns, lengthened the useful life of the equipment and appreciably improved working conditions. The following figures throw interesting light on the effects of the introduction of automation in a workshop producing ammonium nitrate: the staff was reduced by 35 per cent, wastage of ammonium by 9 per cent, and losses of nitric acid by 42 per cent; with these savings, the capital outlay on automation could be recovered in eighteen months to two years. The automation of the carbonization columns at the Donetz soda factory raised the annual production of soda by 42,000 tons, and the automation costs were recovered in six months. In the sugar industry, integrated automation produced a 7 per cent rise in the output of granulated sugar, with a 23 per cent reduction of manpower; in addition, labour productivity rose by 40 per cent, and the cost of automation was recovered in less than two seasons.

Automation in mechanical engineering produces particularly striking economic results. Thus, multi-spindle automatic lathes are twenty times more productive than ordinary universal lathes, and the extra cost of installing the automatic lathes can be recovered within two years from savings on wages alone. In many cases, moreover, the automatic multi-spindle lathes, the machine-tool units and the combined machine tools are cheaper to buy than the ordinary lathes. Automation is particularly effective when accompanied by the introduction of more advanced technological methods. For example, production can be stepped up scores or even hundreds of times in the manufacture of bolts by using cool-drawing automatic machines instead of lathes. On some of the best automated lines, productivity is increased from ten to thirty times with a 20-30 per cent reduction of labour costs.

The above data are, however, not complete, and do not give any idea of the real economic impact of automation. In this connexion, the working out of a sound scientific method for assessment of the economic advantages of automation now represents one of the major technical and economic problems of the national economy of the U.S.S.R. A precise definition of the indices, criteria and methods to be used for assessment of the effectiveness of automation must be found before the future general trends of the automation process can be mapped out. In practice, the most important index of the economic efficacy of automation is the reduction of production costs; but it is not the only one.

The main factors to be taken into account in determining

the efficacy of automation are labour productivity, production costs and the speed of recovery of the capital outlay. But there are a number of others, as automation also results in an appreciably lower accident rate, a rise in the cultural level of the workers, and an improvement of health and hygienic conditions, all of which most certainly cannot be ignored. Another advantage of automation is to reduce faulty workmanship, whether due to defects in the machinery, equipment and materials worked, or to unskilled labour or breaches of technological discipline. In addition, automation usually improves the quality of output, speeds up the production cycle and makes for better use of production capacity.

How does automation affect the work and lives of Soviet workers? Experience to date vindicates the theory of Soviet economists, which is that the automation of production processes will require not less skilled workers but on the contrary, in the majority of cases, more skilled ones, especially when the degree of automation is high. And this is an essential prerequisite for the introduction and effective utilization of new technical methods. Thus automation emphasizes the fact that the raising of the level of professional training is a basic condition without which the application of new, progressive technological processes cannot be extended. In a socialist economy, where both unemployment and over-production are unknown, automation helps to create conditions such that the cultural and technical level of the workers can gradually be raised to that of technicians and engineers. Automation, in fact, needs workers for technical supervision of the machinery, regulating and setting it, and doing repairs—all operations calling for a high degree of technical skill.

In the Soviet Union, the problem of the training of highly qualified personnel is a matter of national importance, and it is being solved at the national level. In the first place, universal and free seven-year schooling has existed for a long time. Under the present five-year plan (1956-60), the principle of universal and compulsory secondary (10-year) educaton is to be applied in the towns and rural centres. During this five-year period, the number of pupils graduating from the tenth classes of the secondary schools will be 6,300,000, or more than double the number for the previous five-year period. Thus the Soviet labour force will be steadily swelled by the addition of graduates of secondary and technical schools. Already in 1956, many new workers with a full ten-year secondary education took up jobs in industry; and there is no doubt that these new recruits possess the maximum potential ability to acquire intricate skills with rapidity and assimilate new technical processes, such as those connected with automation. A further contributing factor is that the secondary school syllabuses are becoming increasingly technical so that the schoolchildren receive a sound grounding in various technical subjects, besides some experience in manual work.

In addition to the general secondary schools there is also, in the Soviet Union, a vast network of secondary training establishments for young workers; these schools were set up in 1944 for the benefit of young people who, having for one reason or another been unable to obtain a general education at the normal time, now wish to improve their educational standard, without leaving their factory jobs. In 1956 there were 6,637 such schools in operation in the U.S.S.R., catering to about 1,400,000 young workers.

There is also in the Soviet Union a number of vocational training schools, mining schools and railway workers' schools for training skilled workers. The students at these establishments are entirely State-supported. During the fifth five-year plan (1951-56), these schools and the apprenticeship training centres for industry, the building trade and transport turned out 1,736,000 skilled workers.

In conclusion, mention must be made of the vast network of courses and schools organized by Soviet enterprises all over the country. They include advanced vocational training courses, schools teaching the latest production methods, new techniques, etc. All forms of industrial training given in Soviet firms are free of charge. The question arises whether, with so vast a training system, automation may not lead to a surplus of qualified personnel. However, one of the advantages of the socialist economic system is that this danger, like that of unemployment, is precluded.

Automation raises the duties of workers to highly technical supervision over a complicated system of automatic machinery, and gives them scope for important creative work. Under socialist conditions of production, ever broader prospects are unfolding, and the way is open for workers to graduate to technical and engineering work of a scientific and cultural level.

In a socialist society, the need for workers in these two fields is growing steadily. How then is the training of these workers organized in the U.S.S.R.? In 1956, a total of 1,867,000 students were enrolled in Soviet higher educational establishments, almost fifteen times as many as in the corresponding establishments in Tsarist Russia, and 2.3 times as many as in 1940. There were also, in 1956, over 1,960,000 pupils in special secondary schools in the U.S.S.R, or thirty-six times as many as in Tsarist Russia, and twice as many as in 1940. The extension of higher and secondary specialized education in the U.S.S.R. is facilitated by the fact that it is accessible to all. The Soviet State is extending the network of these establishments; since 1957, all instruction has been made absolutely free of charge (before that fees were very low) ; all students making satisfactory progress receive allowances, and there are hostels for those needing them. It is essential to add that there is also, in the U.S.S.R., an extensive system of secondary and higher special education by correspondence; and that many enterprises have organized special evening classes with the help of the higher educational establishments and technical schools. This enables workers to train as engineers, technicians, scientists or cultural workers while continuing in their regular jobs.

Thus the labour force is replenished by young people with a secondary education, and new engineers, technicians, scientists and cultural workers are recruited from former workers by a uniform natural process in accordance with the principles of an over-all plan which covers all branches of the socialist economic and cultural system, and even regulates the distribution of labour as between the various branches of the national economy.

Socialist production, which is based on the public ownership of the tools and means of production and which is subordinated to an over-all State plan, can be automated without harming the interests of the workers, because it serves the interests of the people as a whole, and not merely those of individual employers, as in systems where the equipment and means of production are privately owned.

It is precisely by virtue of the new social pattern of the U.S.S.R. that the automation of production, which constitutes an invaluable means of raising productivity and increasing the rate of industrial development, also leads to the betterment of working conditions by raising the material and cultural standards of the working people. In the U.S.S.R., higher productivity and increased output lead directly to higher wages, cheaper manufactured goods and foodstuffs, and increased demand and expansion of the capacity of the home market.

Thus, as a result of the rise in real wages and frequent reductions in the prices of consumer goods on the state market—which had immediate repercussions on the collective farm market prices—the real wages of factory and office workers employed in the U.S.S.R. were, in 1956, 75 per cent higher than in 1940 (before the war); while the real wages of industrial workers were about 90 per cent higher. Even by 1954, the state retail prices were 2.3 times lower than in 1947 (the prices of foodstuffs 2.6 times lower, and those of manufactured goods, 1.6 times lower).

Thanks to the increase of labour productivity and of output, for which automation is largely responsible, the Soviet State is able, as befits a workers' state, to shorten the working day without reducing wages. It is in accordance with this principle that, during the sixth five-year plan period (1956-60), the working day is to be shortened to seven hours and even, in some branches of industry, to six. The changeover to the shorter working day began in 1956, and has not affected wages in any way. Again, the socialist economic system, based on planned, balanced development in all branches of production, paves the way for unlimited expansion of production. It is no mere chance that, in the U.S.S.R., the employment figures are rising steadily, despite the steadily increasing pace of automation. In 1955, for instance, the main total of persons employed in industry was 48,400,000, as against 38,900,000 in 1950.

Under the socialist production system, where all the workers have a personal interest in results, automation of manufacturing processes makes work more systematic and more creative, improves working conditions, and raises the general cultural level. It also cuts down considerably the number of industrial accidents. It is true that some of the emulsions and oils used at present cause skin diseases among the workers. The changeover to different cooling methods is complicated by the fact that it is difficult to separate completely the lubricating system of automatic machines from the cooling liquid. But, as the experience of the Kiev automatic machine tool factory shows, there are good prospects of being able to use an impoved emulsion as a cooling agent; and further work in this field will doubtless produce useful results.

Such, in broad outline, are the social factors involved in automating production in the U.S.S.R. socialist economy. A large number of studies are at present being published in the U.S.S.R. on the economic problems of automation and on the organization of research into many complex aspects of this subject, which is regarded as an extremely important part of the general problem of the development of industrial and economic research.

AUTOMATION DICTIONARY*

A

AC. A suffix meaning "automatic computer" as in ORDVAC, EDVAC, ENIAC, etc.

ACCESS, RANDOM. Access to storage under conditions in which the next position from which information is to be obtained is in no way dependent on the previous one.

ACCESS TIME. (1) The time interval between the instant at which information is: (a) called for from storage and the instant at which delivery is completed, i.e., the read time; or (b) ready for storage and the instant at which storage is completed, i.e., the write time. (2) the latency plus the word-time.

ACCUMULATOR. The zero-access register (and associated equipment) in the arithmetic unit in which are formed sums and other arithmetical and logical results; a unit in a digital computer where numbers are totaled, i.e., accumulated. Often the accumulator stores one quantity and upon receipt of any second quantity, it forms and stores the sum of the first and second quantities.

ACCURACY. Freedom from error. Accuracy contrasts with precision; e.g., a four-place table, correctly computed, is accurate; a six-place table containing an error is more precise, but not accurate.

ACTION. In the automatic field it refers specifically to control action. It is that which is done to regulate the controlling element in a process or operation. The action ranges from the

---

The dictionary represents the combined efforts of C. L. Peterson, Vice-President, Brown Instruments Division, Minneapolis Honeywell Regulator Company (as it appeared in Advanced Management, July, 1956, published by the Society for the Advancement of Management, 74 Fifth Ave., New York 11, N. Y.) and the Automation Committee and the Technical Division of the National Office Management Association, Willow Grove, Pa., which has published a glossary of Automation terms.

simple familiar "on" and "off" movements to not so familiar derivative and rate types of action.

ADDER. A device capable of forming the sum of two or more quantities.

ADDRESS. A label such as an integer or other set of characters which identifies a register, location, or device in which information is stored.

ADDRESS, ABSOLUTE. The label (s) assigned by the machine designer to a particular storage location; specific address.

ADDRESS, RELATIVE. A label used to identify a word in a routine or subroutine with respect to its position in that routine or subroutine. Relative addresses are translated into absolute addresses by the addition of some specific "reference" address, usually that at which the first word of the routine is stored, e.g., if a relative address instruction specifies an address n and the address of the first word of the routine is k, then the absolute address is n plus k.

ADDRESS, SYMBOLIC. A label chosen to identify a particular word, function or other information in a routine, independent of the location of the information within the routine; floating address.

ALLOCATE. To assign storage locations to the main routines and subroutines, thereby fixing the absolute values of any symbolic addresses. In some cases allocation may require segmentation.

AMPLIFIER, BUFFER. An amplifier used to isolate the output of any device, e.g., oscillator, from the effects produced by changes in voltage or loading in subsequent circuits.

AMPLIFIER, TORQUE. A device which produces an output turning moment in proportion to the input moment, wherein the output moment and associated power is supplied by the device, and the device requires an input moment and power smaller than the output moment and power.

ANALOG. The representation of numerical quantities by means of physical variables, e.g., translation, rotation, voltage, resistance; contrasted with "digital."

ANALYZER, DIFFERENTIAL. An analog computer designed and used primarily for solving many types of differential equations.

AND. A logical operator which has the property such that if P and Q are two statements, then the statement "P and Q" is true or false precisely according to the following table of possible combinations:

P	Q	P and Q
false	false	false
false	true	false
true	false	false
true	true	true

The "and" operator is often represented by a centered dot (.), or by no sign as in P. Q or PQ.

AND-GATE. A signal circuit with two or more input wires which has the property that the output wire gives a signal only if all input wires receive co-incident signals.

AQUADAG. A graphite coating on the inside of certain cathode ray tubes for collecting secondary electrons emitted by the screen.

ARITHMETIC UNIT. That portion of the hardware of an automatic computer in which arithmetical and logical operations are performed.

ASSEMBLE. To integrate subroutines (supplied, selected, or generated) into the main routine, by adapting, or specializing to the task at hand by means of preset parameters, by adapting, or changing relative and symbolic addresses to absolute form, or incorporating, or placing in storage.

ATTENUATE. To obtain a fractional part or reduce in amplitude an action or signal. Measurement may be made as percentage, per unit, or in decibels, which is 10 times $\log_{10}$ of power ratio; contrasted with amplify.

AUTOMATIC CONTROLLER. A device or instrument which is capable of measuring and regulating anything from color to chemical make-up.

AUTOMATION. The modern-day engineer's word for the state of being automatic. Once referred to machine tool applications, but has come to mean the act or method of making a manufacturing—or processing—system partially or fully automatic.

AVAILABLE-TIME, MACHINE. Time during which a computer has the power turned on, is not under maintenance, and is known or believed to be operating correctly.

AZIMUTH. The angular measurement in an horizontal plane

and in a clockwise direction from a specific reference direction, usually a form of North, i.e., true azimuth is measured from true north, grid-azimuth from grid north or thrust or base line.

B

BAND. A group of recording tracks on a magnetic drum.

BASE. A number base; a quantity used implicitly to define some system of representing numbers by positional notation; radix.

BEAM, HOLDING. A diffused beam of electrons used for regenerating the charges stored on the screen of a cathode ray storage tube.

BIAS. The average D.C. voltage maintained between the cathode and control grid of a vacuum tube.

BINARY. A characteristic or property involving a selection, choice or condition in which there are but two possible alternatives.

BINARY, NUMBER. A single digit or group of characters or symbols representing the total, aggregate or amount of units utilizing the base two; usually using only "0" and "1" digits to express quantity.

BIQUINARY. A form of notation utilizing a mixed base; see Notation, Biquinary.

BIT. See Digit, Binary, a contraction of binary digit.

BLOCK. A group of words considered or transported as a unit; an item; a message; in flow charts, an assembly of boxes, each box representing a logical unit of programming, usually requiring transfer to and from the high speed storage; in circuitry, a group of electrical circuits performing a specific function, as in a "block" diagram, in which unit, e.g. oscillator, is represented as a block (symbol).

BLOCK, INPUT. A section of internal storage of a computer reserved for the receiving and processing of input information.

BOOTSTRAP. The special coded instructions at the beginning of an input tape, together with one or two instructions inserted by switches or buttons into the computer; in circuitry, a positive feedback or regenerative circuit.

BRANCH. A conditional jump.

BREAKPOINT. A point in a routine at which the computer may, under the control of a manually-set switch, be stopped for a visual check of progress.

BUFFER. An isolating circuit used to avoid any reaction of a driven circuit upon the corresponding driving circuit, e.g. a circuit having an output and a multiplicity of inputs so designed that the output is energized whenever one or more inputs are energized. Thus, a buffer performs the circuit function which is equivalent to the logical "OR."

BUS. A path over which information is transferred; a trunk; an electrical conductor, channel or line; a heavy wire or heavy lead.

C

CABLE. An electrical conductor designed to provide common electric potential between two or more points.

CABLE, COAXIAL. A transmission line consisting of two conductors concentric with and insulated from each other.

CALL-NUMBER. A set of characters identifying a subroutine and containing information concerning parameters to be inserted in the subroutine, information to be used in generating the subroutine, or information related to the operands; a call-word when exactly one word is filled.

CAPACITANCE. The property of two or more bodies which enables them to store electrical energy in an electrostatic field between the bodies.

CAPACITY. No mystery, simply the measure of the maximum amount of a material (or energy) which can be stored.

CARD. Heavy, stiff paper of uniform size and shape, adapted for being punched in an intelligent array of holes. The punched holes are sensed electrically by wire brushes or mechanically by metal feelers. One standard card is $7\frac{3}{8}$ inches long by $3\frac{1}{4}$ inches wide and contains 20 columns in each of which any one of 12 positions may be punched.

CARRIAGE, AUTOMATIC. A typewriting paper guiding or holding device which is automatically controlled by information and program so as to feed forms or continuous paper to a set of impression keys and to provide the necessary space, skip, eject, tabulate, or performing operations.

CARRY. (1) A signal, or expression, produced as a result of an arithmetic operation on one digit place of two or more numbers expressed in Positional Notation and transferred to the next higher place for processing there.

(2) Usually a signal or expression as defined in (1) above

which arises in adding when the sum of two digits in the same digit equals or exceeds the Base of the number system in use. If a carry into a digit place will result in a carry out of the same digit place, and if the normal adding circuit is bypassed when generating this new carry, it is called a High-Speed Carry, or Standing-on-Nines Carry. If the normal adding circuit is used in such a case, the carry is called a Cascaded Carry. If a carry resulting from the addition of carries is not allowed to propagate (e.g., when forming the partial product in one step of a multiplication process), the process is called a Partial Carry. If it is allowed to propagate, the process is called a Complete Carry. If a carry generated in the most significant digit place is sent directly to the least significant place (e.g., when adding two negative numbers using the nine complements) that carry is called an End-Around Carry.

(3) In direct subtraction, a signal or expression as defined in (1) above which arises when the difference between the digits is less than zero. Such a carry is frequently called a Borrow.

(4) The action of forwarding a carry.

(5) The command directing a carry to be forwarded.

CASCADE CONTROL. An automatic control system in which the control units, linked in chain fashion, feed into one another in succession, each regulating the operation of the next in line. (Sometimes called "piggy-back" control.)

CATHODE-FOLLOWER. A vacuum-tube circuit in which the input signal is applied to the control grid and the output is taken from the cathode, possessing high input impedance and low output impedance characteristics.

CELL. Storage for one unit of information, usually one character or one word; usually a location specified by whole or part of the address and possessed of the faculty of store; specific terms as column, field, location and block are preferable when appropriate.

CELL, BINARY. An element that can have one or the other of two stable states or conditions and thus can store a unit of information.

CHANNEL. A path along which information, particularly a series of digits or characters, may flow. In storage which is serial by character and parallel by bit (e.g., a magnetic tape or drum in some coded-decimal computers, a channel comprises several parallel tracks. In a circulating storage a channel is one recirculating path containing a fixed number of words stored serially by word.

CHARACTER. One of a set of elementary symbols such as those corresponding to the keys on a typewriter. The symbols may include the decimal digits 0 through 9, the letters A through Z, punctuation marks, operation symbols, and any other single symbols which a computer may read, store, or write; a pulse code representation of such a symbol.

CHECK. A means of verification of information during or after an operation.

CHECK, BUILT-IN OR AUTOMATIC. Any provision constructed in hardware for verifying the accuracy of information transmitted, manipulated, or stored by any unit or device in a computer. Extent of automatic checking is the relative proportion of machine processes which are checked or the relative proportion of machine hardware devoted to checking.

CHECK, DUPLICATION. A check which requires that the results of two independent performances (either concurrently on duplicate equipment or at a later time on the same equipment) of the same operation be identical.

CHECK, FORBIDDEN-COMBINATION. A Check (usually an Automatic Check) which tests for the occurrence of a nonpermissible code expression. A self-checking code (or error-detecting code) uses code expressions such that one (or more) error(s) in a code expression produces forbidden combination. A parity check makes use of a self-checking code employing binary digits in which the total number of 1's (or 0's) in each permissible code expression is always even or always odd. A check may be made for either even parity or odd parity. A redundancy check employs a self-checking code which makes use of redundant digits called check digits.

CHECK, MATHEMATICAL OR ARITHMETICAL. A check making use of mathematical identities or other properties, frequently with some degree of discrepancy being acceptable; e.g., checking multiplication by verifying that $A \cdot B = B \cdot A$, checking a tabulated function by differencing, etc.

CHECK, MODULO N. A form of check digits, such that the number of ones in each number A operated upon is compared with a check number B, carried along with A and equal to the remainder of A when divided by N, e.g., in a "modulo 4 check," the check number will be 0, 1, 2, or 3 and the remainder of A when divided by 4 must equal the reported check number B, or else an error or malfunction has occurred; a method of verification by congruences, e.g. casting out nines.

CHECK, ODD-EVEN. A check system in which a one or zero is carried along in a word depending on whether the total number of ones (or zeros) in a word is odd or even.

CHECK, PARITY. A summation check in which the binary digits, in a character or word, are added (modulo 2) and the sum checked against a single, previously computed parity digit; i.e., a check which tests whether the number of ones is odd or even.

CHECK, PROGRAMMED. A system of determining the correct program and machine functioning either by running a sample problem with similar programming and known answer, including mathematical or logical checks such as comparing A times B with B times A and usually where reliance is placed on a high probability of correctness rather than built-in error-detection circuits or by building a checking system into the actual program being run and utilized for checking during the actual running of the problem.

CHECK, REDUNDANT. A check which uses extra digits, short of complete duplication, to help detect malfunctions and mistakes.

CHECK, SUMMATION. A redundant check in which groups of digits are summed, usually without regard for overflow, and that sum checked against a previously computed sum to verify accuracy.

CHECK, TRANSFER. Verification of transmitted information by temporary storing, re-transmitting and comparing.

CHECK, TWIN. A continuous duplication check achieved by duplication of hardware and automatic comparison.

CHECKING, MARGINAL. A system or method of determining computer circuit weaknesses and incipient malfunctions by varying the power applied to various circuits, usually by a lowering of the D.C. supply or filament voltages.

CLAMPING-CIRCUIT. A circuit which maintains either amplitude extreme of a waveform at a given voltage level, or potential.

CLEAR. To replace all information in a storage device by ones or zeros as expressed in the number system employed.

CLOCK, MASTER. The source of standard signals required for sequencing computer operation, usually consisting of a timing pulse generator, a cycling unit and sets of special pulses that occur at given intervals of time. Usually in synchronous machines, the basic frequency utilized is the clocking pulse.

CLOSED LOOP. A family of automatic control units linked together with a process to form an endless chain. The effects of control action are constantly measured so that if the process goes off the beam, the control units pitch in to bring it back into line.

CLOSED-SHOP. This is intended to mean that mode of computing machine support wherein the applied programs and utility routines are written by members of a specialized group whose only professional concern is the use of computers.

CODE. A system of symbols and their use in representing rules for handling the flow or processing of information; to actually prepare problems for solution on a specific computer.

CODE, COMPUTER. The code representing the operations built into the hardware of the computer.

CODE, EXCESS-THREE. A coded decimal notation for decimal digits which represents each decimal digit as the corresponding binary number plus three, e.g., the decimal digits 0, 1, 7, 9 are represented as 0011, 0100, 1010, 1100, respectively. In this notation, the nines complement of the decimal digit is equal to the ones complement of the corresponding four binary digits.

CODE, INSTRUCTION. An artificial language for describing or expressing the instructions which can be carried out by a digital computer. In automatically sequenced computers, the instruction code is used when describing or expressing sequences of instruction, and each instruction word usually contains a part specifying the operation to be performed and one or more addresses which identify a particular location in storage. Sometimes an address part of an instruction is not intended to specify a location in storage but is used for some other purpose.

If more than one address is used, the code is called a multiple-address code.

CODE, INTERPRETER. A code which is acceptable to an interpretive routine.

CODE, MULTIPLE-ADDRESS. An instruction or code in which more than one address or storage location is utilized. In a typical instruction of a Four-Address Code the addresses specify the location of two operands, the destination of the result, and the location of the next instruction in the sequence. In a typical Three-Address Code, the fourth address specifying the location of the next instruction is dispensed with, the instructions are taken from storage in a preassigned order. In a typical Two-Address Code, the addresses may specify the locations of the operands. The results may be placed at one of the addresses or the destination of the results may be specified by another instruction.

CODE, OPERATIONAL. That part of an instruction which designates the operation to be performed.

CODING. The list, in computer code or in pseudo-code, of the successive computer operations required to solve a given problem.

CODING, ABSOLUTE, RELATIVE or SYMBOLIC. Coding in which one uses absolute, relative, or symbolic addresses, respectively; coding in which all addresses refer to an arbitrarily selected position, or in which all addresses are represented symbolically.

CODING, ALPHABETIC. A system of abbreviation used in preparing information for input into a computer such that information is reported in the form of letters, e.g., New York as NY, carriage return as CN, etc.

CODING, AUTOMATIC. Any technique in which a computer is used to help bridge the gap between some "easiest" form, intellectually and manually, of describing the steps to be followed in solving a given problem and some "most efficient" final coding of the same problem for a given computer; two basic forms are Routine, compilation and Routine, interpretation.

CODING, NUMERIC. A system of abbreviation used in the preparation of information for machine acceptance by reducing all information to numerical quantities; in contrast to alphabetic coding.

COLLATE. To combine two or more similarly ordered sets of items to produce another ordered set composed of information from the original sets. Both the number of items and the size of the individual items in the resulting set may differ from those of either of the original sets and of their sums, sequence 23, 24, 48 may be collated into 12, 23, 24, 29, 42, 48; to combine two or more sequences of items according to a prescribed rule such that all items appear in the final sequence.

COLLATOR. A machine which has two card feeds, four card pockets and three stations at which a card may be compared or sequenced with regard to other cards so as to select a pocket in which it is to be placed, e.g., the machine is suitable for matching detail cards with master cards, merging cards in proper sequence, etc.

COLUMN. One of the character or digit positions in a positional notation representation of a unit of information, columns are usually numbered from right to left column, zero being the right-most column if there is no point, or the column immediately to the left of the point if there is one; a position or place in a number in which the position designates the power of the base and the digit is the coefficient, e.g., in 3876, the 8 is the coefficient of $_{10}2$, the position of the 8 designating the 2.

COMMAND. A pulse, signal, or set of signals initiating one step in the performance of a computer operation. See instruction and order.

COMPARATOR. A device for comparing two different transcriptions of the same information to verify the accuracy of transcription, storage, arithmetic operation or other process, in which a signal is given dependent upon the relative state of two items, i.e. larger, smaller, equal, difference, etc.

COMPARE. To examine the representation of a quantity for the purpose of discovering its relationship to zero, or of two quantities for the purpose of discovering identity or relative magnitude.

COMPARISON. Determining the identity, relative magnitude and relative sign of two quantities and thereby initiating an action.

COMPARISON, LOGICAL. The operation concerned with the determination of similarity or dissimilarity of two items, e.g., if A and B are alike, the result shall be "1" or yes, if A and B are not alike or equal, the result shall be "0" or no, signifying "not alike."

COMPILER. A program making routine, which produces a specific program for a particular problem by determining the intended meaning of an element of information expressed in pseudo-code, selecting or generating the required subroutine, transforming the subroutine into specific coding for the specific problem, assigning specific storage registers, etc. and entering it as an element of the problem program, maintaining a record of the subroutines used and their position in the problem program and continuing to the next element of information in pseudo-code.

COMPLEMENT. A quantity which is derived from a given quantity, expressed to the base n, by one of the following rules and which is frequently used to represent the negative of the given quantity. (a) *Complement on n:* subtract each digit of the given quantity from $n—1$, add unity to the least significant digit, and perform all resultant carrys. For example, the *twos complement* of binary 11010 is 00110; *the tens complement* of decimal 456 is 544. (b) *Complement of n—1:* subtract each digit of the given quantity from n—1. For example, the *ones complement* of binary 11010 is 00101; the *nines complement* of decimal 456 is 543.

COMPUTER. A term applied to calculating machines ranging from the Chinese abacus to electronic "brains." In automation, it refers to machines which, once set up, perform a series of individual computations without outside tutoring. More specifically, a device for performing sequences of arithmetic and logical operations; sometimes, still more specifically, a stored-program digital computer capable of performing sequences of internally-stored instructions, as opposed to *calculators* on which the sequence is impressed manually (desk calculator) or from tape or cards (card programmed calculator).

COMPUTER, ANALOG. A calculating machine which solves problems by translating physical conditions like flow, temperature or pressure into electrical quantities and using electrical equivalent circuits for the physical phenomenon.

COMPUTER, ASYNCHRONOUS. A calculating device in which the performance of any operation starts as a result of a signal that the previous operation has been completed; contrasted with synchronous computer.

COMPUTER, AUTOMATIC. A calculating device which handles long sequences of operations without human intervention.

COMPUTER, DIGITAL. The numbers are usually expressed as a space-time distribution of punched holes, electrical pulses, sonic pulses, etc.

COMPUTER, SYNCHRONOUS. A calculating device in which the performance of all operations is controlled with equally spaced signals from a master clock.

CONDITIONAL. Subject to the result of a comparison made during computation; subject to human intervention.

CONTENTS. The information stored in any storage medium. Quite prevalently, the symbol () is used to indicate "the contents of"; e.g., (m) indicates the contents of the storage location whose address is m; (A) indicates the contents of register A; (T_2) may indicate the contents of the tape on input-output unit two, etc.

CONTROL. (1) Usually, those parts of a digital computer

which affect the carrying out of instructions in proper sequence, the interpretation of each instruction, and the application of the proper signals to the arithmetic unit and other parts in accordance with this interpretation. (2) Frequently, one or more of the components in any mechanism responsible for interpreting and carrying out manually-initiated directions. Sometimes called manual control. (3) In some business applications of mathematics, a mathematical check.

CONTROL AGENT. The middleman, or force, in a control system. Generally means the energy or fuel which is regulated to affect the value of the conditions or factor being controlled.

CONTROL MEANS. Part of an automatic control device which makes the corrective action.

CONTROL POINT. The value you actually get as the result of some control action. Oddly enough it may not be just what you were hoping for.

CONTROL-SEQUENCE. The normal order of selection of instructions for execution. In some computers, one of the addresses in each instruction specifies the control sequence. In most other computers, the sequence is consecutive except where a jump occurs.

CONTROL, SEQUENTIAL. A manner of operation of a computer such that instructions are fed in a given order to the computer during the solution of a problem.

CONTROL SIGNAL. The energy applied to the device which makes corrective changes.

CONTROL-UNIT. That portion of the hardware of an automatic digital computer which directs the sequence of operations, interprets the coded instructions, and initiates the proper commands to the computer circuits to execute the instructions.

CONVERT. To change numerical information from one number base to another (e.g., decimal to binary) and/or from some form of fixed point to some form of floating-point representation, or vice versa.

CONVERTER. A unit which changes the language of information from one form to another so as to make it available or acceptable to another machine, e.g., a unit which takes information punched on cards to information recorded on magnetic tape, possibly including editing facilities.

COPY. To reproduce information in a new location replacing whatever was previously stored there and leaving the source of the information unchanged.

CORE, MAGNETIC. A magnetic material capable of assuming and remaining at one or two or more conditions of magnetization, thus capable of providing storage, gating or switching functions, usually of toroidal shape and pulsed or polarized by electric currents carried on wide wound around the material.

COUNTER. A device, register, or storage location for storing integers, permitting these integers to be increased or decreased by unity or by an arbitrary integer, and capable of being reset to zero or to an arbitrary integer.

COUNTER, CONTROL. A device which records the storage location of the instructed word, which is to be operated upon following the instruction word in current use. The control counter may select storage locations in sequence, thus obtaining the next instruction word from the following storage location, unless a transfer of special instruction is encountered.

COUNTER, RING. A loop of interconnected bistable elements such that one and only one is in a specified state at any given time and such that, as input signals are counted, the position of the one specified state moves in an ordered sequence around the loop.

COUPLING. The means by which energy is transferred from one circuit to another; the common impedance necessary for coupling.

COUPLING, CAPACITIVE. A method of transferring energy from one circuit to another by means of a capacitor that is common to both circuits.

COUPLING, DIRECT. A method of transferring energy from one circuit to another by means of resistors common to both circuits.

CRT. Cathode ray tube; a device yielding a visual plot of the variation of several parameters by means of a proportionally deflected beam of electrons.

CYBERNETICS. n. A new field of science which attempts to relate the operation of automatic devices to the automatic functioning of the human body's nervous system. Once accomplished, it hopes to evolve a theory blanketing the field of control and communication—both in machines and men.

CYCLE. A set of operations repeated as a unit; a non-arithmetic shift in which the digits dropped off at one end of a word are returned at the other end in circular fashion; cycle right and cycle left. To repeat a set of operations a prescribed number of times including, when required, supplying necessary address changes by arithmetic processes or by means of a hardware device such as a *B-box* or *cycle-counter*.

CYCLE COUNT. To increase or decrease the cycle index by unity or by an arbitrary integer.

CYCLE-CRITERION. The total number of times the cycle is to be repeated; the register which stores that number.

CYCLE-INDEX. The number of times a cycle has been executed; or the difference, or the negative of the difference, between that number and the number of repetitions desired.

CYCLE, MAJOR. The maximum access time of a recirculating serial storage element; the time for one rotation, e.g., of a magnetic drum or of pulses in an acoustic delay line; a whole number of minor cycles.

CYCLE, MINOR. The word time of a serial computer, including the spacing between words.

CYCLE, RESET. To return a cycle index to its initial value.

CYCLING. A rhythmic change of the factor under control at or near the desired value.

D

DAMPING. A characteristic built into control systems which diminishes the system's natural enthusiasm for making excessive corrections when it detects something gone awry. Not dampening; no water involved.

DATA HANDLING SYSTEM. To the control engineer, automatically-operated equipment engineered to simplify the use and interpretation of the bewildering mass of data gathered by modern instrument installations. Can, for example, automatically handle information fed to it from thousands of widely scattered points in a plant. (Engineers, with their tendency for "two-for-one" terminology, also refer to this as a data reduction system.)

DATA-REDUCTION. The art or process of transforming masses of raw test or experimentally obtained data, usually gathered by instrumentation, into useful, ordered, or simplified intelligence.

DATA-REDUCTION, ON-LINE. The processing of information as rapidly as the information is received by the computing system.

DEAD BAND. A specific range of values in which the incoming signal can be altered without also changing the outgoing response. Sometimes called "dead zone."

DEAD TIME. Any definite delay between two related actions. It is measured, obviously, in units of time.

DEBUG. To isolate and remove all malfunctions from a computer or all mistakes from a routine.

DECADE. A group or assembly of ten units, e.g., a decade counter counts to ten in one column; a decade resistor box inserts resistance quantities in multiples of powers of 10.

DECIMAL, CODED, BINARY. Decimal notation in which the individual decimal digits are represented by some binary code, e.g., in the 8-4-2-1 coded decimal notation, the number twelve is represented as 0001 0010 for 1 and 2, respectively. Whereas in pure binary notation, it is represented as 1100. Other coded decimal notations are known as: 5-4-2-1, excess three, 2-4-2-1, etc.

DECODE. To ascertain the intended meaning of the individual characters or groups of characters in the pseudo-coded program.

DECODER. A device capable of ascertaining the significance or meaning of a group of signals and initiating a computer event based thereon; matrix.

DEFLECTION-SENSITIVITY. The quotient of the displacement of the electron beam at the place of impact by the change in deflecting field. It is usually expressed in millimeters per volt applied between the deflection electrodes, or in millimeters per gauss of the deflecting magnetic field.

DELAY-LINE, ELECTRIC. A transmission line of lumped or distributed capacitive and inductive elements in which the velocity of propagation of electromagnetic energy is small compared with the velocity of light. Storage is accomplished by recirculation of wave patterns containing information, usually in binary form.

DELAY-LINE, MAGNETIC. A metallic medium along which the velocity of propagation of magnetic energy is small relative to the speed of light. Storage is accomplished by recirculation of wave patterns containing information, usually in binary form.

DELAY-LINE, MERCURY or QUARTZ. A sonic or acoustic delay-line in which mercury or quartz is used as the medium of sound transmission. See Delay-line, Sonic or Acoustic.

DELAY-LINE, SONIC or ACOUSTIC. A device capable of transmitting retarded sound pulses, transmission being accomplished by wave patterns of elastic deformation. Storage is accomplished by re-circulation of wave patterns containing information, usually in binary form.

DENSITY PACKING. The number of units of useful information contained within a given linear dimension, usually expressed in units per inch, e.g., the number binary digit magnetic pulses stored on tape or drum per linear inch on a single track by a single head.

DERIVATIVE ACTION. Control operation in which the speed of a correction is made according to how fast a condition is going off-the-beam (same as *rate action*).

DESIGN, LOGICAL. (1) The planning of a computer or data-processing system prior to its detailed engineering design. (2) The synthesizing of a network of logical elements to perform a specified function. (3) The result of (1) and (2) above, frequently called the logic of the system, machine, or network.

DEVIATION. No change from the layman's definition—means the difference between the actual value of a condition and the one at which it is supposed to be controlled.

DIAGRAM. A schematic representation of a sequence of subroutines designed to solve a problem; a coarser and less symbolic representation than a flow chart, frequently including descriptions in English words; a schematic or logical drawing showing the electrical circuit or logical arrangements within a component.

DIAGRAM, LOGICAL. In logical design, a diagram representing the logical elements and their interconnections without necessarily expressing construction or engineering details.

DIAPHRAGM MOTOR VALVE. A pneumatic-powered valve which regulates fluid flows in response to a pneumatic signal.

DIFFERENTIAL GAP. The difference between two target values, one of which applies to an upswing to conditions, the other to a downswing. In other words, suppose the temperature in your office falls and the thermostat is set at 70°F. If the differential gap is 4°, then the temperature will drop to 68°F. before the heating system moves into action, raising the temperature to 72°F. Thus, the two degree difference on either side of the 70°F makes up the 4° of differential gap.

DIFFERENTIATOR. A device whose output function is proportional to a derivative of its input function with respect to one or more variables.

DIGIT. One of the *n* symbols of integral value ranging from 0 to *n—1* inclusive in a scale of numbering of base n, e.g., one of the ten decimal digits, 0, 1, 2, 3, 4, 5, 6, 7, 8, 9.

DIGIT, BINARY. A whole number in the binary scale of notation; this digit may be only 0 (zero) or 1 (one). It may be equivalent to an "on" or "off" condition, a "yes" or a "no," etc.

DIGIT, DECIMAL, CODED. One of ten arbitrarily-selected patterns of ones and zeros used to represent the decimal digits.

DIGITAL. The quality of utilizing numbers in a given scale of notation to represent all the qualities that occur in a problem or a calculation.

DIGITIZE. To render an analog measurement of a physical variable into a numerical value, expressing the quantity in digital form.

DIGITS, CHECK. One or more redundant digits in a character or word, which depend upon the remaining digits in such a fashion that if a digit changes, the malfunction can be detected, e.g., a given digit may be zero if the sum of other digits in the word is odd, and this (check) digit may be one if the sum of other digits in the word is even.

DIGITS, EQUIVALENT BINARY. The number of binary digits required to express a number in another base with the same precision, e.g., approximately $3\frac{1}{3}$ times the number of decimal digits is required to express a decimal number in binary form. For the case of coded decimal notation, the number of binary digits required is 4 times the number of decimal digits.

DISCRETE UNITS. Distinct or individual units. For example, automobile crank shafts or bathtubs would be discrete units as contrasted to petroleum or orange juice, which is produced in one continuous flow.

DOWN-TIME. The period during which a computer is malfunctioning or not operating correctly due to machine failures; contrasted with available time, idle time or standby time.

DRUM, MAGNETIC. A rotating cylinder on whose magnetic-material coating information is stored in the form of magnetized dipoles, the orientation of polarity of which is used to store binary information.

DUMMY. An artificial address, instruction, or other unit of information inserted solely to fulfill prescribed conditions (such as word-length or block-length) without affecting operations.

DUMP, A.C. The removal of all A.C. power, intentionally,

accidentally or conditionally from a system or component. An A.C. dump usually results in the removal of all power.

DUMP, D.C. The removal of all D.C. power, intentionally, accidentally, or conditionally, from a system or component.

DUMP, POWER. The removal of all power accidentally or intentionally.

DYNAMIC BEHAVIOR. Describes how a control system or an individual unit carries on with respect to time.

E

ECCLES-JORDON (TRIGGER). A direct coupled multi-vibrator circuit possessing two conditions of stable equilibrium. Also known as a flip-flop circuit or "toggle."

ECHO CHECKING. A system of assuring accuracy by reflecting the transmitted information back to the transmitter and comparing the reflected information with that which was transmitted.

EDIT. To rearrange information. Editing may involve the deletion of unwanted data, the selection of pertinent data, the insertion of invariant symbols such as page numbers and typewriter characters, and the application of standard processes such as zero-suppression.

ELECTRONIC. Pertaining to the application of that branch of science which deals with the motion, emission and behavior of currents of free electrons, especially in vacuum, gas or photo-tubes and special conductors or semi-conductors. Contrasted with electric which pertains to the flow of large currents in wires only.

ELEMENT, LOGICAL. In a computer or data-processing system, the smallest building blocks which can be represented by operators in an appropriate system of symbolic logic. Typical logical elements are the and-gate and the flip-flop, which can be represented as operators in a suitable symbolic logic.

ELEVATION. The angular measurement in a vertical plane from a specific reference, usually the horizontal plane.

ENCODER. A network or system in which only one input is excited at a time and each input produces a combination of outputs. Sometimes called a matrix.

END-POINT CONTROL. Simply—quality control through continuous, automatic analysis. In highly automatic operations, final product is analyzed; if there are any undesirable variations, the controller automatically brings about the necessary changes.

EQUILIBRIUM. For control systems, as for humans, this means balance.

ERASE. To replace all the binary digits in a storage device by binary zeros. In a binary computer, *erasing* is equivalent to clearing, while in a coded decimal computer where the pulse code for decimal zero may contain binary ones, *clearing* leaves decimal zero while *erasing* leaves all-zero pulse codes.

ERROR. Not a mistake by an automatic controller but rather the margin by which it missed its target value. *Errors* occur in numerical methods; *mistakes* occur in programming, coding, data transcription, and operating; *malfunctions* occur in computers and are due to physical limitations on the properties of materials; the differential margin by which a control-led unit deviates from its target value.

ERROR, INHERITED. The error in the initial values; especially the error inherited from the previous steps in the step-by-step integration.

ERROR, ROUNDING. The error resulting from deleting the less significant digits of a quantity and applying some rule of correction to the part retained. A common round-off rule is to take the quantity to the nearest digit. Thus, pi, 3.14169265 . . . , rounded to four decimals is 3.1416. Note; Alston S. Householder suggests the following terms: "initial errors," "generated errors," "propagated errors" and "residual errors." If x is the true value of the argument, and x* the quantity used in computation, then, assuming one wishes f (x) , x—x* is the initial error; f (x) —f (x*) is the propagated error. If fa is the Taylor, or other, approximation utilized, then f (x*) —fa (x*) is the residual error. If f* is the actual result, then fa—f* is the generated error, and this is what builds up as a result of rounding.

ERROR, TRUNCATION. The error resulting from the use of only a finite number of terms of an infinite series, or from the approximation of operations in the infinitesimal calculus by operations in the calculus of finite differences.

EXCHANGE. To interchange the contents of two storage devices or locations.

EXTRACT. To remove from a set of items of information all those items that meet some arbitrary criterion; to replace the contents of specific parts of a quantity (as indicated by some other quantity called an extractor) by the contents of specific parts of a third quantity, e.g., if the number 01101 is stored, the machine can remove and act upon or according to the third digit, in this case a 1.

F

FACTOR, SCALE. One or more coefficients used to multiply or divide quantities in a problem in order to convert them so as to have them lie in a given range of magnitude, e.g., plus one to minus one.

FEED, CARD. A mechanism which moves cards serially into a machine.

FEEDBACK. Part of a closed loop system which brings back information about the condition under control for comparison to the target value.

FERROELECTRIC. A phenomenon exhibited by materials within which permanent electric dipoles exist and a residual displacement in the D-E plane occurs.

FERROMAGNETICS. In computer technology, the science that deals with the storage of information and the logical control of pulse sequences through the utilization of the magnetic polarization properties of materials to store binary information.

FIELD. A set of one or more characters (not necessarily all lying on the same word) which is treated as a whole; a set of one or more columns on a punched card consistently used to record similar information.

FIELD, CARD. A set of card columns fixed as to number and position into which the same unit of information is regularly entered.

FILE. A sequential set of items.

FINAL CONTROL ELEMENT. Unit of a control system (such as a valve) which directly changes the amount of energy or fuel to the process.

FIXED-POINT. A notation or system of arithmetic in which all numerical quantities are expressed by a predetermined number of digits with the point implicitly located at some predetermined position; contrasted with floating-point.

FLIP-FLOP. A bi-stable device; a device capable of assuming two stable states; a bi-stable device which may assume a given stable state depending upon the pulse history of one or more input points and having one or more output points. The device is capable of storing a bit of information; controlling gates, etc.; a toggle. See Eccles Jordan.

FLOATING-POINT. A notation in which a number x is represented by a pair of numbers y and z (and two integers n and m which are understood parameters in any given representation) with y and z chosen so that x = y.nz where z is an integer, ordinarily either m>|y|>m/nfi or y = 0 (where n is usually 2 or 10 and m is usually 1). The quantity y is called the fraction or mantissa; the integer z is called the exponent or characteristic, e.g. a decimal number 241,000,000 might be shown as 2.41, 8, since it is equal to 2.41×10^8.

FLOW-CHART. A graphical representation of a sequence of operations, using symbols to represent the operations such as compute, substitute, compare, jump, copy, read, write, etc. A *flow chart* is a more detailed representation than a *diagram*.

FORCE. To intervene manually in a routine and cause the computer to execute a jump instruction.

FOUR-ADDRESS. See code, Multiple-address.

FREQUENCY RESPONSE ANALYSIS. One of the most frequently misunderstood terms. Simply refers to a method of putting a control system through its paces. Does this by introducing a varying rhythmic change (like alternating current) into a process or control unit to see what effect, if any, these changes will have on the process or control unit. Since the information determines how a system or control unit will react, it is possible to use this method of analysis to predict what the addition of new equipment will mean to an operation.

FUNCTION-TABLE. Two or more sets of information so arranged that an entry in one set selects one or more entries in the remaining sets; a dictionary; a device constructed or hardware, or a subroutine, which can either (a) decode multiple inputs into a single output or (b) encode a single input into multiple outputs; a tabulation of the values of a function for a set of values of the variable.

FUNCTOR. A logical element which performs a specific function or provides a linkage between variables.

G

GAIN. Amount of increase in a signal (or measurement) as it passes through a control system or a specific control element. If a signal gets smaller, it is said to be *attenuated*. To further complicate things, gain can also mean the sensitivity of a device to changes.

GATE. A circuit which has the ability to produce an output which is dependent upon a specified type or the coincidence nature of the input, e.g. an "and" gate has an output pulse when there is time coincidence at all inputs; an "or" gate has an output when any one or any combination of input pulses occur in time coincidence; any gate may contain a number of "inhibits," in which there is no output under any condition of input if there is time coincidence of an inhibit or except signal.

GENERATE. To produce a needed subroutine from parameters and skeletal coding.

GENERATOR. A program for a computer which generates the coding of a problem; a mechanical device which produces an electrical output.

H

HALF-ADDER. A circuit having two output points, S and C, and two input points, A and B, such that the output is related to the input according to the following table:

INPUT		OUTPUT	
A	B	S	C
0	0	0	0
0	1	1	0
1	0	1	0
1	1	0	1

If A and B are arbitrary input pulses, and S and C are "sum without carry" and carry, respectively, it may be seen that two half-adders, properly connected, may be used for performing binary addition.

HARDWARE. The mechanical, magnetic, electronic and electrical devices from which a computer is fabricated; the assembly of material forming a computer.

HEAD. A device which reads, records or erases information in a storage medium, usually a small electro-magnet used to read, write or erase information on a magnetic drum or tape or the set of perforating or reading fingers and block assembly for punching or reading holes in paper tape.

HOLD. The function of retaining information in one storage device after transferring it to another device; in contrast to clear.

HUNTING. Even control or measuring instruments sometimes have trouble finding the target. When an instrument wanders around the target without success, engineers appropriately claim it is "hunting."

HYSTERESIS. The difference between the response of a unit or system to an increasing signal and the response to a decreasing signal.

I

IGNORE. A typewriter character indicating that no action whatsoever be taken. (In Teletype or Flexowriter code, all holes punched is an ignore); an instruction requiring non-performance of what normally might be executed; not to be executed.

IMPEDANCE, CHARACTERISTIC. The ratio of voltage to current at every point along a transmission line on which there are no standing waves; the square root of the product of the open and short circuit impedance of the line.

INFORMATION. An aggregation of data.

INPUT. A sort of cable-ese meaning incoming signal to a control unit or system, the information which is transferred from external storage into the internal storage; a modifier designating the device performing this function.

INSTRUCTION. A set of characters which defines an operation together with one or more addresses (or no address) and which, as a unit, causes the computer to operate accordingly on the indicated quantities. The term "instruction" is preferable to the terms "command" and "order"; command is reserved for electronic signals; order is reserved for "the order of the characters" (implying sequence) or "the order of the interpolation," etc.

INSTRUCTION, BREAKPOINT. An instruction which, if some specified switch is set, will cause the computer to stop.

INSTRUCTION, BREAKPOINT, CONDITIONAL. A conditional jump instruction which, if some specified switch is set, will cause the computer to stop, after which either the routine may be continued as coded or a jump may be forced.

INSTRUCTION, MULTIPLE-ADDRESS. See code, Multiple-address.

INSTRUCTION, ONE-ADDRESS. An instruction consisting of an operation and exactly one address. The instruction code of a single-address computer may include both zero and multi-address instructions as special cases.

INSTRUCTION, ONE-PLUS-ONE or THREE-PLUS-ONE ADDRESS. A two- or four-address instruction, respectively, in which one of the addresses always specifies the location of the next instruction to be performed.

INSTRUCTION, TRANSFER. A computer operational step in which a signal or set of signals specifies the location of the next operation to be performed and directs the computer to that operation (or instruction).

INSTRUCTION, TWO, THREE or FOUR ADDRESS. An instruction consisting of an operation and 2, 3 or 4 addresses, respectively.

INSTRUCTION, ZERO-ADDRESS. An instruction specifying an operation in which the location of the operands are defined by the computer code, so that no address need be given explicitly.

INSTRUMENT. As used industrially, definitely not a French horn or surgeon forceps. Used broadly to connote a device incorporating measuring, recording, and/or controlling abilities.

INSTRUMENTATION. Used to describe the application of industrial instruments to a process or manufacturing operation. Also describes the instruments themselves.

INTEGRATOR. A device which continuously adds up a quantity being measured over a period of time. Similar in use to your electric meter at home.

INTERLACE. To assign successive storage locations to physically separated storage positions, e.g., on a magnetic drum or tape, usually for the express purpose of reducing access time.

ITEM. A set of one or more fields containing related information; a unit of correlated information relating to a single person or object; the contents of a single message.

INTERPRETER. An interpretive routine.

J

JUMP. An instruction or signal which, conditionally or unconditionally, specifies the location of the next instruction and directs the computer to that instruction. A jump is used to alter the normal sequence control of the computer. Under certain special conditions, a jump may be forced by manual intervention, in other words a transfer of control is made to a specified instruction.

JUMP, CONDITIONAL. An instruction which will cause

the proper one of two (or more) addresses to be used in obtaining the next instruction, depending upon some property of one or more numerical expressions or other conditions.

K

KEY. A group of characters usually forming a field, utilized in the identification or location of an item; a marked lever manually operated for copying a character, e.g., typewriter, paper tape perforator, card punch manual keyboard, digitizer or manual word generator.

L

LAG. Preferred engineering terms for delay in response, or a relative measure of the time delay between two events, states, or mechanisms.

LANGUAGE, MACHINE. Information recorded in a form which may be made available to a computer, e.g., punched paper tape may contain information available to a machine, whereas the same information in the form of printed characters on a page is not available to a machine; information which can be sensed by a machine.

LATENCY. In a serial storage system, the access time less the word time, e.g., the time spent waiting for the desired location to appear under the drum heads or at the end of an acoustic tank.

LIBRARY, ROUTINE. An ordered set or collection of standard and proven routines and subroutines by which problems and parts of problems may be solved, usually stored in relative or symbolic coding. (A library may be subdivided into various *volumes,* such as floating decimal, double-precision, or complex, according to the type of arithmetic employed by the subroutines.)

LINE, DELAY. A device capable of causing an energy impulse to be retarded in time from point to point, thus providing a means of storage by circulating intelligence bearing-pulse configurations and patterns. Examples of delay lines are material media such as mercury, in which sonic patterns may be propagated in time; lumped constant electrical lines; coaxial cables, transmission lines and recirculating magnetic drum loops.

LINE-PRINTING. Printing an entire line of characters

across a page as the paper feeds in one direction past a type bar or cylinder bearing all characters on a single element.

LINE TRANSMISSION. Any conductor or systems of conductors used to carry electrical energy from its source to a load.

LOAD. What the process calls for in fuel or energy input.

LOCATION. A unit storage position in the main internal storage, storing one computer word; a storage register.

LOCATION, STORAGE. A storage position holding one computer word, usually designated by a specific address or a specific register.

LOGGER. Not a woodsman but an instrument which automatically scans conditions (temperature, pressure, humidity) and records—or logs—findings on a chart. Can come equipped with lights or alarms to signal danger points.

LOGIC. The science that deals with the canons and criteria of validity in thought and demonstration; the science of the formal principles of reasoning; the basic principles and applications of truth tables, gating, interconnection, etc. required for arithmetical computation in a computer.

LOGIC, SYMBOLIC. Exact reasoning about relations using symbols that are efficient in calculation. A branch of this subject known as Boolean algebra has been of considerable assistance in the logical design of computing circuits.

LOGICAL. See operation, logical.

LOOP. The repetition of a group of instructions in a routine.

LOOP, CLOSED. Repetition of a group of instructions indefinitely.

M

MALFUNCTION. A failure in the operation of the hardware of a computer.

MANIPULATED VARIABLE. A quantity—or a condition—which is altered by the automatic units to set off a change in the value of the chief condition under regulation.

MATRIX. In mathematics, an array of quantities in a prescribed form, usually capable of being subject to a mathematical operation by means of an operator or another matrix according to prescribed rules; an array of circuit elements, e.g., diodes, wires, magnetic cores, relays, etc. which are capable of performing a specific function, e.g., conversion from one numerical system to another.

MEASURING MEANS. Whatever is used to measure a condition. A thermometer is a measuring means for room temperatures.

MEMORY. The term "storage" is preferred.

MERGE. To produce a single sequence of items, ordered according to some rule (i.e., arranged in some orderly sequence), from two or more sequences previously ordered according to the same rule, without changing the items in size, structure, or total number. Merging is a special case of collation.

MESSAGE. A group of words, variable in length, transported as a unit; a transported item of information.

MICROSECOND. A millionth part of a second.

MILLISECOND. A thousandth part of a second.

MINIATURIZATION. Method of reducing the size of instruments to minimize the space requirements.

MISTAKE. A human blunder which results in an incorrect instruction in a program or in coding, an incorrect element of information, or an incorrect manual operation.

MNEMONIC. Assisting, or intended to assist, memory; of or pertaining to memory; mnemonics is the art of improving the efficiency of the memory (in computers, storage).

MODIFIER. A quantity used to alter the address of an operand, e.g., the cycle index.

MODIFY. To alter in an instruction the address of the operand; to alter a subroutine according to a defined parameter.

MULTIVIBRATOR. A type of relaxation oscillator used for the generation of non-sinusoidal waves in which the output of each of its two tubes is coupled to the input of the other to sustain oscillations.

MULTIVIBRATOR, ASTABLE. A free-running type of relaxation oscillator used for the generation of non-sinusoidal waves.

MULTIVIBRATOR, MONOSTABLE. A type of relaxation oscillator used to sustain a trigger pulse for a specified time, since the device assumes another state for a specified length of time at the end of which it returns to its original state, after being pulsed or forced into another state.

N

NEUTRAL ZONE. An automatic control engineer's version of No Man's Land—a range of values in which no control action occurs.

NOISE. Similiar to radio static. Meaningless stray signals in a control system that do not require correction.

NORMALIZE. To adjust the exponent and mantissa of a floating-point result so that the mantissa lies in the prescribed standard (normal) range; standardize.

NOTATION. See "NUMBER-SYSTEM."

NOTATION, BIQUINARY. One of any number of mixed-base notations in which the term ni in the definition of number system is replaced by the product $\prod_{j=0}^{i-1} m_j$. In the biquinary system, mj is two for j odd, five for j even; a scale of notation in the base is alternately 2 and 5, e.g., the decimal number 3671 is biquinary 03 11 12 01, the first of each pair of digits counting 0 or 1 units of five and the second counts 0, 1, 2, 3 or 4 units. For comparison, the same number in Roman numerals is MMM-DCLXXI. Biquinary notation expresses the representation of numbers by the abacus, and by the two hands and five fingers of man and is used in some computers.

NOTATION, CODED-DECIMAL. Decimal notation in which the individual decimal digits are represented by some code.

NOTATION, MIXED-BASE. A number system in which the term ni in the definition of number-system is replaced by the product $\prod_{j=0}^{i=1} m$, e.g., in the biquinary system mj is two for j odd and five for j even.

NUMBER, BINARY. A numerical value written in the base-two system of notation.

NUMBER, OPERATION. A number indicating the position of an operation or its equivalent subroutine in the sequence forming a problem routine. When a problem is stated in pseudo-code, each step is sometimes assigned an operation number.

NUMBER, RANDOM. A set of digits constructed of such

a sequence that each successive digit is equally likely to be any of n digits to the base n of the number.

NUMBER-SYSTEM. Numerical notation; positional notation; a systematic method for representing numerical quantities in which any quantity is represented approximately by the factors needed to equate it to a sum of multiples of powers of some chosen base n. In writing numbers, the base is sometimes indicated as a subscript (itself always in decimal notation) whenever there is any doubt about what base is being employed; *Binary, Ternary, Quatenary, Quinary, Octal (Octonary), Decimal, Duodecimal, Sexadecimal (Hexadecimal)* or Duotricenary Notation—notation using the base 2, 3, 4, 5, 8, 10, 12, 16 or 32 respectively.

O

OCTAL. Pertaining to the number base of eight, e.g. in octal notation, octal 214 is 2 times 64 plus 1 times 8 plus 4 times 1 equals decimal 140; octal 214 is binary 010, 001, 100.

OFFSET. Describes the difference between the value or condition desired and that actually attained.

ONE-ADDRESS. Single address; a system of machine instruction such that each complete instruction explicitly describes one operation and one storage location.

ON-LINE OPERATION. A type of system application in which the input data to the system is fed directly from the measuring devices and the computer results obtained during the progress of the event, e.g. a computer receives data from wind tunnel measurements during a run, and the computations of dependent variables are performed during the run enabling a change in the conditions so as to produce particularly desirable results.

OPEN LOOP. A control system in which there is no self-correcting action for "misses" of the target value, as there is in a closed loop system. Might be likened to a hunter firing a rifle at a deer; the bullet goes where it's aimed, and if it misses the target, no deer!

OPERAND. Any one of the quantities entering or arising in an operation. An operand may be an argument, a result, a parameter, or an indication of the location of the next instruction.

OPERATION. A defined action; the action specified by a

single computer instruction or pseudo-instruction; an arithmetical, logical, or transferal unit of a problem, usually executed under the direction of a subroutine.

OPERATION, ARITHMETICAL. An operation in which numerical quantities form the elements of the calculation (e.g., addition, subtraction, multiplication, division).

OPERATION, AVERAGE-CALCULATING. A common or typical calculating operation longer than an addition and shorter than a multiplication; often taken as the mean of nine addition and one multiplication time.

OPERATION, COMPLETE. An operation which includes (a) obtaining all operands from storage, (b) performing the operation, (c) returning resulting operands to storage, and (d) obtaining the next instruction.

OPERATION, COMPUTER. The electronic action of hardware resulting from an instruction; in general, computer manipulation required to secure computed results.

OPERATION, FIXED-CYCLE. A type of computer performance whereby a fixed amount of time is allocated to an operation; synchronous or clocked type arrangement within a computer in which events occur as a function of measured time.

OPERATION, LOGICAL. An operation in which logical (yes-or-no) quantities form the elements being operated on (e.g., comparison, extraction). A usual requirement is that the value appearing in a given column of the result shall not depend on the values appearing in more than one given column of each of the arguments.

OPERATION, REAL-TIME, ON-LINE, SIMULATED. The processing of data in synchronism with a physical process in such a fashion that the results of the data-processing are useful to the physical operation.

OPERATION, RED-TAPE. An operation which does not directly contribute to the result; i.e., arithmetical, logical, and transfer operations used in modifying the address section of other instructions in counting cycles, in rearranging data, etc.

OPERATION, SERIAL. The flow of information through a computer in time sequence, using only one digit, word, line or channel at a time. Contrasted with parallel operation.

OPERATION, TRANSFER. An operation which moves information from one storage location or one storage medium to another (e.g., read, record, copy, transmit, exchange). *Transfer* is sometimes taken to refer specifically to movement between different media; *storage* to movement within the same medium.

OPERATION, VARIABLE CYCLE. Computer action in which any cycle of action or operation may be of different lengths. This kind of action takes place in an asynchronous computer.

OPERATOR. The person who actually manipulates the computer controls, places information media into the input devices, removes the output, presses the start button, etc.; a mathematical symbol which represents a mathematical process to be performed on an associated function.

OPTIMALIZATION. The approach to economically perfect plant operation accomplished, primarily, by analytical rather than hit-or-miss methods.

OR-CIRCUIT. An electrical or mechanical device which will yield an output signal whenever there are one or more inputs on a multichannel input, e.g. an OR gate is one in which a pulse output occurs whenever one or more inputs are pulsed; forward merging of pulses simultaneously providing reverse isolation.

ORDER. A defined successive arrangement of elements or events. The word order is losing favor as a synonym for instruction, command or operation partly due to ambiguity.

OR-OPERATOR. A logical operator which has the property such that if P or Q are two statements, then the statement "P or Q" is true or false precisely according to the following table of possible combinations:

P	Q	P or Q
False	True	True
True	False	True
True	True	True
False	False	False

OSCILLATIONS, FREE. Oscillating currents which continue to flow in a tuned circuit after the impressed voltage has been removed. Their frequency is the resonant frequency of the circuit.

OSCILLOGRAPH RECORDER. A device capable of charting high speed variations in measured quantities, such as temperature or pressure, as found in aircraft testing, for example.

OUTPUT. Outgoing signal of a control unit or operation, or information transferred from the internal storage of a computer to secondary or external storage; information transferred to any device exterior to the computer.

OUTPUT-BLOCK. A portion of the internal storage re-

The mixed sand is conveyed through pneumatic tubes to the places where it is needed.

Core-moulding is done on highly mechanized rotary core blowers and in single-station core-blowing machines. Some 5 per cent of the cores required to produce the crankchamber and holes for the cylinder bores were being shell moulded on a five-station rotary-table shell-moulding machine designed and produced by the Company. The metal pattern is heated by gas jets at one station; the table then indexes round and a release-agent is sprayed on to the hot pattern; at the next station a hot core box comes down over the pattern, and the resin-sand mix is blown vertically downwards into the space so formed; at the next stage or two the shell mould passes through baking ovens and is then indexed round to an automatic mould-release station. The sand is not at present being reclaimed but it is intended to do so later. Ford has little doubt that the process is much more economical than sand moulding at American labour costs.

The drag moulds are carried automatically past two successive stations, at each of which a set of cores is loaded into them. The cores are hand-loaded into a fixture which locates all of them accurately in relation to dowel holes, and then clamps them into position in a pneumatic clamp. The set of cores so clamped is loaded with the fixture into the mould, and the dowels carried in the mould engage with the holes in the fixture. The pneumatic clamp is then released and the fixture is withdrawn, leaving the cores accurately placed. At the second station the remaining cores are placed in a similar manner.

In this foundry all the iron is melted in acid cupolas. About 220 pairs of samples are taken each 16-hour day and are spectrographically analysed, two samples being tested on each occasion as a precaution. The machine used is of the normal spark spectrograph type.

The metal is conveyed from the cupolas to the pouring ladles in large receivers, which are driven along overhead monorails. In this respect the foundry seemed to be a much safer work place than that at the Dearborn plant. The pouring ladles are themselves carried on another overhead monorail and their trolleys have electric clutches, which can be engaged with an overhead conveyor and synchronized with the mould conveyor below. When one mould has been poured the operator withdraws his clutch and holds his ladle at rest until the next mould in the line overtakes it. Thus there is no relative motion between pouring ladles and moulds during the pouring operation. The filled moulds are automatically transferred to a cooling conveyor and thence to an automatic sand-knockout machine.

Unlike many of the American foundries, the Ford Cleveland foundry removes the flash from the joint faces by automatic transfer through a large grinding machine. The castings are then passed to an automatic pressure-testing machine, which applies air pressure to the water-jacket space, cuts off the supply and measures the internal pressure again after a short interval. As at Dearborn, defective castings are welded up after being pre-heated to 1100°F, and are given a final annealing.

Engine plant

This plant has recently begun to produce cold-extruded gudgeon pins; they are in fact double-reverse-flow extruded from a slug about 1 in. diameter and 2 in. long. The pins are automatically conveyed from the extrusion press through six centreless grinders with wheels of successively finer grit. One operator looks after the six machines and ocasionally checks the size of a sample from each with an air gauge. There is no automatic feed-back correction to the machines. The finished pins are passed through an automatic gauging centre, which checks them for diameter, roundness and taper and sorts them accordingly. The machine has provision for an automatic hardness test, but this is not at present being used. All the gauging in this machine is done by electrical contact heads.

At this plant, all joint faces on the blocks and cylinder heads are broached in enormous Cincinnati machines. The whole V-8 line had been shut down the afternoon of the visit because of trouble with the automatic clamping device on the broaching machine. Whenever a serious failure of this kind occurs the company sends home at once all the men affected and stops their pay for the rest of the day provided it has met the minimum laid down in the company-union contract—four hours pay for any day on which the men have been asked to attend. On this occasion the company hoped to have the broaching machine running by the next morning and so asked the men to report as usual. If during the night it had been found that the machine would not be ready, telegrams would have been sent to all the men telling them not to report.

The block machining line is highly automatic. '*Toolometers*' (The Cross Company, Detroit, Mich.) are everywhere and at several stations there is automatic pressure-testing of the water spaces. A bell rings if the feed to any machine jams. If taps and drills have broken inside castings, they are removed by an electric-spark process with the help of an '*Electron*' drill, made by the Elox Corporation, Clawson, Detroit. Defects uncovered during machining are frequently remedied by welding and subsequent annealing. Finally the cylinder bores of each block are measured by hand with air gauges, and the sizes are marked with a wax pencil on the joint faces near each bore concerned. At the same time the inspector writes the engine number and the cylinder-bore sizes on a telewriter, which passes the information to the piston store. The store-keeper selects pistons of appropriate sizes and puts them on a conveyor, which automatically delivers them to the main engine assembly line at a point next to the block for which they have been tagged.

The connecting-rod small and big ends are bored simultaneously in a special machine. The big-end bolts and nuts are fitted by hand and are then tightened simultaneously by a double headed pneumatic spanner of special design whilst the rod is held in a fixture. The small-end bushes are inserted in the rod and their oil holes are drilled automatically.

The automatic piston line receives bought-out die castings and passes them through all turning and grinding processes and thence on to a store, where they are graded by diameter.

On the engine assembly line the crankshafts are assembled in the blocks on a roller conveyor, all the bearing shells, caps and bolts being inserted by hand. All the main-bearing caps bolts are tightened simultaneously to a predetermined torque by a hand-welded, power-driven, multihead spanner which is suspended from an overhead-balance device. The assemblers do not even bother to check the tightness of the crankshafts in their bearings at this stage. After this operation, further assembly is carried out with the block mounted on a suspended hanger and processed as at Dearborn.

One large machine, in a bay near the assembly line, was apparently unused. It had been designed for the automatic assembly of valves, valve springs, cotters, and caps in the cylinder heads. The Company, having tried for a long while to make the machine work reliably, has now abandoned it and still has all valve assembly done by hand. It thinks that the operation is a little too complicated to be done economically at present by automatic means.

When assembly is complete, the engines are run whilst still suspended from their hangers. There are quick-acting clips on all water, oil and power connections, but there are no elaborate automatic loading and connecting devices as at the Chrysler engine plant.

Some of these engines seemed rough when on test, but the company intends to provide for a final balancing of the engine on the conveyor after assembly. This will presumably be done by removing metal at the fan pulley and flywheel and measuring the unbalance by electric vibration pick-ups mounted on the freely suspended engine.

There is little electrical trouble with the automatic transfer equipment; but the tools require a copious flow of coolant. The switches work quite reliably until their contacts begin to wear—after four or five years' operation. It is then best to replace the lot.

FORD MOTOR COMPANY
Dearborn, Detroit

This plant, also known as the River Rouge, occupies a site of 1200 acres, roughly two miles long by a mile wide. There are 55,000 employees, most of whom work one of two 2-hour shifts; there is parking space for 20,000 cars.

Engine plant

The building in which this plant is housed was erected during World War II for the manufacture of Pratt and Witney R 2800 aircraft engines. It was converted to the manufacture of car engines soon after the war. In it are built V8 engines for the Ford, Lincoln and Continental cars, also truck and industrial engines.

Iron crankshafts are moulded so acurately that no machining is necessary anywhere on the webs or on the balance weights, which are not even skimmed to a true radius. No holes are cored in the crankpins, which are cast solid and later drilled blind from one end. The machining processes are:

(1) milling the ends to correct overall length;

(2) centring the ends, locating from the cast outside diameter of the balance weights;

(3) miling the locations on balance weights for subsequent turning of pins and journals;

(4) turning the journals and pins on Le Blond lathes (R. K. Le Blond Machine Tool Company, Cincinnati, Ohio) ;

(5) cutting keyways for timing-chain sprocket and fan pulley;

(6) drilling out crankpin centres and oil holes;

(7) grinding all journal and nose diameters (plunge-grinding at one machine setting) ;

(8) grinding all crankpins (individually) ;

(9) polishing pins and journals by machine; and

(10) dynamic balancing the crankshaft on an automatic machine which drills radial holes in the end balance weights to a depth and in a position calculated by the machine.

All joint faces on crankcases are machined by broaching on Cincinnati machines. The main-bearing housing caps are cast in sets of five, and separated by mechanical saws. 'Toolometer' boards control tool life throughout the plant. Engines are assembled on hangers suspended from an overhead monorail. Pneumatic nut-runners are used for all major assembly operations: these are suspended units which are guided by hand on to groups of nuts or bolts, the torque reaction being taken by the hanger which carries the assembly. The method of engine transfer leaves the operating floor much clearer than does a roller-conveyor assembly line. Very many more workers are employed here than at the Chrysler engine plant.

Foundry

Melting is done in fourteen cupolas, seven being in use at a time, and seven being shut down for lining repairs—each cupola working alternate days. All but one have acid linings. The exception is water-cooled and has a basic lining; it runs fourteen days between shutdowns. The maximum daily output of engine blocks (mostly V-8s) is 6400.

At the time of the visit, the valve chest cores were being finished with a silica wash instead of the normal plumbago. The cylinder-head cores were placed by hand in a jig, which correctly spaced them. Movement of a lever on the jig caused the cores to be clamped in position pneumatically. They were then lowered with the jig into the mould, where the jig engaged with accurately placed dowels. The pneumatic clamp was then released and the jig lifted away, leaving the complete set of cores in the correct positions.

A surprising number of defective castings were being reclaimed by arc or oxy-acetylene welding, there being eight welding booths in continuous operation. It was said that cast-iron elctrodes were used for the oxy-acetylene welding, but no information was obtained on the composition of the arc-welding electrodes. The castings were heated to 1100°F before welding.

FORD MOTOR COMPANY, AUTOMATIC TRANSMISSION PLANT
Livonia, Michigan

The plant at Livonia has 2 million square feet under one roof. It was a tank factory during the Korean war, but two years ago the Government decided that the tank-manufacturing equipment should be removed and placed in storage. The Company now employs between 4000 and 5000 people here, producing automatic transmissions for Mercury and Lincoln cars. It also makes those parts that are common to the Ford automatic transmission, the remainder being made at a plant in Cincinnati.

The chief difference between the production of automatic transmissions and other motor-manufacturing operations is the very high precision required. Each transmission unit contains 22 piston valves of the type used in all hydraulic servo-mechanisms. These valves have to be fitted to their housings with very small clearances, which involves working to limits of 0·0002 in. The hardened steel piston valves work in die-cast housings of high-silicon aluminium alloy.

The Ford automatic transmission uses, not engine oil but a very thin, synthetic hydraulic fluid, with a viscosity apparently little higher than that of water. Even so, no jointing compound and not a single paper gasket is used anywhere in the assembly. Most of the joints are faced on specially adapted precision-boring machines using tungsten-carbide cutters. Throughout the plant there are numerous automatic lapping machines for removing very fine scratches from the die-cast housings.

All the main transmission shafts that carry the drive from the automatic-transmission unit to the propeller shaft are profile-turned on high-speed tracer-controlled lathes.

Some interesting processes were seen for de-burring the various components. For instance, valve bodies and housings, all of light alloy, had their surfaces cleaned and their oil holes de-burred by blasting with finely ground walnut shell. Steel components were de-burred by tumbling with rounded granite chips of appropriate size, the major diameter being about ¾ in. Many burrs on the piston valves (and much consequent trouble due to the jamming of the valves in service) have been avoided by running out the sides of the grooves at an angle of 85° instead of the usual 90°, i.e., by giving a slightly obtuse angle to the corners of the lands.

The die-cast light-alloy rotor, which forms the first stage of the hydraulic transmission, is machined only on the blade tips and in the bore. It carries a steel shroud, which is bound tightly around the blade tips and welded to form a complete circle.

At the time of the visit the Company was developing the manufacture of ⅜ in. diameter splindles for the planet gears from 0.85 per cent carbon-steel wire by shearing and cold heading from the coil.

The gears in this transmission are not ground. Most are cut on 12-station shapers or on hobbing machines.

The splines on the ends of the output shafts are slotted in the normal manner, but the power-input shaft splines are cold-rolled between parallel dies.

Air-gauges are very widely used throughout the plant.

The output turbine rotors and stators are built up from pressed-steel components, with tabs on the radial blades passing through punched slots and rolled by machine after assembly by hand. The main rotor casing is pressed from steel plate about 0.2 in. thick. The transmission sump is pressed from strip in six operations on one machine; transfer is automatic from the coil through each operation to the storage bin, so that there is no handling whatsoever from the coil to the finished product.

Most of the oil holes and bolt holes are drilled on multi-station drilling machines, which give automatic transfer around a rotary table. But there is no automatic transfer from one machine to another as yet: this does not seem to be worth while since the parts can be very easily handled in trays. The only automatic transfer-machine lines in the plant are those working on large iron castings for the clutch casing and the transmission housing. These are heavy components, which would require a lot of man-handling if not transferred automatically.

The assembly of the transmission is carried out in typical Ford manner; hangers are suspended from an overhead monorail and the floor area is kept clear. Even the performance checks on the complete assembly are made while it is still on the hanger. Very few men appear to be employed on rectification.

It was said that more than 75 per cent of Ford cars are now sold with automatic transmission.

FORD MOTOR COMPANY,
ASSEMBLY PLANT
Milpitas, San Jose, California

The original Ford Northwest California Assembly Plant was built at Richmond near San Francisco in 1933, but by 1954 it was hopelessly inadequate. The Richmond premises were closed and the plant moved to Milpitas 18 months ago. The Milpitas output goes to Northern California, Nevada and Utah, while the cars assembled in the Long Beach plant go to Southern California, Arizona and New Mexico.

The Milpitas plant is 855 feet wide and 1700 feet long; 2800 men work there, on one eight-hour shift only, five days a week, producing 700 cars and trucks a day. All components and sub-assemblies are moved to Milpitas from other Ford plants (e.g., Detroit, Cleveland and Buffalo) by rail, but if some items run short consignments are usually supplied by air.

The plant-layout office was found to be exceptionally good. The complete plant layout for the whole works was reproduced on aluminium photo-litho sheet to a scale of $\frac{1}{4}$ in. to a foot, each separate sheet being 3 ft × 4 ft and engraved with $\frac{1}{4}$-in. squares. The sheets were laid out on a very large table and were split down the middle so that one could conveniently reach across from the centre or outside gangways to any part of the plan. It seemed to be Ford practice to make this kind of layout available in every plant, and it was said that the sheets were produced in the central Ford planning office in Detroit. All conveyors, both overhead and floor type, were shown on the plan—over seven miles of overhead conveyor alone. Electric trolleys were available outside the plant manager's office to take officials and foremen wherever they had to go; others went by bicycle.

The most interesting processes, so far as automation goes, are listed below:

(1) In the assembly of the front suspension system to the chassis, a large pneumatic clamp straddles both wishbone assemblies and compresses the suspension springs, whilst the assembly is being completed.

(2) A teletyping system is used throughout the plant to ensure that the right types of component in the correct colour schemes come together at the right time and at the right place. Every car assembled at the plant has been ordered by a customer through his dealer. A coded card is made out listing all his special requirements and these instructions are transmitted through the teletyping system. The customer can even state which type and make of tyre he wants fitted. The only place in the assembly line where human judgment has to be used, to ensure that parts match up correctly, is between the body- and chassis-assembly lines. About five body-assembly lines run parallel to each other and at right angles to the main floor-type conveyor carrying the chassis. Bodies are switched into the main chassis assembly conveyor by an operator, who selects the bodies he requires from their individual lines. The appropriate bodies must be there because they have been produced on an assembly line controlled by the same teletyping machine.

(3) Many of the paint-spraying booths are fully automatic, but no electrostatic spraying is employed anywhere. The working conditions in the non-automatic booths seem to be rather unpleasant, and the operators have to wear breathing apparatus.

(4) Throughout the body assembly lines there are literally hundreds of spot-welding guns and seam welders carried on balanced suspensions (Thor Power Tool Company, Aurora, Ill.). The dexterity of oper-

ators with this portable welding equipment is remarkable, but the percise location of many of the
spot welds does not seem to be important.

(5) Some very specialized spot-welding machines—collo-
quially referred to as 'piano machines' from their
general appearance—are used to make about 200
simultaneous spot welds on curved and irregular
shaped joints between the firewall and the side members, and between the front and rear members of
the car bottom.

(6) In one fully automatic operation tyres are conveyed
through a machine which applies soap solution to
the beads, rolls them on to the whee rims, inflates
them to the correct pressure and discharges the assemblies into a conveyor which takes them to the
assembly line. Neither tyre nor wheel is touched by
hand between the time each item is placed on its
conveyor and the time the complete assembly is
delivered to the appropriate point in the final assembly line (two wheels to each side of the assembly
line, and one to the centre for the spare, all of the
correct colour, and fitted with the customer's choice
of tyre). This automatic tyre and rim assembly has,
of course, been very greatly simplified by the use of
tubeless tyres, for about 10 men were required previously to fit the tubes inside the tyres and make sure
they were not trapped during inflation. Obviously
tyres with tubes are only supplied on Ford cars to
special order, and the cost is probably prohibitive
because the rims have to be specially drilled to accommodate the valves, and the tyre assembly has to
be done in the old-fashioned manner.

(7) Specially designed C-shaped suspension brackets permit the whole seat assembly to be carried into the
body by a hoist running on an overhead conveyor
and to be lowered into its final position without
lifting and tugging by hand. The seat assemblies
carry their own captive bolts, which are lowered
straight through holes in the floor and quickly secured by nuts fitted from underneath the car.

(8) On 'station wagon' bodies, much of the decoration,
either coloured flashes or imitation wooden com-

ponents, is given by securing coloured plastic tapes with adhesives, instead of masking and paint spraying. The results are very attractive and are obviously much cheaper than would be possible by the alternative process.

FORD MOTOR COMPANY, STAMPING PLANT
Walton Hills, Cleveland, Ohio

This plant has an Automation Department containing a design and development section, and a manufacture and maintenance section; the latter employs fitters, welders, die, makers, electricians, etc. The design staff consists of six mechanical engineers, and one electrical engineer. The Department Manager is a one-time Master-Mechanic who graduated in mechanical engineering by part-time study. The Department is responsible for everything connected with automation in the plant. It designs automatic machinery; it then develops experimental models, tests them, installs them and maintains them in service. While some automatic equipment is made at the plant, most of it is made by outside companies. The engineers responsible for the Automation Department are running evening courses at the plant, teaching electrical theory and applied mechanics to their tradesmen. They teach electrical maintenance to mechanical fitters, welding to die makers, etc., so that the skill of their men is broadened.

The Department Manager has generally abandoned hydraulic actuating equipment for press loaders and unloaders ('mechanical hands') , because the viscosity of the oil tends to make operation too slow and too variable with temperature. He now employs all-pneumatic actuation, using air at 40 lb. per square inch gauge. To increase operating speeds, he employs various link and rack-and-pinion mechanisms, increasing the speed of the mechanical hand up to about four times the ram speed. An ingenious chain-and-sprocket mechanism with a link connection to the mechanical hand reduces the acceleration at the ends of the stroke. Linear ball bearings (with raceways locally manufactured) are used for all high-speed operating slides. The construction of the automatic equipment is very rugged throughout, as any breakdown stops large sections of the plant and is extremely expensive. Long endurance tests are run on each new item of transfer equipment before it is installed. Most of the equipment is built up by fabrication from solid-drawn tubes of square section, which are very easily prepared and welded in position.

Hourly rated employees are paid the normal overtime premiums of time-and-a-half and double time. Salaried employees are paid overtime in accordance with Federal statutes and Company policies.

During the tour of the works many items of Ford-designed transfer equipment were seen in operation. The operating speed of the mechanical hands was remarkable; they also made it unnecessary for the workers to approach the danger area near the dies. Automatic transfer from one press to another was accomplished by slotted tables, whose rising, traversing and falling motions moved the pressings forward step by step on to chutes, which in turn guided them straight into the next pair of dies. Most of the presses were of Danly or Bliss manufacture (Danly Machine Specialties, Chicago, Ill. and the E. W. Bliss Company, Canton, Ohio) and Ford had made them automatic simply by mounting micro-switches on the press frames and striker bars on the rams.

It was easy to be impressed by the way in which automation is being tackled in this plant, and by the amount of resources made available to the engineer in charge.

THE INTERNATIONAL HARVESTER COMPANY
5556 Brookville Road, Indianapolis, Indiana

This visit was made primarily to see automation in truck-engine production. The plant at Indianapolis employs about 4000 men and includes a large, very well organized iron foundry producing 600 tons of iron castings per day. The daily output of cylinder-block castings is 1400, of which 500 are dispatched to subsidiary engine manufacturers and 900 are machined at the Indianapolis works and built up into five main engine types. These are all either 6-cylinder in-line engines or V-8 engines. The foundry is equipped with the latest Osborn and Sutter core-blowing machines. (Osborn Manufacturing Company, Cleveland, Ohio, and Sutter Products Company, Dearborn, Mich.). It is experimenting with shell-moulded cores, which are being produced by Sutter, and some excellent examples are used for cylinder-block water spaces and steering boxes. A shell-moulding served primarily for receiving, processing and transmitting data which is to be transferred out.

OVERFLOW. In an arithmetic operation, the generation of a quantity beyond the capacity of the register or location which is to receive the result; over capacity; the information contained in an item of information which is in excess of a given amount.

OVERSHOOT. Occurs when the process exceeds the target value as operating conditions change.

P

PACK. To include several brief or minor items of information into one machine item or word by utilizing different sets of digits for the specification of each brief or minor item.

PARALLEL. Handled simultaneously in separate facilities; operating on two or more parts of a word or item simultaneously; contrasted with serial.

PARAMETER. In a subroutine, a quantity which may be given different values when the subroutine is used in different main routines or in different parts of one main routine, but which usually remains unchanged throughout any one such use; in a generator, a quantity used to specify input-output devices, to designate subroutines to be included, or otherwise to describe the desired routine to be generated.

PARAMETER, PRESET. A parameter incorporated into a subroutine during input.

PARAMETER, PROGRAM. A parameter incorporated into a subroutine during computation. A program parameter frequently comprises a word stored relative to either the subroutine or the entry point and dealt with by the subroutine during each reference. It may be altered by the routine and/or may vary from one point of entry to another.

PATCH. Section of coding inserted into a routine to correct a mistake or alter the routine; explicitly transferring control from a routine to a section of coding and back again.

PENTODE. A five-electrode vacuum tube containing a cathode, control grid, suppressor grid, screen grid, and plate.

PERFORATION, RATE OF. Number of characters, rows or words punched in a paper tape by a device per unit of time.

PHASE SHIFT. A time difference between the input and output signal of a control unit or system.

PHOSPHORESCENCE. The property of emitting light for some time after excitation.

PIEZOELECTRIC. The effect of producing a voltage by placing a stress, either by compression, by expansion, or by twisting, on a crystal; and, conversely, the effect of producing a stress in a crystal by applying a voltage to it.

PLOTTING-BOARD. A unit capable of graphically presenting information, usually as curves of one or more variables; analogue curve or point tracer.

PLUG-BOARD. A removable panel containing an ordered array of terminals which may be interconnected by short electrical leads according to a prescribed pattern and hence designating a specific program. The entire panel, pre-wired, may be inserted for different programs.

POINT. The dot that marks the separation between the integral and fractional parts of a quantity; i.e., between the coefficients of the zero and the minus one powers of the number base. It is usually called, for a number system using base two, a *binary point;* for base ten, a *decimal point,* etc.; a base point; radix.

POST MORTEM. A routine which, either automatically or on demand, prints information concerning the contents of the registers and storage locations at the time the routine stopped, in order to assist in the location of a mistake in coding.

POTENTIOMETER. Probably the most versatile of all measuring instruments. It comes in 36 million variations. It can measure anything from bubble gum mix to nuclear energy generation by comparing the difference between known and unknown electrical potentials.

PRECISION. The degree of exactness with which a quantity is stated; a relative term often based on the number of significant digits in a measurement. See also Accuracy.

PRECISION, DOUBLE. Retention of twice as many digits of a quantity as the computer normally handles, e.g., a computer whose basic word consists of 10 decimal digits is called upon to handle 20 decimal digit quantities by keeping track of the 10-place fragments.

PRE-STORE. To set an initial value for the address of an operand or a cycle index; to restore; to store a quantity in an available or convenient location before it is required in a routine.

PROCESS. Actually, the system under control. It does not include the automatic control equipment.

PROGRAM. A plan for the solution of a problem. A complete program includes plans for the transcription of data, coding for the computer and plans for the absorption of the results into the system. The list of coded instructions is called a *routine;* to plan a computation or process from the asking of a question to the delivery of the results, including the integration of the operation into an existing system. Thus programming consists of planning and coding, including numerical analysis, systems analysis, specification of printing formats, and any other functions necessary to the integration of a computer in a system.

PROGRAM CONTROL. A control system which automatically holds or changes its target value on the basis of time to follow a prescribed "program" for the process. Setting the timer and thermostat in your oven at home to bake a cake is a simple example of this type of control.

PROGRAM SENSITIVE MALFUNCTION. A malfunction which occurs only when some unusual combination of program steps occur.

PROGRAMMER. A person who prepares instruction sequences without necessarily converting them into the detailed codes.

PROGRAMMING, AUTOMATIC. Any technique in which the computer is used to help plan as well as to help code a problem; e.g., compiling routines, interpretive routines.

PROGRAMMING, OPTIMUM. Improper terminology for minimal latency coding, i.e., for producing a minimal latency routine.

PROGRAMMING, RANDOM ACCESS. Programming without regard for the time required for access to the storage positions called for in the program; contrast with minimum access programming.

PROPORTIONAL BAND. The range of values of the condition being regulated which will cause the controller to operate over its full range. Usually expressed by engineers in terms of percentage of instrument full scale range.

PROPORTIONAL CONTROL. Control action related to the extent a condition being regulated is off-the-beam.

PSEUDO-CODE. An arbitrary code, independent of the hardware of a computer, which must be translated into computer code.

PSEUDO-RANDOM. Having the property of satisfying one or more of the standard criteria for statistical randomness but being produced by a definite calculation process.

PULSE. A change in the intensity or level of some medium, usually over a relatively short period of time, e.g., a shift in electric potential of a point for a short period of time compared to the time period, i.e., if the voltage level of a point shifts from −10 to +20 volts with respect to ground for a period of 2 microseconds, one says that the point received a 30 volt 2 microsecond pulse.

PULSE-CODE. Sets of pulses to which particular meanings have been assigned; the binary representations of characters.

PUNCH, CALCULATING, ELECTRONIC. A card handling machine which reads a punched card, performs a number of sequential operations and punches the result on a card.

PUNCH, CARD. A device which perforates or places holes in card in specific locations designated by a program.

PUNCH-POSITION. The location of the row in a columniated card, e.g., in an 80-column card the rows or "punch position" may be 0 to 9 or "X" and "Y" corresponding to position 11 and 12.

PUNCH, SUMMARY. A card handling machine which may be electrically connected to another machine, e.g., tabulator and which will punch out on a card the information produced, calculated or summarized by the other machine.

PUNCHING, RATE of. Number of cards, characters, blocks, fields or words of information placed in the form of holes distribution on cards, or tape per unit of time.

PYROMETER. *Pyro* is the Greek word for fire, and meter, a device for measuring. Thus, a pyrometer is an instrument for taking the temperature of a process. Not confined to measuring high temperature.

QUANTITY. A positive or negative real number in the mathematical sense. The term quantity is preferred to the term number in referring to numerical data; the term number is used in the sense of natural number and reserved for "the number of digits," the "number of operations," etc.

QUANTITY, DOUBLE-PRECISION. A quantity having twice as many digits as are normally carried in a specific computer.

R

RANDOM-ACCESS. Access to storage under conditions in which the next position from which information is to be obtained is in no way dependent on the previous one.

RANGE. All the values which a function may have.

RATE ACTION. A type of control action in which the rate of correction is made in proportion of how fast the condition has gone awry. Also called *derivative action*.

RATIO, OPERATING. The ratio obtained by dividing the number of hours of correct machine operation by the total hours of scheduled operation, e.g., on a 168 hour week scheduled operation, if 12 hours of preventive maintenance is required and 4.8 hours of unscheduled down time occurs, then the operating ratio is (168-16.8)/168, which is equavalent to a 90% operating ratio.

READ. To copy, usually from one form of storage to another, particularly from external or secondary storage to internal storage; to sense the meaning of arrangements of hardware; to sense the presence of information on a recording medium.

READ-AROUND-RATIO. In electrostatic storage tubes, the number of times a specific spot (digit or location) may be consulted before "spill over" will cause a loss of information stored in surrounding spots, immediately prior to which the surrounding information must be restored; read-around number.

READER, CARD. A mechanism that permits the sensing of information punched on cards by means of wire brushes or metal feelers.

READER, TAPE, MAGNETIC. A device capable of restoring to a train or sequence of electrical pulses, information recorded on a magnetic tape in the form of a series of magnetized spots, usually for the purpose of transferring the information to some other storage medium.

READER, TAPE, PAPER. A device capable of restoring to a train or sequence of electrical pulses, information punched on a paper tape in the form of a series of holes, usually for the purpose of transferring the information to some other storage medium.

READING, RATE of. Number of characters, words, fields, block or cards sensed by an input sensing device per unit of time.

REAL-TIME. The performance of a computation during the actual time that the related physical process transpires in order that results of the computations are useful in guiding the physical process.

RECORD. A listing of information, usually in printed or printable form; one output on a compiler consisting of a list of the operations and their positions in the final specific routine and containing information describing the segmentation and storage allocation of the routine; to copy or set down information in reusable form for future reference; to make a transcription of data by a systematic alteration of the condition, property or configuration of a physical medium, e.g., placing information on magnetic tape or a drum by means of magnetized spots.

REGENERATION. The process of returning a part of the output signal of an amplifier to its input circuit in such a manner that it reinforces the grid excitation and thereby increases the total amplification; periodic restoration of stored information.

REGISTER. The hardware for storing one or more computer words. Registers are usually zero-access storage devices.

REGISTER, CIRCULATING or MEMORY. A register (or memory) consisting of a means for delaying information and a means for regenerating and reinserting the information into the delaying means.

REGISTER, CONTROL. The accumulator, register or storage unit which stores the current instruction governing a computer operation; an instruction register.

REGISTER, PROGRAM. A register in the control unit which stores the current instruction of the program and controls computer operation during the execution of the instruction; control register; program counter.

REGULATION, VOLTAGE. A measure of the degree to which power source maintains its output-voltage stability under varying load conditions.

REPETITION, RATE of PULSE. The number of electric pulses per unit of time experienced by a point in a computer, usually the maximum, normal, or standard rate of pulses.

REPRESENTATIVE-CIRCULATING-TIME. A method of evaluating the speed performance of a computer. One method is to use one-tenth of the time required to perform nine complete additions and one complete multiplication. A complete addition or a complete multiplication time includes the time required to procure two operands from high speed storage, perform the operation, and store the result and the time required to select and execute the required number of instructions to do this.

RERUN. To repeat all or part of a program on a computer.

RERUN-POINT. That stage of a computer run at which all information pertinent to the running of the routine is available either to the routine itself or to a rerun routine in order that a run may be reconstituted.

RESET. To return a device to zero or to an initial or arbitrarily selected condition.

RESET ACTION. A type of control action in which the corrections are made in proportion to the length of time a condition has been off-the-beam and the amount of deviation.

RESET RATE. The number of corrections per minute made by the control system. Usually expressed as X number of repeats per minute.

RESISTANCE. A characteristic of an industrial process which opposes flow, either fluid or electrical. A partially open faucet resists the full flow of water.

REPRODUCIBILITY. Nothing at all to do with the spawning of new little robots. In instrument work it means the exactness with which measurement of a given value can be duplicated.

RESOLVER. A device which separates or breaks up a quantity, particularly a vector, into constituent parts or elements, e.g., to form three mutually perpendicular components of a space vector.

RESPONSE, FREQUENCY. A measure of the ability of a device to take into account, follow or act upon a rapidly varying condition, e.g., as applied to amplifiers, the frequency at which the gain has fallen to the one half power point or to 0.707 of the voltage gain factor; as applied to a mechanical controller, the maximum rate at which changes in condition can be followed and acted upon.

RESTORE. To return a cycle index, a variable address, or other computer word to its initial or pre-selected value; periodic regeneration of charge, especially in volatile, condenser-action storage systems.

RETURN. To go back to a specific, planned point in a program, usually when an error is detected, for the purpose of re-running the program. Rerun points are usually three to five minutes apart to avoid long periods of lost computer time. Information pertinent to a rerun is available in standby registers from point to point.

REWIND. To return a film or magnetic tape to its beginning.

ROLLBACK. Equivalent to rerun when referring to tape-sequenced computers; to recapture tape-inscribed data.

ROLL-OUT. To read a register or counter by adding ones

to the respective digits simultaneously obtaining a signal as each column returns to zero, until all columns have returned to zero, usually requiring n additions, where n is the number base.

ROUND-OFF. To change a more precise quantity to a less precise one, according to some rule.

ROUTINE. A set of coded instructions arranged in proper sequence to direct the computer to perform a desired operation or series of operations.

ROUTINE, COMPILING. An executive routine which, before th desired computation is started, translates a program expressed in pseudo-code into machine code (or into another pseudo-code for further translation by an interpreter). In accomplishing the translation, the compiler is required to decode, convert, select, generate, allocate, adapt, orient, incorporate, or record.

ROUTINE, DIAGNOSTIC. A specific routine designed to locate either a malfunction in the computer or a mistake in coding.

ROUTINE, EXECUTIVE. A set of coded instructions designed to process and control other sets of coded instructions; a set of coded instructions used in realizing "automatic coding"; a master set of coded instructions.

ROUTINE, FLOATING-POINT. A set of coded instructions arranged in proper sequence to direct the computer to perform a specific set of operations which will permit floating-point operation, e.g., enable the use of a fixed-point machine to handle information on a floating-point basis from an external point of view, floating-point routines are usually used in computers which do not have built in floating-point circuitry, in which case floating-point operation must be programmed.

ROUTINE, GENERAL. A routine expressed in computer coding designed to solve a class of problems, specializing to a specific problem when appropriate parametric values are supplied.

ROUTINE, INTERPRETIVE. An executive routine which, as the computation progresses, translates a stored program expressed in some machine-like pseudo-code into machine code and performs the indicated operations, by means of subroutines as they are translated. An interpretive routine is essentially a closed subroutine which operates successively on an indefinitely-long sequence of program parameters (the pseudo-instructions and operands). It may usually be entered as a closed subroutine and exited by a pseudo-code exit instruction.

ROUTINE, MINIMAL LATENCY. Especially in reference to serial storage systems, a routine so coded, by judicious arrangement of data and instructions in storage, that the actual latency is appreciably less than the expected random-access latency.

ROUTINE, RERUN. A routine designed to be used in the wake of a computer malfunction or a coding or operating mistake to reconstitute a routine from the last previous rerun point; roll back routine.

ROUTINE, SEQUENCE CHECKING. A routine which checks every instruction executed, printing certain data, e.g., to print out the coded instruction with addresses, and the contents of each of several registers, or it may be designed to print out only selected data, such as transfer instructions and the quantity actually transferred.

ROUTINE, SERVICE. A routine designed to assist in the actual operation of the computer. Tape comparison, block location, certain post mortems, and correction routines fall in this class.

ROUTINE, SPECIFIC. A routine expressed in computer coding designed to solve a particular mathematical, logical, or data-handling problem in which each address refers to explicitly stated registers and locations.

ROUTINE, TEST. A routine designed to show whether a computer is functioning properly or not.

RUN. One performance of a program on a computer; performance of one routine, or several routines automatically linked so that they form an operating unit, during which manual manipulations are not required of the computer operator.

S

SCALE. To alter the units in which all variables are expressed so as to bring all magnitudes within the capacity of the computer or routine at hand.

SCANNER. An instrument which automatically checks a number of measuring points and indicates which have wandered too far from their desired values.

SEGMENT. To divide a routine in parts each consisting of an integral number of subroutines, each part capable of being completely stored in the internal storage and containing the necessary instructions to jump to other segments; in a routine too long to fit into internal storage, a part short enough to be stored entirely in the internal storage and containing the coding necessary to call in and jump automatically to other segments. Routines which exceed internal storage capacity may be automatically divided into segments by a compiler.

SELECT. To take the alternative A if the report on a condition is of one state, and alternative B if the report on the condition is of another state; to choose a needed subroutine from a file of subroutines.

SELECTOR. A device which interrogates a condition and initiates a particular operation according to the interrogation report.

SENSE. To examine, particularly relative to a criterion; to determine the present arrangement of some element of hardware, especially a manually-set switch; to read holes punched in paper.

SENSITIVITY. The degree of response of an instrument or control unit to a change in the incoming signal.

SENTINEL. A symbol marking the beginning or the end of some element of information such as a field, item, block, tape, etc.; a tag.

SEQUENCE, PSEUDO-RANDOM. An order of numbers produced by a definite recursive rule but satisfying one or more of the standard tests for randomness.

SEQUENCER. A machine which puts items of information into a particular order, e.g., it will determine whether A is greater than, equal to, or less than B, and sort or order accordingly.

SERIAL. Handle one after the other in a single facility, such as transfer or store in a digit by digit time sequence.

SERVOMECHANISM. A type of closed loop control system in which mechanical *position* is the controlled variable. For example an anti-aircraft gun positioning system is a servomechanism. This term often used incorrectly with reference to all types of automatic control systems.

SERVO TECHNIQUES. Methods devised by engineers to study performance of servomechanisms or control systems.

SET POINT. The target value which the automatic unit strives to reach, or hold.

SHIFT. To move the characters to a unit of information column-wise right or left. For a number, this is equivalent to multiplying or dividing by a power of the base of notation.

SHIFT, ARITHMETIC. To multiply or divide a quantity by a power of the number base, e.g. binary 1011 represents decimal 11, therefore two arithmetic shifts to the left is binary 101100, which represents decimal 44.

SHIFT, CYCLIC. A shift in which the digits dropped off at one end of a word are returned at the other in a circular fashion; logical, non-arithmetical or circular shift.

SIGNAL. Information relayed from one point in the control system to another.

SIGNIFICANCE. The arbitrary rank, priority, or order of relative magnitude assigned to a given position or column in a number; the significant digits of a number are a set of digits, usually from consecutive columns beginning with the most significant digit different from zero and ending with the least significant digit whose value is known and assumed relevant, e.g., 2300.0 has five significant digits, whereas 2300 probably has two significant digits.

SIMULATION. The representation of physical systems and phenomena by computers, models or other equipment.

SINUSOIDAL. An adjective used to describe a type of signal which varies with time. The electricity in your house is a common example of a sinusoidal variation. In the language of the automatic control engineer, this term is used to describe the type of input signal introduced into a control system to study its control characteristics (see FREQUENCY RESPONSE).

SKIP. An instruction to proceed to the next instruction; a "blank" instruction.

SOLVER, EQUATION. A calculating device, usually analog, which arrives at the solution to systems of linear simultaneous non-differential equations or determines the roots of polynomials or both.

SORT. To arrange items of information according to rules dependent upon a key or field contained in the items.

STACKER, CARD. A mechanism that accumulates cards in a bin after they have passed through a machine operation; a hopper.

STANDARDIZE. To adjust the exponent and mantissa of a floating-point result so that the mantissa lies in the prescribed normal range; normalize; see Floating-point Representation.

STATIC BEHAVIOR. Describes how a control system, or an individual unit, carries on under fixed conditions (as contrasted to dynamic behavior which refers to behavior under changing conditions).

STEP CHANGE. Simply the change from one value to another in a single step.

STORAGE. Preferred to memory, any device into which units of information can be copied, which will hold this information, and from which the information can be obtained at a later time; devices, such as plugboards, which hold information in the form of arrangements of physical elements, hardware, or equipment; the erasable storage in any given computer.

STORAGE, BUFFER. A synchronizing element between two different forms of storage, usually between internal and external; an input device in which information is assembled from external or secondary storage and stored ready for transfer to internal storage; an output device into which information is copied from internal storage and held for transfer to secondary or external storage. Computation continues while transfers between buffer storage and secondary or internal storage or vice versa take place.

STORAGE, CIRCULATING. A device using a delay line, or unit which stores information in a train or pattern of pulses, where the pattern of pulses issuing at the final end are sensed, amplified, reshaped, and re-inserted in the delay line at the beginning end.

STORAGE, DYNAMIC. Storage such that information at a certain position is moving in time and so is not always available instantly; e.g., acoustic delay line, magnetic drum; circulating or re-circulating of information in a medium.

STORAGE, ELECTROSTATIC. A device possessing the capability of storing changeable information in the form of charged or uncharged areas on the screen of a cathode ray tube.

STORAGE, ERASABLE. Media which may hold information that can be changed; i.e., the media can be re-used; e.g., magnetic tape, drum or core.

STORAGE, EXTERNAL. Storage facilities divorced from the computer itself but holding information in the form prescribed for the computer; e.g., magnetic tapes, magnetic wire, punched cards, etc.

STORAGE, INTERNAL. Storage facilities forming an integral physical part of the computer and directly controlled by the computer; the total storage automatically accessible to the computer.

STORAGE, MAGNETIC. Any storage system which utilizes the magnetic properties of materials to store information.

STORAGE, MERCURY. Columns of a liquid mercury medium used as a storage element by the delaying action or time of travel of sonic pulses which are circulated by having electrical amplifier, shaper, and timer circuits complete the loop.

STORAGE, NON-ERASABLE. Media used for containing information which cannot be erased and reused, such as punched paper tapes, and punched cards.

STORAGE, NON-VOLATILE. Storage media which retain information in the absence of power and which may be made available upon restoration of power; e.g., magnetic tapes, drums, or cores.

STORAGE, PARALLEL. Storage in which all bits, or characters, or (especially) words are essentially equally available in space, without time being one of the coordinates. Parallel storage contrasts with serial storage. When words are in parallel, the storage is said to be *parallel by words;* when characters within words (or binary digits within words or characters) are dealt with simultaneously, not one after the other, the storage is *parallel by characters* (or *parallel by bit* respectively) .

STORAGE, SECONDARY. Storage facilities not an integral part of the computer but directly connected to and controlled by the computer; e.g., magnetic drum, magnetic tapes, etc.

STORAGE, SERIAL. Storage in which time is one of the coordinates used to locate any given bit, character, or (especially) word. Storage in which words, within given groups of several words, appear one after the other in time sequence, and in which access time therefore includes a variable latency or waiting time of zero to many word-times, is said to be *serial by word.* Storage in which the individual bits comprising a word appear in time sequence is *serial by bit.* Storage for coded-decimal or other non-binary numbers in which the characters appear in time sequence is *serial by character;* e.g., magnetic drums are usually serial by word but may be serial by bit, or parallel by bit, or serial by character and parallel by bit, etc.

STORAGE, STATIC. Storage such that information is fixed in space and available at any time; e.g., flip-flop, electrostatic, or coincident-current magnetic-core storage.

STORAGE, TEMPORARY. Internal storage locations reserved for intermediate and partial results.

STORAGE, VOLATILE. Storage media such that if the applied power is cut off, the stored information is lost; e.g., acoustic delay lines, electrostatic tubes.

STORAGE, WORKING. A portion of the internal storage reserved for the data upon which operations are being performed.

STORAGE, ZERO-ACCESS. Storage for which the latency (waiting time) is negligible at all times.

STORE. To transfer an element of information to a device from which the unaltered information can be obtained at a later time.

STRAIN GAGE. A measuring element (transducer) which can be used to convert a force, pressure, tension, etc., into an electrical signal. The signal is then fed to an instrument for measurement and, if desired, control.

SUBROUTINE. The set of instructions necessary to direct the computer to carry out a well defined mathematical or logical operation; a subunit of a routine. A subroutine is often written in relative or symbolic coding even when the routine to which it belongs is not.

SUBROUTINE, CLOSED. A subroutine not stored in its proper place in the linear operational sequence, but stored away from the routine which refers to it. Such a subroutine is entered by a jump, and provision is made to return, i.e., to jump back to the proper point in the main routine at the end of the subroutine.

SUBROUTINE, DYNAMIC. A subroutine which involves parameters, such as decimal point position or item size, from which a relatively coded subroutine is derived. The computer itself is expected to adjust or generate the subroutine according to the parametric values chosen.

SUBROUTINE, OPEN. A subroutine inserted directly into the linear operational sequence, not entered by a jump. Such a subroutine must be recopied at each point that it is needed in a routine.

SUBROUTINE, STATIC. A subroutine which involves no parameters other than the addresses of the operands.

SUBSTITUTE. To replace an element of information by some other element of information.

SUPERVISORY COUNCIL. A control system which furnishes intelligence, usually to a centralized location, to be used by an operator to supervise the control of a process or operation.

SWITCH, ELECTRONIC. A circuit which causes a start-and-stop action or a switching action by electronic means.

SWITCH, FUNCTION. A circuit having a fixed number of inputs and outputs designed such that the output information is a function of the input information, each expressed in a certain code or signal configuration or pattern.

SYSTEM ENGINEERING. A method of engineering approach which takes into consideration all of the elements in the control system, down to the smallest valve, and the process itself. It is believed to have the most promise as an intelligent approach leading toward fuller industrial automation.

SYMBOL, LOGICAL. A symbol used to represent a logical element graphically.

SYSTEM. An assembly of components united by some form of regulated interaction; an organized whole.

T

TABULATOR. A machine which reads information from one medium, e.g., cards, paper tape, magnetic tape, etc. and produces lists, tables, and totals on separate forms or continuous paper.

TAG. A unit of information, whose composition differs from that of other members of the set so that it can be used as a marker or label; a sentinel.

TANK. A unit of acoustic delay line storage, containing a set of channels each forming a separate recirculation path; a circuit consisting of inductance and capacitance used for the purpose of sustaining electrical oscillations.

TAPE, MAGNETIC. A tape or ribbon of any material impregnated or coated with magnetic material on which information may be placed in the form of magnetically polarized spots.

TAPE, PROGRAM. A tape which contains the sequence of instructions required for solving a problem and which may be read by the computer.

TELEMETERING. Transmission of a measurement over long distances, usually by electrical means. A receiving instrument converts the transmitted electrical signals into units of whatever is being measured.

TERNARY. Pertaining to the system of notation utilizing the base of 3, employing the characters 0, 1, and 2.

TEST, CRIPPLED-LEAP-FROG. A variation of the leap-frog test, modified so that it repeats its tests from a single set of storage locations rather than a changing set of locations.

TEST, LEAP-FROG. A program designed to discover computer malfunction, characterized by the property that it performs a series of arithmetical or logical operations on one group of storage locations, transfers itself to another group of storage locations, checks the correctness of the transfer, then begins the series of operations over again. Eventually, all storage positions will have been occupied and the test will be repeated.

TETRAD. A group of four, usually four pulses, in particular, a group of four pulses used to express a decimal digit, or a sexadecimal digit by means of four (binary) pulses.

TETRODE. A four-electrode vacuum tube containing a cathode, control grid, screen grid, and plate.

THERMISTOR. A special type of temperature sensing element. Its extreme sensitivity permits it to transmit a strong signal from a very tiny temperature change, or it is made in many shapes, such as beads, disks, flakes, washers, and rods, to which contact wires are attached. As its temperature is changed, the electrical resistance of the thermistor varies. The associated temperature coefficient of resistance is extremely high, nonlinear, and negative.

THERMOCOUPLE. A temperature sensing element that creates an electrical signal in proportion to the temperature at the element.

THREE-ADDRESS. See Code, Multiple-address.

THYRATRON. A hot-cathode, gas-discharge tube in which one or more electrodes are used to control electrostatically the starting of an unidirectional flow of current.

TIME, CODE CHECKING. All time spent checking out a problem on the machine making sure that the problem is set up correctly, and that the code is correct.

TIME, ENGINEERING or SERVICING. All machine down time necessary for routine testing (good or bad), for machine servicing due to breakdowns, or for preventing servicing measures, e.g., block tube changes. Includes all test time (good or bad) following breakdown and subsequent repair or preventive servicing.

TIME, IDLE. Time in which machine is believed to be in good operating condition and attended by service engineers but not in use on problems. To verify that the machine is in good operating condition, machine tests of the leap-frog variety may be run.

TIME, NO CHARGE MACHINE-FAULT. Unproductive time due to a computer fault such as the following: (1) non-duplication, (2) transcribing error, (3) input-output malfunction, (4) machine malfunction resulting in an incomplete run.

TIME, NO CHARGE NON-MACHINE-FAULT. Unproductive time due to no fault of the computer such as the following: (1) good duplication, (2) error in preparation of input data, (3) error in arranging the program deck, (4) error in operating instructions or misinterpretation of instructions, (5) unscheduled good testing time, run during normal production period when machine malfunction is suspected but is demonstrated not to exist.

TIME, PRODUCTION. Good computing time, including occasional duplication of one case for a check of rerunning of the test run. Also, duplication requested by the sponsor; any reruns caused by misinformation or bad data supplied by sponsor. Error studies using different intervals, convergence criteria, etc.

TIME, STANDBY UNATTENDED. Time in which the machine is in an unknown condition and not in use of problems. Includes time in which machine is known to be defective and work is not being done to restore it to operating condition. Includes breakdowns which render it unavailable due to outside conditions (power failure, etc.).

TIME, SYSTEM, IMPROVEMENT. All machine down time needed for the installation and testing of new components, large or small, and machine down time necessary for modification of existing components. Includes all programmed tests following the above actions to prove machine is operating properly.

TRACK. In a serial magnetic storage element, a single path containing a set of pulses.

TRANSCRIBE. To copy, with or without translating, from one external storage medium to another.

TRANSDUCER. An element which converts one form of energy into another. Usually refers to an element creating a signal in relation to a condition being measured, like a thermocouple, or strain gage, e.g., a quartz crystal imbedded in mercury can change electrical energy to sound energy as is done in sonic delay lines in computer storage systems.

TRANSFER, CONDITIONALLY. To copy, exchange, read, record, store, transmit, or write data or to change control or jump to another location according to a certain specified rule or in accordance with a certain criterion.

TRANSFER FUNCTION. Simply a mathematical expression of the control engineer which expresses the relationship between the outgoing and incoming signals of a process, or control element. Useful in studies of control problems.

TRANSFER, PARALLEL. A system of data transfer in which the characters of an element of information are transferred simultaneously over a set of paths.

TRANSFER, SERIAL. A system of data transfer in which the characters of an element of information are transferred in sequence over a single path in consecutive time positions.

TRANSFER, UNCONDITIONAL. An instruction which causes the subsequent instruction to be taken from an address which is not the next one in the sequence in a digital computer which ordinarily obtains its instructions serially from an ordered sequence at all other times.

TRANSFORM. To change information in structure or composition without altering the meaning or value; to normalize, edit, or substitute.

TRANSIENT STATE. Generally implies a temporarily abnormal condition of a variable like speed, temperature, or pressure, which is changing erratically. Contrasted to this is "steady-state" which means that the variable is either held at a constant value, or else changes uniformly with time.

TRANSISTOR. Tiny element in an electronic circuit that does much the same job as a vacuum tube. It is light, practically unbreakable, long-lived, and highly efficient.

TRANSLATE. To change information (e.g., problem statements in pseudo-code, data, or coding) from one language to another without significantly affecting the meaning.

TRANSMIT. To reproduce information in a new location replacing whatever was previously stored and clearing or erasing the source of the information.

TRANSPORT. To convey as a whole from one storage device to another.

TROUBLE-SHOOT. To search for a coding mistake or the cause of a computer malfunction in order to remove same.

TRUNCATE. To drop digits of a number of terms of a series thus lessening precision, e.g. the number 3.14159265 is truncated to five figures in 3.1415, whereas one may round off to 3.1416.

TRUNK. A path over which information is transferred; a bus.

TUBE, ACORN. A small vacuum tube designed for ultra-high-frequency circuits. The tube has short electron transit time and low interelectrode capacity.

TUBE, CATHODE-RAY. An electronic vacuum tube containing a screen on which information may be stored by means of a multigrid modulated beam of electrons from the thermionic emitter, storage effected by means of charged or uncharged spots; a storage tube; a Williams tube; an oscilloscope tube; a picture tube.

TUBE, WILLIAMS. A cathode-ray tube used as an electrostatic storage device of the type designed by F. C. Williams, University of Manchester, England.

TWO-ADDRESS. See Code, Multiple-address.

TYPEWRITER, ELECTRIC. A hand operated electric powered individual character printing device having the property that almost every operation of the machine after the keys are touched by human fingers is performed by electric power instead of manual power; a typewriter powered by electricity, in all other respects the same as a manually powered typewriter.

U

ULTRASONICS. A new term to describe a range of vibration frequencies well above that which can be heard by the human ear. No limit has yet been put on the upper limits. Vibrations in this range cause unusual and helpful phenomena. For example, the manner in which ultrasonic vibrations penetrate solids permits a new method of testing for internal flaws (also called "black sound").

UNCONDITIONAL. Not subject to conditions external to the specific instruction.

UNPACK. To decompose packed information into a sequence of separate words or elements.

UNWIND. To code explicitly, at length and in full all the operations of a cycle thus eliminating all red-tape operations in the final problem coding. Unwinding may be performed automatically by the computer during assembly, generation, or compilation.

V

VALIDITY. Correctness; especially the degree of the closeness by which iterated results approach the correct result.

VARIABLE. A factor or condition which can be measured, altered or controlled, i.e., temperature, pressure, flow, liquid.

VARISTOR. A passive resistor-like circuit element whose resistance is a function of the current through it or voltage across its terminals, i.e. the current through it is a non-linear function of the voltage across its terminals, hence the linear form of Ohm's Law is not obeyed; a self-varying resistance.

VERIFIER. A device on which a manual transcription can be verified by comparing a retranscription with it character-by-character as it is being retranscribed.

VERIFY. To check a data transfer or transcription, especially those involving manual processes.

W

WIRE, MAGNETIC. Wire made of a magnetic material along small incremental lengths of which magnetic dipoles are placed in accordance with binary information.

WORD. A set of characters which occupies one storage location and is treated by the computer circuits as a unit and transported as such. Ordinarily a word is treated by the control unit as an instruction, and by the arithmetic unit as a quantity. Word lengths are fixed or variable depending on the particular computer.

WORD, INFORMATION. An ordered set of characters bearing at least one meaning and handled by a computer as a unit, which may be contrasted with instruction words.

WORD-TIME. Especially in reference to words stored serially, the time required to transport one word from one storage device to another. See also Access Time.

WRITE. To transfer information to an output medium; to copy, usually from internal storage to external storage; to record information in a register, location, or other storage device or medium.

Z

ZERO. Nothing; positive binary zero is usually indicated by the absence of digits or pulses in a word; negative binary zero in a computer operating on one's complements by a pulse in every pulse position in a word; in a coded decimal machine, decimal zero and binary zero may not have the same representation. In most computers, there exist distinct and valid representation both for plus and for minus zero.

ZERO-SUPPRESSION. The editing or elimination of non-significant zeros to the left of the integral part of a quantity before printing operations are initiated; a part of editing.

ZONE. A portion of internal storage allocated for a particular function or purpose; any of the three top positions of 12, 11 and 0 on a punch card. In these zone positions, a second punch can be inserted so that with punches in the remaining positions 1 to 9, alphabetic characters may be represented.

THIRTY-SEVEN SHORT CASE HISTORIES OF AUTOMATION IN THE U. S. AND CANADA*

AIRCRAFT AND AIRCRAFT ENGINES

CONVAIR DIVISION OF THE GENERAL DYNAMICS CORPORATION
San Diego, California

This visit was made because it was understood** that Convair had taken over from the Massachusetts Institute of Technology the development of a practical digitally controlled machine-tool for the United States Air Force, Air Materiel Command. The purpose was to see what progress had been made.

Digitally controlled machine-tools

It appears that the U.S. Air Force originally gave the Institute a direct contract to develop an electronic digitally controlled universal miler because no one else would undertake the job. When the Institute showed that it could be done, the Air Force gave Convair a contract to apply the machine-tool to the manufacture of airframe components, and suggested that the firm convert a Bridgewater universal milling machine (manufactured by the Bridgewater Machine Tool Company, Akron, Ohio) to elec-

---

* This appendix presents the technical findings of S. B. Bailey, M. Sc. (Eng.), H. I. Mech. E., a Department of Scientific and Industrial Research staff engineer in the United Kingdom who visited factories, universities and research laboratories in the U. S. A. and Canada concerned with one or more aspects of automation, during a six months visit in North America in 1955-56. The excerpts reproduced here from his report *Automation in North America* (Overseas Technical Reports No. 3), Department of Scientific and Industrial Research (London, 1958), are reprinted with the special permission of the Controller of Her Britannic Majesty's Stationery Office, London, and offer a representative picture of automation in North American industry today.

** Convair Division plans its automatic milling machine, *Control Engineering*, June 15, 1955.

tronic control. This milling machine has only one cutter head, and Convair are convinced that it is not the right answer to the aircraft industry's problems, which are:

(1) milling the forged light-alloy bulkheads, which transmit load from the main spars to the fuselage and which serve as the main fuselage joints. This involves extremely complex contour milling; also slotting to accommodate the tongues on the spar ends; and

(2) milling forged light-alloy wing spars. With present methods, the spar booms are straight and flat and their use results in flats on the wing profile, which spoil the correct aerofoil section. At modern flight speeds this causes serious aerodynamic troubles, which aeronautical engineers wish to avoid by milling the booms with the correct curvature. This means double curvature, because of the effect of wing sweep-back.

Convair are of the opinion that at least three cutting heads are desirable, and they now have to convince the U.S. Air Force that this is so. For this purpose they have carried out a complete analysis of the operations necessary to produce bulkheads and spars on a Bridgewater-type machine and on their own projected machine, and they find the latter should do the job in about one-third the time of the former.

If the U.S. Air Force accepts these results, the firm will probably get a contract to produce a three- or four-head milling machine, and they will sub-contract the construction of this to a suitable firm or firms. There is nothing to prevent them from dividing the contract between a machine-tool firm and an electronic-computer firm, but the machine-tool company will have to provide the guarantees for the latter. British firms which approach Convair regarding the use of electronic analogue or digital computers for machine-tool control will be considered along with U.S. firms, but they will have to be able to offer automatic control in at least seven channels.

It seems evident that the U.S. aircraft industry is almost as dissatisfied with the forward thinking of the U.S. machine-tool industry as the British electronics industry is with the British machine-tool industry—and for the same reason. Both machine-tool industries can sell all the catalogue-type tools they produce and they evidently do not consider there is sufficient demand for electronically controlled machine-tools to justify the capital risked in their development; hence the U.S. Air Force contracts with M.I.T. and Convair.

Automatic riveting machines

Convair have two 'Drivmatic' automatic riveting machines (General Riveters, Buffalo, N.Y.). In their original form, these machines clamped the plates to be riveted, drilled the hole, inserted the rivet and then squeezed it. More recently, they have become available with automatic workpiece transfer using punched-tape control. A 35-mm. film is run through a film reader, which is mechanically coupled to the work table so that one inch of film movement corresponds to four inches of table movement. A template, on which the rivet positions have been marked out, is clamped to the table, which is then traversed until the first centrepop comes under the drill point. A hole about two mm. square is then punched in the film, the table is moved on to the next rivet position, another hole is punched, and so on. When the tape is fully punched it is wound back and the workpieces are clamped to the table.

In operation, an electropneumatic hoist motor is used to traverse the machine table and, with the table, to move the film. A contact on the film reader falls through the square hole in the film and stops the table motion. The plates are then clamped, the hole is drilled, and the rivet is inserted and squeezed, all automatically. In practice, there are two reading heads on the tape reader, the first of which anticipates the coming instruction from the second and slows the table movement down so that it will not overshoot on receiving the second instruction. The table motion is effected quite crudely through a system of roller chains and sprockets, and inevitably the system has considerable backlash. Convair said that it could not rely on the system to control the rivet spacing more accurately than ± 1/32 in. even if traversing was arranged to be always in the same direction. They have modified one machine so that automatic transverse movement is provided in addition to the normal longitudinal movement. On later models, this provision is made by the manufacturer.

Convair said that they have on order a punched-tape-controlled drilling machine from the McKay Company of Youngstown, Ohio. The controller for this machine is being made by the Automatic Temperature Controller Company of Philadelphia,

Pa. the specification calls for a hole-spacing accuracy of ± 0.0001 in., but it seems unlikely that this will be achieved. Convair will be satisfied if the hole positions are within the ± 0.005 in. normally specified for the spacing of drilled holes.

DOUGLAS AIRCRAFT COMPANY
Long Beach, California

This visit was arranged for a discussion of the firm's experience with a 'Drivmatic' tape-controlled automatic riveter. Douglas have three of these machines, all equipped by the makers with automatic transverse as well as longitudinal table motion; the largest has a worktable 54 ft long × 12 ft wide, which runs on a track 120 ft long, and all three machines will place rivets up to ⅜ in. in diameter. Douglas are very pleased with the machines and have found that their rivet-spacing accuracy can be held to ± 0.005 in. The ratio of film speed to table speed was about 1:1; perhaps that is the reason why they are more accurate than those operated by Convair.

The first machines that Douglas bought were equipped for feeding conventional countersunk-head rivets, but it is now common practice to use I-section extruded stringers, and it is often impossible to find room to insert a headed rivet mechanically in the space available between the flanges of the stringer. Consequently, the machines have now been modified by the makers to use slugs instead of rivets. Each slug carries along the middle portion of its length a soft aluminium sleeve 0·004 in. thick, which acts as a seal when the slug is squeezed in position. In action, the machine moves the table to the desired position, clamps the plate to the stringer, drills the hole, inserts the slug, squeeze-forms the lower head, hammer-forms the upper (countersunk) head and mills it flush with the surface. It then moves on to the next position automatically. It drives up to five rivets a minute with much greater consistency than is possible by hand riveting or by a hand-held rivet squeezer. On one large aircraft recently constructed, 'Drivmatic' riveters placed over 35 000 rivets in the integral fuel-tank bays and not a single one leaked under test.

Douglas are very interested in electronic machine-tool control, particularly for profiling very thick wing skins (on modern fighter aircraft the skin is now as much as 0·5 in. thick), and such components as bulkheads and wing-root brackets. They are closely in touch with recent developments at the Cincinnati Milling Machine Co., the Bendix Aviation Corporation, Electronic Control System, Transducers, and Giddings and Lewis in the U.S.A., and at Ferranti and Electrical and Musical Industries (E.M.I.) in the U.K.

ORENDA ENGINES
Malton, Ontario

About 5000 people are employed at this very modern plant. Little can be done in the way of automation because modifications change the design of jet engines so frequently that not enough of one design is usually put through to justify automation. However, two automatic machine tools are worth describing:

(1) The Pratt and Whitney blade-profile grinding machine (Pratt and Whitney Company, West Hartford, Conn.), which grinds finished turbine blades from the rough forging, using an endless grinding tape about 0·5 in. wide. This machine has three master cams: one to produce the blade profile, another to twist the axes of the rollers carrying the grinding tape so that the tape is always tangential to the surface desired, and the third to vary the speed of rotation as the blade profile demands.

(2) The 'Man-au-trol' automatic turret-boring mill (Bullard Machine Tool Company, Bridgeport, Conn.), which has up to forty motions automatically controlled by sets of cam drums and longitudinal positioning controls, a pair for each motion. The longitudinal controls are by means of sliders, which can be pre-set on taut wires and which actuate micro-switches. Machines of this type are being used for boring the grooves and other internal profiles of compressor stator casings. Their accuracy was stated to be ± 0.002 in. on a diameter of some 30 in.

Throughout the shops, the system of line flow had been adopted wherever possible, with a separate machine for each operation. As a result, some machine tools were only in use for an hour per day, but the production time for each component was the minimum. With this system, production could be stepped up very greatly in time of need simply by increasing the manpower available.

One or two special-purpose machines have been made up to the firm's special requirements by the Modern Tool Company, Toronto, Ontario.

Orenda Engines make a very favourable impression. It must have been a difficult task to build up a firm of this kind in a country with no previous experience of building aircraft engines. The result reflects the greatest credit on those responsible.

MOTOR VEHICLES AND ACCESSORIES

CHRYSLER CORPORATION, MOUND ROAD
ENGINE PLANT
Detroit, Michigan

This is probably the most modern engine plant in the American motor-car industry, as it was only completed in October 1955. It is more highly automatic than any other mechanical engineering plant seen in the U.K. or in North America, and is an excellent example of what can be achieved when the engine designer and production engineer work together from the start, and the plant engineer is given a completely free hand.

The whole of the manufacturing processes at the plant has been described in American journals.* The following points are of particular interest.

Product design for automatic production

Hardly any nuts or bolts on the engine had to be tightened by hand because they were inaccessible to mechanical devices.

Machining of joint faces

All joint faces are broached. The two mutually perpendicular faces on the block are broached in one operation by an enormous Cincinnati machine, which removes up to 0·25 in. of metal from the rough casting in one stroke. The broach carriage is mechanically driven (as in a planing machine) to avoid the vibration troubles sometimes experienced with hydraulic drive.

---

* *American Machinist* (August 15, 1955), and *Tooling for Production*, (October, 1955).

Arrangement of tool spindles

All multiple-head drilling and tapping machines are arranged with their spindles horizontal instead of being inclined at 45°, as is usual in the production of V-8 engines. Arranged in this manner the drills are accessible and can be changed very easily by the operator, but the transfer machine has to rotate the block when necessary during its passage down the line.

Concentricity of valve seats and valve guide bores

This is ensured by performing the valve-seat cutting and valve guide housing boring operations on the same machine tool at the same setting.

Assembly of connecting rods and caps

Before a connecting rod big-end is bored, the cap and big-end bolts must be fitted, and the nuts fully tightened. At Mound Road, the whole operation is done automatically on an *'Indexomatic' machine* (Russell T. Gilman, Janesville, Wis.). The rods and caps are brought to the machine by conveyor and automatically placed on an indexing turret, so that they face each other correctly. The big-end bolts are fed from a hopper, correctly oriented, and pressed into position. Nuts are brought from a hopper by a *'Syntron'* conveyor (Syntron Company, Homer City, Pa.), correctly oriented, and then fed into pneumatic wrenches, which screw them on to the bolts and apply the correct tightening torque. The assembled rod is then ejected through a chute on to a conveyor, which carries it off for boring. The whole of this operation is done automatically—with three men watching to clear stoppages and fill the hoppers with nuts and bolts.

Removal of machining swarf

Despite the enormous output from this plant, working conditions were very much better than those found in the average British motor factory. There were no fumes from hot cutting fluids; no piles of machining swarf on the floor; no oil and grease underfoot. The reasons are first, that in the excellently designed plant all swarf and cutting fluids fall freely into large ducts below the transfer machines; and second, that there is a system of forced ventilation, which extracts over half a million cubic feet of air a minute through the ducts already mentioned, passes them through a highly efficient filtering plant, and then returns them. This forced ventilation carries away the fine swarf from the cast-iron machining operations as quickly as it is produced.

The Chrysler Corporation describes the plant as follows:

'The plant engineer did not particularly care for the fine dust sheen that would be evident on plant surfaces after production would get underway. He was interested in a collector that would operate at 95 per cent efficiency and remove dust down to 0·3 microns in lieu of, for example, 98 per cent efficiency in the removal of dust particles in the 3·5 to 5 micron range.

'A 10 000 cfm unit was selected, with face velocities at the heads to run about 1500 fpm. A representative collector installation uses a tunnel for its main inlet trunk line. The tunnel, which has a conveyor for disposing of cast iron chips, runs directly below a line of machines which straddle it. This dual purpose trough is covered with relatively air-tight steel panels for ease of inspection and maintenance. There are roughly 5500 ft. of the drag chain conveyor for carrying chips to outside overhead bins.

'The dust-laden air flows through a tunnel on the way to the aerodyne collector. This passageway serves as a plenum chamber with relatively low air velocity, allowing heavy particles to settle out.

'After flowing through the fan to the collector, the dust-laden air enters a metal cone, whose surface has been perforated to provide many air paths as well as change the direction of air flow. The heavier particles have a tendency to fall back into the stream inside the cone and go on into a cyclone. The lighter air from the cyclone exhaust opening at the top is returned to the fan and again enters the cone. In the meantime, the heavy dust particles drop through an air valve into the tunnel. The duct through which the particles fall from the collector is extended to a few inches above the drag chain conveyor to prevent re-entry of the particles into the collection system.

'Air that has passed through the cone continues through

the filter media where the very fine dust is removed, and is returned to the room in an exceptionally clean state.'

Automatic hopper-loading of trays with small parts

Each engine built requires a trayful of small components like bolts, nuts, washers and plugs. In the Mound Road engine plant, the trays are filled from an automatic store. They move past the store to the same time-cycle as the engine-assembly transfer line (16 seconds dwell—8 seconds move). At each dwell period *en route* through the store, the twenty or so compartments in the tray are filled from hoppers employing a '*Syntron*' feed and simple counting mechanisms. The twenty or more hoppers are replenished when necessary by a couple of unskilled labourers who need not even be able to count.

Tool and cutter life

All tool lives are determined by '*Toolometers*' (The Cross Company, Detroit, Mich.) which shut the transfer machine down when any tool has performed its scheduled number of operations. The operator can see at a glance which other tools are due for replacement soon and he can act accordingly.

Transfer-machine control panels

These are arranged on bridges over the lines they control. Each panel carries a circut diagram with a wandering test prod, which can be used to locate faults very rapidly. This arrangement keeps the panels, relays, etc., right away from cutting fluids, chips, etc., which might impede their operation, and ensures that the faults can be located and rectified in about a quarter of the time required.

Automatic engine-assembly line

A machine takes the five camshaft bushes, turns them round until their oil holes are facing in the right direction, then presses them into the bores in the crankcase—all this is completely automatic. Then the eight sets of big-end bolts are tightened automatically at four separate transfer-machine stations. Subsequently, other bolts are tightened automatically, each group at one station;

first all those that secure the sump, then in turn all those that
secure each cylinder head, the rocker-shaft pedestals and the induction manifold. This is hardly automatic assembly, but all the
hard physical work is done by automatic machines. It would not
be possible if the engine were not designed so that all bolts were
directly approachable with a box spanner of indefinite length.
All the above operations are performed by compressed-air motors,
with automatic torque control, so that all bolts are uniformly
tightened.

Automatic engine-testing line

After assembly and aluminium spraying (the latter with electrostatic reduction of overspray), the completed engines are automatically moved into a separate bay having 72 test stands. Then
the following operations are done automatically, without human
interference. An empty stand is selected, the engine is moved into
it, all oil, water, fuel and electrical connections are made, and the
engine is started up and run for 15 minutes. After this a signal
lamp summons an inspector who comes and checks the engine for
correct functioning, signs the inspection card, and releases it on
its journey.

Safety

Throughout the length of all transfer machine lines, a thin
steel wire rope runs along each side of the track along which the
product moves. This functions like the communication cord in a
railway carriage, so that pulling the cord stops the machine spindles and the transfer mechanism.

DELCO-REMY DIVISION OF GENERAL MOTORS
2401 Columbus Avenue, Anderson, Indiana

This visit was made in order to study automation in a large
auto-electrical works. The arrangements at the plant were made
by the manufacturing manager, typical of the young energetic
engineer who quickly assumes high responsibility in the United
States. In his early forties, he is in charge of all manufacturing
processes and is responsible for the development of all special
manufacturing machinery.

Delco-Remy has 16 000 employees (nearly a third of them

women, who get equal pay for equal work) in five main plants situated near the centre of the town; no two are more than a mile apart. It is one of the largest firms in the town and makes auto-electrical equipment including horns, switches, ignition coils, distributors, generators, starter motors and voltage regulators. The output is enormous; for instance about 7½ million ignition coils are produced a year.

The manufacturing plants abound with special-purpose machines of the firm's own design and manufacture; they are used for such purposes as coil winding, condenser winding, armature winding and commutator building. There is a very high degree of automation everywhere, but possibly 80 per cent of the components fitted to assemblies on conveyors are fitted by hand, often with the assistance of powered screwdrivers, nut-runners, etc. 'Syntron' conveyors are widely used for lifting small parts on to chutes, and for feeding automatic screwdrivers. Most screws are automatically fitted and driven, but all nuts are placed on their mating threads by hand and are only run down and tightened mechanically. One ingenious device enables all condenser cans fed down a chute to arrive at the bottom right way up.

A new product at the time of the visit was an ignition distributor, whose moulded cover has a side-access door through which the contact points can be adjusted, while the engine is running, without removing the distributor cap.

The iron foundry, which makes all the housings and pole pieces used in Delco-Remy electrical equipment, did not seem to be quite as well-organized as the assembly plants, although there was evidently a very high rate of production, partly owing to the use of rotary-table machines for continuous moulding and casting.

FORD MOTOR COMPANY ENGINE PLANT AND FOUNDRY
Brookpark, Cleveland, Ohio

Foundry

The foundry serves Nos. 1 and 2 Engine Plants, which are placed next to it. It has impressive facilities for storing sand. The incoming sand is passed through a rotary drying kiln on its way to a covered store, which has capacity for 50 000 tons of dried sand. From the store the sand is conveyed to mullers, which automatically weigh out the correct ingredients for the sand-oil mix.

machine was on order from Sutter at the time of the visit but
had not been delivered.

One interesting example of the saving of space and manpower by more up-to-date machinery was a *'Rotoblast'* machine
(The Pangborn Corporation, Hagerstown, Md.) which had displaced 16 tumbling mills and two *'Wheelabrator'* machines
(American Wheelabrator & Equipment Corporation, Mishawaka,
Ind.) for removing foundry sand and scale from completed castings.

The melting at the Indianapolis works is done in acid cupolas. Oily steel machining swarf is made into briquettes with a
small percentage of cement, which obtains water (for hydration)
from the soluble oil coolant remaining on the swarf.

There is no automation in the placing of cores in moulds.

The works has been planned in such a way that the main
assembly line is continuous and the various machine lines for
components come into it at right-angles just where the components are required. Nearly all the flat faces on the castings are
still milled, rather than broached. The only broaching seen was
for the main-bearing cap housings. Very few boring bars are
used: nearly all in-line holes for the camshaft and crankshaft
bearings are bored in multi-head machines using overhung cutters.

No attempt is made to trim the flash off castings before machining. It has been found cheaper to machine it off, taking very
heavy initial cuts with carbide tools.

The four bores in each side of the V-8 cylinder blocks are
honed simultaneously on a *'Barnesdril'** honing machine incorporating automatic gauging and feed-back to the hones. On this
machine, expansion of the individual hones is regulated by tapered lead sizing plugs, uniformly spaced around a ring located
above the honing stones. On every stroke each of the four honing
tools enters its cylinder bore and its tapered plug strikes the top
face of the crankcase. As the bore diameter gradually increases
with honing, the tapered plugs enter further and further into the
bore until, when the bore diameter reaches the mean dimension
required, the ring and plugs enter completely. As soon as this
occurs, the feeds to the individual hone expanders are automati-

---

* Made by the Barnes Drill Company of Rockford, Ill. Another automatic honing machine which incorporates the feed-back principle is the
'Microhoner,' made by the Micromat Hone Corporation of Detroit, Mich.

cally cut off. In this way, the bore diameters are automatically held to ±0.0002 in.; afterwards they are individually checked by air gauges.

The tool feeds on the drilling heads of transfer machines are effected hydraulically so that the total thrust loads are indicated by pressure gauges. The complete set of drills in a particular head are changed when the gauge reading exceeds a pre-determined limit. The firm does not believe in relating the drill life directly to the number of pieces drilled because of variations in the machineability of the castings and in the endurance of the drills.

Crankshafts are automatically balanced on a Gisholt machine (Gisholt Machine Company, Madison, Wis.) operated by two men. The first operator loads the machine and sets the shaft spinning. The machine feeds its out-of-balance signals direct to a special-purpose drilling machine incorporating a memory device. The second operator loads the shaft into this special-purpose drilling machine, which then drills the shaft automatically in four places to give correct dynamic balance. Neither of the operators concerned is rated as a skilled man.

All the crankshafts are plunge-ground on Landis machines (Landis Tool Company, Waynesboro, Pa.) incorporating automatic wheel-trimming devices. One machine grinds all the journals at one setting and another grinds all the crankpins. The wheel radii still have to be dressed by hand with the help of a gauge. The larger shafts are induction hardened on the pins and journals.

Unless crankcase castings are to stand outside for any length of time, the International Harvester Company no longer paints their interiors, as some firms do, to prevent sand and scale working off in service. They consider that the paint hides more sand than it sticks on and that it makes inspection more difficult. They do not worry about light rust on crankwebs and crankcase walls as it is not abrasive and they have proved that it does not harm the bearings in any way. They are trying to get away altogether from painting crankcases and hope to use a petroleum-base oil spray instead.

One point noticed on the assembly line is that the light alloy pistons are free to expand relative to the gudgeon pins. To this end, one gudgeon pin boss is bored 0.0003 in. larger than the other and the gudgeon pin is always inserted from the same side of the piston.

Torque-loading spanners are used on the main bearing nuts, the big-end nuts and the cylinder-head retaining nuts. The spanners in use are of very simple cantilever design and are not foolproof. All male threads on the engine are coated with 'Parkerluberite' (Parker Rust Proof Company, Detroit, Mich.) and no other lubricant is added before assembly.

In this plant the company has experimented with nodular-graphite cast iron for crankshafts and other components, but has come to the conclusion that the magnesium-nickel process is not economic with irons from acid cupolas. They are very interested to hear of British and German work on the direct magnesium process. There seems little doubt that the company will turn over to nodular cast-iron shafts as soon as they can be made economically, because it will then be able to produce them itself in the Indianapolis foundry and will become independent of external sources of supply.

The foundry and machine shop seemed to be very well-planned. There was as much automation as would be economical in a works of this size, producing engines of five main types. The speed and dexterity of the foundry and machine-shop workers was remarkable.

PONTIAC DIVISION OF GENERAL MOTORS
Pontiac, Michigan

This plant employs 17,000 men and produces 1200 cars per day. The visit consisted of an organized tour, lasting an hour, on trucks provided for general visitors. The running commentary was provided by the guide who drove the trucks. He was employed solely as a guide and was unable to answer really technical questions. There was no time for detailed study, but the following points were noted:

(1) The joint faces on the cylinder blocks were milled.
(2) The joint faces on the cylinder heads were broached.
(3) The bodies were supplied direct to the plant on a conveyor from the adjacent Fisher Body Works in the exact order of finish required. There was no need, therefore, to keep a stock of bodies at the plant.
(4) Metal-cutting automation was about as advanced as at the largest British car plants.

(5) There were a few semi-automatic assembly operations, *e.g.*, the use of pneumatic multi-head spanners
for tightening nuts on cylinder heads, and the assembly of gudgeon pins in pistons and connecting rods.

(6) When the finished cars left the conveyor they were
driven on to a pair of rollers so that inspectors could
check the engine, transmission and brakes without
taking the car on to the road.

WHITE MOTOR COMPANY, TRUCK DIVISION
842 East 79th Street, Cleveland, Ohio

This firm is one of the largest independent truck manufacturers in the U.S.A.; its 6000 employees turn out about 10,000
trucks a year. The White Company used to make motor buses as
well as trucks, but it dropped out of this field when the market
for city transport buses contracted. About 75 per cent of the
trucks it makes have side-valve petrol engines designed and manufactured by the Company. Although not perhaps so economical
as an overhead valve design they have quite a big advantage on
the White truck because their low overall height enables them
to be more easily accommodated beneath the cab than others.
The remaining 25 per cent of the trucks use other diesel engines,
but the company hope to replace these in due course with engines
of their own manufacture. To this end they have bought a large
diesel-engine firm at Springfield, Ohio, and they intend to build
their own truck engines there under license from the German
firm, Krupp-Maffei. The diesel trucks are easy to sell in the less
populous West where gross weights up to 100,000 lb and axle
loads up to 18,000 lb are permitted. Petrol engines are favoured
in the East because they are more flexible in heavy traffic. The
maximum speed limit for big trucks is 55 m.p.h. in most states,
but the trucks are designed for a continuous cruising speed of
65 m.p.h. Air brakes are universal on the larger trucks and hydraulic brakes on the smaller ones.

The plant is very similar to several in the U.K. which also
produce a great variety of trucks, including many made to special
order. It has no foundry and buys out all its castings, gearboxes
and road springs. Assembly is done by typical production-line
methods, the output being too small to justify much automation.
It seems that the firm could use a good deal of automatic equip-

ment in the office, e.g., in ordering spare parts and in production scheduling and accounting, but it has not yet done anything about it.

The factory works two shifts, the first being from 7 a.m. to 3:30 p.m. and the second (their smaller shift) from 3:30 to midnight, with a half-hour break for lunch. The manual workers and staff seem to be very enthusiastic.

WHITE MOTOR COMPANY, DIESEL ENGINE DIVISION
1401 Sheridan, Springfield, Ohio

This plant has only recently been acquired by the White Motor Company from the National Supply Company, and its new owners had no time to reorganize it before this visit. Under its former ownership, it manufactured large stationary engines of up to 2000 BHP; there had up to eight cylinders in-line, ran on fuel oil or natural gas, and were sold mainly to the oil industry which used them to drive pumps and generate electric power at oil wells. These engines are still made, but there is also the development of small (424 and 636 cu. in.) Krupp-Maffei licensed V-6 and V-8 truck engines (see above).

There is a large iron foundry which produces the crankcases, bed-plates, pistons and cylinder liners but most of the equipment is very old and the ventilation is poor by modern U.S. standards. No novel techniques were seen either there or in the machine shops and assembly shops. The firm has for many years produced large engines in small quantities, and one would not expect to find any very remarkable materials handling equipment or automation. There will probably be big changes there once the small diesel engine has been developed and is ready for mass production.

DISTRIBUTION

CANADIAN TIRE CORPORATION
837 Yonge Street, Toronto

This company specializes in the supply of spares for motor-cars of all types, and it works in direct competition with the approved agents of motor-car manufacturers. Parts are often supplied to the company by the firms that actually make them under sub-contract for the vehicle producer, but the company can sell them more cheaply because its overheads are lower than those of the big manufacturers.

It was said that the Canadian Tire Corporation was the first retail firm to have an IBM punched-card machine installed for sales direct to customers. Each article stocked is exhibited in the store, and near it is a pocket containing a supply of IBM cards bearing the reference number of the part and appropriate coded holes. The customer chooses the goods he wishes to buy, picks up a card for each item, and takes the cards to a sales clerk, who inserts them one after another into the IBM machine. This prints out an order, a receipt, and a stock adjustment card. The order is placed in a pneumatic tube, which passes it direct to the stores, where a storekeeper can quickly pick up the items required from bins, serially numbered to agree with the references in the IBM cards, and hand them to the customer across the counter. A customer can place his order and have it delivered to him in half a minute, without having to look up part numbers or write invoices, receipts, etc. Carried to its logical conclusion, the selection of parts in the stores could be made automatic, and parts could be delivered along a conveyor belt to the counter.

At the time of the visit, the system had not been installed very long, and the firm was undecided whether to pursue it further. It was invaluable at rush hours but extremely inefficient in slack periods because one operator had to attend the IBM machine and another the air tube in the stores, even when no orders were being received. It seemed that this disadvantage could have been very easily overcome by a simple set of bells, which would warn the people concerned whenever a customer approached the counter.

Since the visit, two more departments have been added to the store and there is more continuous work for both machine and operators. Management and staff have become accustomed to the system and now consider it indispensable.

DOMESTIC APPLIANCES

CANADIAN GENERAL ELECTRIC COMPANY
5781 Notre Dame E., Montreal, Quebec

At this plant, 750 employees produce about 100,000 major appliances (refrigerators, cookers and washing machines) a year.

There is very little automatic equipment apart from continuous overhead monorail conveyors. There is no press-tool automation, but there are several special-purpose machines, such as those for roll flanging, stretch forming, and seam welding. One of them is operated by one man, who loads it with the sides, top and bottom of a refrigerator. The machine seam welds the four pairs of line contacts, automatically turning the assembly four times as it goes.

The chromium-plating bath is completely automatic. Degreasing and dip-nickel finishing prior to vitreous enamelling are done automatically on a conveyor. Leaks in refrigerators are detected by evacuation and passing through a helium atmosphere; later the gas content of the refrigerant circuit is tested. Much spot welding is done with the aid of hand-operated machines suspended from balancing devices.

Progress with automation at this plant is obviously restricted by the small size of the Canadian market. Specialized tooling for mass production cannot often be justified, even though Canadian wage rates are high.

THE CANADIAN WESTINGHOUSE COMPANY
Hamilton, Ontario

The first visit was to the company's No. 2 Plant, one division of which manufactures 50-60 different types of thermionic valve. The total output is 5-6 million valves per annum, but the maximum number of any one type is only 400,000. Almost all operatives are female. All cage-assembly work is done by hand. Bases and connector pins are heat-sealed in automatic machines with 'Syntron' or rotating finger feeds. The parts are then transferred by hand to machines that bend the connector wires in appropriate directions. The bases are spot welded to the cages by hand. The complete assemblies are sealed in bulbs in automatic lamp-manufacturing machines. There is no automatic transfer of valves at any stage. On completion, each valve is fitted by hand into ageing and testing fixtures. There is no automatic rejection of defective valves. Only one machine has automatic unloading; the valves are discharged on to a small belt, and from there into a cardboard box. Grid and filament winding is automatic, but all parts are moved to and from machines by hand.

There is a very simple vacuum chuck with a number of equally spaced transversed slots. When it is not held by the operator, the slots quickly pick up cathodes from a tray to place them at the desired uniform spacing in a frame. This frame carries them while they are sprayed with coating to increase electron emission.

Valves are marked for identification in a rotary printing machine, which has to be loaded by hand to ensure that the valves are the right way round. All valves are packed by hand.

Substantial production increases to meet a greater market demand would undoubtedly justify more automation in this plant, particularly in cage assembly, transfer between machines, ageing, testing, hopper feeding to marking machines, and packing.

Television tubes were also manufactured in the No. 2 plant. After coating (by sedimentation from a water suspension of phosphors), lacquering, and aluminizing, the electron guns were assembled in the tubes and the final closure, and the evacuation and sealing were done on a fully automatic conveyor line, each trolley on which carried its own vacuum pump. Overhead conveyors were used whenever possible, but tubes were loaded on and off conveyors by hand because they are very fragile.

That part of the plant where domestic washing machines, drying machines, refrigerators and cookers were made was shut down for the lunch break when visited; but it evidently had very little automation. It was explained that the existing facilities were adequate and economic for the present market demand for the products. There was a conveyorised assembly line and an overhead conveyor for components and insulating materials. All press loading was done by hand and all spot welding by balanced suspended single-shot spot welders.

In another part of Plant 2 was the 'Cypak' system of plant control by logic functions (Westinghouse Electric Corporation, Pittsburgh, Pa.) which uses magnetic amplifiers instead of relays. It was said that many control functions in U.S. industry have been taken over by 'Cypak' and that Westinghouse engineers in Canada were also planning many applications. 'Cypak' is fundamentally more reliable than any system which relies upon contacts. Westinghouse run an annual 'Machine Tool Electrification Forum' to which they invite hundreds of potential customers. The last was held at Buffalo, New York, on April 19-20, 1955, and was largely devoted to the 'Cypak' System.

Later a visit was made to the Brantford television plant, about 35 miles from Hamilton. At that time, about 450 sets were

assembled a day almost entirely by female labour, the total
'woman' hours a set being 4.3. Each set incorporates two printed-
circuit assemblies, on which are mounted all the valves. A conveyor carries the printed-circuit boards past operators, who plug
in the various components; then the boards are dip fluxed and
soldered automatically. The sets are assembled, wired, tested, and
fitted into cabinets on another conveyorized line. This television
plant was said to be the most 'automatic' in Canada.

Canadian Westinghouse faces a severe challenge in competing with U.K. and other European firms in products like switchgear, generators, and heavy industrial equipment of all kinds in
which the labour content of the work is great. The reason is that
very low wages are paid in Europe for equivalent work. However,
the company has been able to meet the challenge through aggressive cost-reduction and product-improvement programmes.

CROSLEY-BENDIX DIVISION OF AVCO
Berry Field, Nashville, Tennessee

This firm employs 3300 people, who produce 2100 units a day
(Bendix washing machines and Crosley electric cookers) in
about ten different designs. It also does much sub-contract work
for the U.S. Air Force.

Crosley-Bendix are beginning to buy their sheet steel in coils
rather than in cut sheet because they can get it much cheaper
that way. They cut and level it themselves on a typical roller
leveller equipped with flying shears and automatic strip-length
control (by the General Electric Company) accurate to $\pm$ 1/32
in. at a setting of 8 ft. The operator of this machine was once a
foreman in the guillotine shop, where panels were cut from rectangular sheet; but the new machine has displaced the guillotine
operators and the one-time foreman is now doing a more highly
skilled job. In this machine electronic control keeps the speed of
the strip constant at the flying shear. Strip which has passed
through the roller leveller has to be used within 24 hours, otherwise it tends to curl.

Throughout the works all material and parts were moved by
overhead monorail conveyors which also acted as floating stores,
holding sufficient stock for about 8 hours' work. The Ransberg
No. 2 process of electrostatic paint-spraying (Ransberg Electro-
Coating Corporation, Indianapolis, Ind.) was in use everywhere,
automization being effected by the discharge of drogs from the

inside rims of rapidly rotating hollow bells maintained at a high potential. This spraying equipment is fully protected by patents: it cannot be bought but has to be hired from the manufacturer.

In the vitreous-enamelling department there were many automatic frit-spraying machines employing simple cam actions to cut off the spray when the traversing head passed beyond the limits of the object being coated; but no electrostatic devices were being used to reduce overspray.

Despite the very large amount of pressing, there had been no attempt to feed work to, or deliver it from the press tools automatically. This, it seems, was because many different kinds of products were being manufactured and there had to be frequent changes in product design to keep pace with the demands of the Sales Department. Throughout the U.S.A., domestic products of this kind are purposely given a 'new look' every year in order to render the old ones obsolete, and so to persuade one customer to buy a new electric cooker because her neighbour has one.

The only automatic equipment seen in the Aircraft Division was a '*Drivmatic*' riveter (General Riveters, Buffalo, N.Y.). It drilled a rivet hole, inserted the rivet and closed it at one machine setting. The machine was being used at the time for attaching stiffeners to wing ribs. A large amount of spot welding was being done on light-alloy aircraft components.

ELECTROLUX CORPORATION
Old Greenwich, Connecticut

The Electrolux Company of Sweden established a sales organization in the U.S.A. in 1924. All sales were made through agents, paid entirely on commission. When the turnover had become fairly high, the company decided to commence manufacture in the U.S.A., at first by sub-contracting it to the White Sewing Machine Company. In 1933, the Old Greenwich plant was built; it now employs about 1000 people, of whom 40 per cent are female. The company was bought out some years ago by American interests, and now has no connection with either the parent firm in Sweden or the British firm at Luton.

The Old Greenwich works makes everything for vacuum cleaners except plastics mouldings and aluminum and magnesium die castings which it buys out. Two cylinder-type machines are made, both with two-stage fans built up from sheet metal. One model has an automatic device, which throws out the sealed dust bag when the pressure drop across the bag exceeds a certain level. This level can be varied by a control knob, to suit the type of work on which the cleaner is engaged. The purpose of the device is twofold; to prevent the cleaner from being operated with a choked bag, and to make it unnecessary for the housewife to empty it—she merely throws the bag away, and puts in a new one.

Apart from the usual vacuum-cleaner accessories, the firm also markets an air-turbine floor polisher. The rotor for the radial-flow air turbine is an aluminium-alloy die-casting, on which a sheet-aluminium annulus is riveted to shroud the blades. Brush strip for polishers is bought out in continuous lengths, and is fed from a coil into automatic machines where it is shaped and cut to size.

Throughout the plant, there was wide use of overhead mono-rail conveyors and merry-go-rounds (carousels) of the moving-store type, which carried a large stock of small components past benches at which assemblers worked. Powered screw-drivers and nut-runners were in use everywhere.

All presses which could be kept fully employed on one job were automatically fed with continuous strip, and unloaded by an air jet into a hopper feeding the next operation. Many presses were still hand fed, but even they were invariably unloaded by air-jets. Some of the presses that were fed by automatic rotary tables from the operator into the working area could quite easily have been converted to hopper systems. There were no multiple-die presses, with automatic transfer of workpieces between con-secutive pairs of dies, like those found in the automobile industry.

Armature and stator coil winding was highly automatic, with special-purpose machines designed in co-operation with an out-side manufacturer. These machines had been introduced to meet not price competition but the gradual ageing of the labour force, since the very rapid hand operations, which were quite satisfactory when done by young girls, ceased to be possible after the girls became women of 50. The older workers had asked to be taken off this operation, and by then it had become so difficult to attract girls to this kind of work that the special machines were designed. The market was still expanding rapidly, so em-ployees could be absorbed in other work and none had to be stood off.

Paint-spraying operations were done on 'Ionic' machines (Ionic Gun Paint Spray Equipment Electric Company, Garfield, N.J.), in which the parts were conveyed past spray guns carrying a high electrostatic charge. This system replaced an earlier electrostatic system in which a charge was induced on the work piece and the spray gun was earthed. It was said that the 'Ionic' system covered the workpiece more uniformly but did not reduce overspray further.

All electroplating and anodizing was done in automatic plants. Copper plating is no longer used as a base for nickel and chromium.

All flexible leads were cut off to length and bared at the ends automatically.

Armatures and fans were dynamically balanced on Gisholt machines (Gisholt Machine Company, Madison, Wis.), but the operator had to control the amount of stock removed during the balancing operation.

Brush-tag terminals and springs were assembled by hand on a rotary-table machine which transferred them through a small induction-heating coil for soldering. Wire solder was fed automatically to the machine.

Wound armatures were insulated by impregnation with synthetic resin on a conveyorized machine, which incorporated a small resin bath with automatic control of temperature and depth. This avoided the need for a large dip-tank full of flammable resin, and so reduced the risk of fire.

Ball-bearing assemblies were supplied to the motor production line tightly packed in long, vertical metal tubes, each with a slot near the base for the removal of bearings: in this way, contamination by dust was reduced. After assembly, each motor was run under power for two hours; the great majority of defective bearings became noisy within this period and could be weeded out.

The name 'Electrolux' is impressed on both sides of the body of the cleaner during the final pressing operation. The impression is filed with a light-coloured paint (by spraying), and the body then passes through a double line of infra-red lamps for accelerated drying. Just before the overspray is dry, a pair of buffing wheels automatically removes the excess before the body passes on to another conveyor.

Many specialized spot-welding machines are used on parts like the motor end covers. They are frequently arranged so that one machine takes a component through two or more separate welding cycles, reducing the peak electrical load; for example 20 spot welds may be done at one machine loading in two consecutive groups of ten.

A well-equipped general engineering shop was available for the design, development, and production of special-purpose machines. Many of these machines were simple transfer machines doing two or three operations only. Others were simpler still; for instance a machine for drilling and tapping four axial holes in the motor housing had two stations, one above the other; they operated alternately, one station being loaded whilst the other operated, and the operator had no idle time.

Each completed cleaner is tested for power consumption and for the vacuum it attains with a standard restrictor orifice. It can attain a depression of more than 60 in. of water attainable at a very low air mass flow.

Very friendly relations existed between management and operatives throughout this plant. Everyone, no matter how junior, appeared to be proud of the work he or she was doing. All the older people, no matter how humble their position, appeared to be known by name to the vice-president, who would stop and chat with them on the way round the works. This friendly attitude was fostered throughout the plant. There was only one canteen, and no top table was reserved for management.

The works is set in beautiful surroundings in a residential part of Old Greenwich and a very fine Social Centre has been built for the use of the employees a short distance away.

Though there is no trade in this factory, the average hourly rate of the factory floor workers is $2.38—well above the union average for this trade. An eight-hour day is worked (single shift), with half an hour for lunch and (paid) 15 minute breaks in the middle of the morning and afternoon.

FRIGIDAIRE DIVISION OF GENERAL MOTORS
Dayton, Ohio

The Frigidaire Division began as the Guardian Refrigerator Company of Detroit in 1915. This Company was almost bankrupt when it was taken over by General Motors in 1918 and renamed the Frigidaire Company. The works were moved to Dayton in 1921 and the Company now employs 24,000 people in five manufacturing plants. The first thing General Motors did when they took charge was to introduce automotive mass-production techniques and to this they attribute most of the success of the firm. Frigidaire now builds household refrigerators, food freezers, ice-cream cabinets, domestic electric cookers, electric water heaters, automatic washing machines and clothes dryers.

Only Nos. 2 and 3 plants at the Moraine Works were visited; they were built in 1925 and 1940 respectively. No. 2 plant manufactures refrigerators and No. 3 plant cookers, water heaters, &c. Their most impressive feature is the extensive use of monorail conveyors and the almost complete absence of trucks. The refrigerators themselves are built on moving assembly lines similar to those used in the motor-car industry. Almost every component of the refrigerator is made at Dayton with the exception of the electric motors, which come from the General Motors Delco Products Division, and the timers for the cookers which are brought out. Even the heating elements for cookers are made at Moraine—from heat-resisting tube, nichrome wire and magnesium oxide powder.

The facilities for vitreous enamelling at Moraine were larger than any the writer had seen elsewhere. The frit sprayers seemed to work under very difficult conditions but were provided with every possible aid to safety and comfort, including breathing masks supplied with conditioned air under pressure.

In the press shop, gangs of four heavy presses for the production of refrigerator cases had been inter-connected; automatic loading and transfer was provided by conveyors and 'mechanical hands'.

One ingenious machine automatically fitted together the nine metal components of an ice tray before this was passed on for welding.

The welding techniques employed on food compartments and on outer casings were very advanced, seam welders of specialized design being used wherever possible. One machine welded all six seams of the food compartment in 12 seconds at one setting.

The main shift at this works operates from 7 a.m. to 3:30 p.m. with half an hour break for lunch, whilst the staff work from 8 a.m. to 4:30 p.m.

GENERAL ELECTRIC COMPANY,
MAJOR APPLIANCE FACTORY
Appliance Park, Louisville, Kentucky

The plant at Appliance Park was commenced about five years ago, when the Company decided that it would be more efficient to concentrate its manufacturing operations on major domestic appliances. Formerly these operations had been scattered over seven different sites in areas (Bridgeport, Conn.; Erie, Pa.; Scranton, Pa.; Trenton, N.J.; Bloomfield, N.J.; Chicago, Ill.; and White Plains, N.Y.) where there was an acute shortage of labour. They chose the present site, which is about eight miles out of Louisville, because labour was available in that area and also because it is only 40 miles away from what the Company calls "the distribution centre of the U.S.A." and has good rail and river communications. In September, 1954, the plant employed 5000 people, of whom only a very few had been transferred from the original scattered plants. It was the policy to start by employing as many men as possible and then to begin recruiting women, because trade union rules required the lowest in order of seniority to be stood off first if output had to be decreased, and the company wanted to make sure that the men would be the last to be stood off. Since then, employment has risen to 14,000 and there is a projected expansion to 16,000 by the end of September 1956. It is Company policy not to let this plant increase beyond its present size (with a ceiling of 16,000 employees) because they do not wish to dominate the scene at Louisville, and to have the prosperity of the whole area dependent upon one firm. With 16,000 workers, they will employ 10 per cent of the available labour.

There are at present slightly over 4 million sq. ft. of plant at Appliance Park in five similar buildings. Each building has its own office block housing engineering and administrative staff. Buildings 4 and 5 are devoted to the manufacture of domestic refrigerators, the refrigerating units being made in Building 4 and the cabinets and assembly work in Building 5. Building 3 manufactures domestic waste-disposal units and dish-washers, and Buildings 1 and 2 make electric cookers and domestic clothes-washing and clothes-drying machines. The five major manufacturing blocks are connected to an enormous storage warehouse half a mile away by totally enclosed overhead conveyors which operate entirely automatically with 'electric eyes' selecting the four major products by markings on the packing cases, and shunting them into the appropriate sections of the warehouse: this system was not yet in full operation on the date of the visit.

In the refrigerator plant there were many specialized machines, such as automatic seam-welding machines which welded all major joints in the cabinets and in the interior cases after loading by one operator. There was a special-purpose spot-welding machine making refrigerator condensers. The condenser consists of about 10 loops of $\frac{1}{4}$-in. copper pipe, across which 70 steel wires are laid (on each side of the tube assembly). This machine is loaded with bent tubes at one end, and the steel wires are fed down from drums above and below them. Inside the machine, spot-welding electrodes come down each time the tube conveyor indexes forward, making at one operation the 280 welds necessary for both sides of two adjacent passes of the condenser tube.

One unsuccessful transfer machine was shown which had been built to handle four consecutive machining operations on compressor crankshafts for refrigerators. The engineers had tried for about two years to make this machine reliable, but in the end they went back to the four separate machines.

Many of the compressor units were assembled with copper brazing wires placed around the joints and with flux painted on by hand. The entire unit then passed through a furnace and out into a quenching tank. This furnace brazing was considerably quicker than hand brazing, and gave a much more reliable product.

The upper and lower components of the pressed steel casings which house the 'sealed-unit' compressor assembly were welded together in specially designed machines using automatically traversed electric arcs with carbon-dioxide shielding.

The cylinder bores in the compressors were all being honed to size by a trial and error process with manual inspection by air gauges. Pistons were selected from three different sizes for assembly to the bores.

A very modern plastics-moulding plant was used for making inner linings for the cabinets and doors of refrigerators, and also for making refrigerator trays. The polystyrene crumb was extruded into sheet and then passed through calender rolls together with a film of clear polystyrene 0.002 in. thick. This gave a coloured polystyrene sheet, with a very fine finish on one side, which was cut to length automatically by flying shears similar to those used in a steel mill. The lengths of polystyrene sheet were then vacuum-formed over beautifully finished dies to give complete components in one operation.

Throughout the refrigerator and home-laundry plants the only completely automatic operations seen were the deep drawing of refrigerator vegetable bins in steel (2 per unit), and of refrigerator shelves in aluminium alloy, on 7-station presses (E.

W. Bliss Company, Canton, Ohio) with automatic feed to the press from continuous strip and automatic transfer from stage to stage within the press and hence to the delivery chute. Most of the pressed items were hand loaded into individual presses and hand unloaded. In some cases four presses engaged on consecutive deep-drawing operations would have roller conveyors from one to the other, but no attempt was seen at automatic transfer of large components from one press to another, such as one finds in the American motor industry.

In the dish-washing machine plant was one interesting little machine which automatically formed the loops on the ends of connecting wires. The operator pushed the end of the wire into one slot, where the braid was removed; she then transferred it to an adjacent slot where the end was bent into a loop to go on to a terminal screw.

Throughout the plant, electrostatic spray painting was used for the finishing coat. The Ransberg No. 1 process, using an electrostatically charged conventional spray gun, is now considered obsolete and has been replaced by the No. 2 process which uses charged, rapidly rotating, hollow bells to atomize the spray.

In the home-laundry plant, electrostatic spraying was applied to vitreous enamelling. The groundcoat on vitreous enamelled items is no longer applied by dipping, but by a flow-coating process. The components are carried on a conveyor through a compartment in which the slip is pumped through nozzles arranged in a multitude of different attitudes so as to make sure every part of the surface is covered. Before emerging from this compartment the component passes through a thin curtain of slip, which is arranged to fall through a narrow jet running right across the exit. The parts drain during the first twenty odd feet of travel along the conveyor, after emerging from the flow-coating booth, and then they pass through a low-temperature firing oven. After visual inspection, they pass through two booths in which the top coating of slip is applied. In the first booth slip is sprayed through the normal kind of compressed-air gun, hand operated, to reinforce the corners and other positions where maximum resistance to wear is required. The booths in which this operation is performed are extremely well designed and ventilated, and the air control is such that the operators do not need to wear either masks or breathing equipment.

After passing through this booth, the parts enter the electrostatic spraying booth where the Ransberg No. 3 process is in use.

In this process slip ground to about 300 mesh (i.e., about three times finer than that normally used) is sprayed on to the workpiece through a highly charged and rapidly-rotating disc about 10 in. in diameter. The axis of this disc is vertical and the disc carried a shroud about ¼ in. above and rotating with it. The slip is metered into the space between the disc and the shroud by a displacement pump. There were two of these discs in the booth, and the washing-machine components moved up and down, as they passed the discs, on a monorail resembling a scenic railway. This electrostatic spray booth was very efficient; it left hardly a trace of slip on the walls or on the floor. It was said that the only way to make electrostatic spraying successful with material containing water was to exercise rigid control of the humidity of the air in the booth; here it was maintained at 65 per cent relative humidity. It was apparently the Company's intention to replace the existing system of fixed discs and vertically moving workpieces by a system in which the workpieces would move at a constant height past vertically reciprocating discs. This appears to be merely a matter of convenience; the existing conveyor must have been very difficult to make, and looked rather fantastic.

Though the company had done an "impossible job" in assembling seven previously scattered plants in one town, it had less automation than the American motor industry, which handles just as large quantities of bigger and more awkwardly shaped components.

HOOVER COMPANY
North Canton, Ohio

This, the original Hoover works, was started in 1875 to produce leather harness for horse-drawn vehicles. With the coming of the motor car Mr. Hoover realized that there was little future in harness manufacture: he looked around for a new product and found one in the newly invented vacuum cleaner. The company started to produce domestic vacuum cleaners in 1908 and now employs some 2500 people in the North Canton plant.

A tour of the works gave the impression that automation had been carried little further at North Canton than at similar plants in the U.K. An exception was the Motor Department, where some interesting coil-winding machines were in use. One machine, made by the Globe Tool and Engineering Company, Dayton,

Ohio, wound the complete armature at one setting. After being placed in the machine by an operator, the armature remained at rest whilst flying arms wound the coil and brought out the ends near the commutator. These ends were cut off automatically, the armature indexed round to the next slot, and so on until winding was complete. As the winding process was entirely automatic, the operator's attention was not required except during loading and unloading: one operator could easily feed two adjacent machines. Another machine was a stator-winding machine developed by the Fort Wayne Tool Die and Engineering Company, Fort Wayne, Ind., which accommodates the complete stator frame and winds both coils upon the pole pieces simultaneously. The wound stator goes straight from the winding machine to a varnish dip and a baking oven, and no wrapping of the coils is necessary.

The company has developed a new method for making flexible suction hose which yields a much cheaper, lighter and more flexible product than cotton-braided flexible metallic hose. The Hoover hose uses a P.V.C. sheathed carbon-steel helix, which is anchored inside an outer layer of P.V.C. tube. More detailed manufacturing methods cannot be given, as they are confidential.

One of the biggest developments at the plant over the past few years has been the gradual change-over from zinc-alloy die castings to aluminium-alloy die castings, which are very much stronger and lighter. No doubt this change was hastened by the very rapid increase in the price of zinc during the past half dozen years. The main castings and fans of Hoover vacuum cleaners of the bag type are at present made as aluminium-alloy die castings. Magnesium alloys have been tried and are quite satisfactory, but at present they cannot be produced quite so cheaply. The two-stage fans for the simple air-suction cleaners, which do not use the Hoover rotating brush and beater, are built from sheet aluminium alloy because this form of construction suits their layout.

An assembly line which the company has set up (as a subcontractor) to produce a small but complicated mechanism was of special sociological interest. The operation was previously carried out on the normal type of conveyor belt, where each operator was assigned a specific small task. In the new set-up, each operator handles the complete assembly of a unit, with the various components brought up automatically in correctly timed sequence. The operatives are happier than they were because they have a more interesting and varied task; the management is happier because responsibility for a complete assembly can be assigned to a particular worker. Both output per operator and quality have improved noticeably under the new arrangement.

The operatives at this plant worked from 7 a.m. to 3:30 p.m. and the office staff from 8 a.m. to 5 p.m. Where necessary a second shift was worked from 3:30 p.m. to midnight.

The company has a museum containing examples of its own vacuum cleaners, and those of other companies, built since 1910. It also has research and development laboratories which are extremely well equipped to do the kind of applied research that yields such rapid results in the U.S.A. Radioactive tracers are being used to determine rapidly the amount of dust removed from different kinds of carpets by various types of suction cleaners. The laboratories are also very well equipped with injection-moulding machines, on which prototype components can be made and the troubles eliminated from the processes before the parts are released for large-scale production.

ELECTRONICS

RADIO CORPORATION OF AMERICA
Harrison, New Jersey

On this visit the latest automatic machinery for radio-valve manufacture was seen. The company develops many types of electronic products, including valves, and designs special-purpose machinery for valve manufacture. It has a broad licensing policy, under which it makes technical information on its products available to licensees in many countries.

Four main types of automatic machine are currently used in the valve industry. The first type includes all presses used for the production of parts like plates and anodes. These are all very highly developed machines, automatically fed from continuous strip and automatically unloaded. The rate of production is up to 6000 parts an hour, depending on the complexity of the part. Machines like this are essential to all valve production, and are made by specialist firms (such as the Baird Machine Company, Stratford, Conn.) for the whole valve industry. They could hardly be made more automatic.

The second type includes all automatic machines for making the wire components used in the valve assembly, such as filaments, getters and anode supports. Most of these machines appear to have been developed by the valve manufacturers themselves. All have features like automatic feed and automatic spot welding, but many rely on an operator to unload products and transfer them to portable racks, often at a very high rate. Much of this unloading could be mechanized and the possibility is being considered.

The third type includes automatic rotary-table machines for producing the glass stems, sealing-in the electrode assemblies, evacuation, and final sealing (tip-off). These machines have been developed from those that have grown up in the electric lamp industry since 1919. They do quite a good job, but the electrode assemblies still have to be placed in the envelopes by hand, and two rotary-table machines are needed:

No. 1 to make the glass stem which carries the support wires;

No. 2 to seal this stem into the envelope and to evacuate the valve and seal it.

The company has built one very complicated machine with separate rotary tables side by side. On the first of these tables the stem is sealed; on the second the valve is evacuated and sealed. This machine gives automatic transfer of the work-piece between the rotary tables, and there are separate two-stage vacuum pumps for each station on the second table. This is a test machine for determining whether exhaust speeds can be greatly increased. It is not intended to replace existing equipment.

The fourth type are automatic valve-testing machines; each incorporates a conveyor into which an operator plugs every valve after it has been aged. The conveyor takes the valve through a heating cycle, and then through a series of electrical tests; the test for internal short circuit is accompanied by two severe blows by rubber hammers on the glass envelope, to reveal loose assemblies. If a valve fails any of the half-dozen tests it is immediately rejected, the number of rejections at each stage being counted automatically. Valves are still loaded by hand because the loader also acts as inspector and can take immediate steps to report excessive rejections at any stage to the type engineer or foreman. The present machine tests between three and four times as many valves an hour as are tested by hand methods.

Until recently all valve 'cages', i.e., the sub-assemblies consisting of micas anodes, cathodes and grids, were asembled by highly skilled female labour. The company has now developed a transfer machine to carry out this operation. Three of these machines were seen working—one on diodes, one on triodes, and one on double triodes. Another machine for pentodes had been constructed and was undergoing development tests. All these machines were straight-line transfer machines with transfer devices designed to suit the small size and fragility of the various components.

On these valve-cage transfer machines, the cost of the machine base is negligible in comparison with the tooling that is mounted on it. It is not economic therefore to convert such a machine from one type of valve production to another. The machine indexes at several thousand units an hour, when working continuously at the designed operating cycle, and it saves a considerable amount of labour.

GENERAL AND HEAVY ENGINEERING

VULCAN IRON AND ENGINEERING COMPANY
Sutherland and Maple, Winnipeg, Manitoba

This is a medium-sized general engineering firm, which used to incorporate an iron and steel foundry. The foundry, which employs 150 men and produces 200 tons of steel castings per week, has now been moved out to Selkirk, about 30 miles away to the north-west, and is operated by a subsidiary company, Manitoba Foundries and Steel. Deep snow prevented a visit to Selkirk, but according to information obtained it mass-produces cast-steel balls for use in grinding mills and also makes small quantities of nickel-steel and manganese-steel casings, chiefly for the Canadian railways. It does not produce any nodular-graphite cast iron because its customers have not yet called for this material. The managing director was quite interested in the possibility of producing nodular iron because he had a basic-lined electric furnace at Selkirk and this might be ideal for the purpose.

The general engineering shop at Winnipeg employs 400 men, chiefly on the design, fabrication and erection of structural steel for commercial and industrial buildings and conveyors; on the manufacture of water-tube boilers up to 150,000 lb. capacity, and miscellaneous fire-tube boilers; and on the fabrication of heat exchangers, storage tanks, and pressure vessels. Most of the present orders are coming from the oil industry, which has recently been established in Canada, particularly around Edmonton, Alberta.

The works is very well designed for heavy plate work and automatic arc-welding processes, for both of which powerful radiographic test equipment is available. It is a typical jobbing firm and not the kind of place where one would expect to find automatic equipment.

INSTRUMENTS AND CONTROL EQUIPMENT

TAYLOR INSTRUMENT COMPANY
Rochester, New York

This firm was established in 1851 and now employs 2300 people in the manufacture of thermometers, barometers and industrial control instruments of all kinds. The direct labour engaged on production is between 950 and 1000 and the total factory labour—excluding research, sales, engineering and payroll, but including maintenance workers—is 1600. Its annual turnover is $23.2 million, about ten times that of the British subsidiary, Taylor-Short and Mason, London, whose direct labour force is 210 and total factory labour 338.

Although the company's products are very widely used in automatic processes its own operations do not lend themselves to mechanization save in the machine shop, which is well furnished with the usual single and multi-spindle automatic lathes. In glass-thermometer production, it has developed simple rotary-table transfer machines to carry the bulbs and stems under a series of gas jets of increasing size, prior to joining. Another machine automatically sealed the tops of thermometer stems without any manual operation except the feeding of blanks into the machine and the removal of the finished product; this machine did work previously performed by four girls.

Many applications of automatic control were to be seen in the calibration and testing of thermometers, industrial instrument components and sub-assemblies. Precision floating-scale manometers of Taylor design and manufacture were widely used for adjusting and calibrating aneroids, flowmeters and pneumatic transmitters.

The intricate and individual instrument assembly and testing done in this plant does not lend itself to conveyors, so none are to be seen in the plant. One man assembles each controller complete from pre-tested sub-assemblies produced by individual craftsmen. The only conveyor in the factory was between final packaging and boxing for shipment in the shipping room.

It was said that the chief respect in which the larger output at Rochester simplified operations as compared with the London factory was that it enabled light-alloy die castings to be used economically instead of more expensive fabrications. Because wages are very much lower in Britain the simpler types of instrument or controller tend to be cheaper to manufacture here.

Nearly all the company's design, research and development work is done at Rochester. The British company alters these designs when necessary to suit British applications (e.g., pipe threads and materials).

MACHINE TOOLS AND MACHINE-TOOL CONTROL SYSTEMS

AUTOMATION CORPORATION OF AMERICA
5546 Satsuma Avenue, North Hollywood, California

Magnasync Manufacturing Company, the parent firm, is a small development and manufacturing concern with a total staff of 35. It was formed eight years ago to make magnetic sound-recording equipment for the motion-picture industry. The Automation Corporation of America is an offshoot of Magnasync; it has developed a numerical system of magnetic programming.

At the time it was seen it was controlling a 6 in. turret lathe (South Bend Lathe Works, South Bend, Ind.) to which the company had fitted a very neat electro-pneumatically operated turret, and an automatic indexing control (with magnetic brake) to the mandrel, so that it could also be used as an automatic dividing head. They were building this automatic lathe for a manufacturer who wanted a tape-controlled machine for producing prototypes and short-runs of various cylindrical components. One such component was $2\frac{1}{2}$ in. long × $1\frac{1}{2}$ in. diameter, with a helical groove around the outside and radial holes of two different diameters drilled at the centre line of the groove in various angular positions. Most of the turning and grooving operations on this piece could have been done on an ordinary turret lathe, but the company's automatic dividing head enabled the radial hole drilling also to be accomplished at the same setting: the one machine carried out the complete process automatically. Programming of the machine to produce this component had taken about sixteen hours. It reduced the production time of the previous manual operation from 30 to 6 minutes a piece.

Another component that was being programmed for the same manufacturer was shaped like a frustrum of a cone, about 4 in. long × 2 in. base diameter, with five V-grooves of large root radius around the outside. Programming involved splitting the outside profile into a series of small tangents (on the drawing board), measuring the co-ordinates of the intersections of these tangents, and working out the appropriate 'x' and 'y' rates. and distances for the cutting tool. Once again programming took about sixteen hours, but it avoided the need to manufacture special-form tools and templates.

More recently, the Automation Corporation of America has concluded negotiations with the True Trace Corporation of El Monte, California, for the application of its magnetic-tape system to hydraulic machine-tool tracing controls manufactured by that firm. In this application, miniature synchronous differentials turn a precision leadscrew which activates the spool in a conventional hydraulic servo driving the machine carriage. It is claimed that carriage movements can be controlled to ± 0.0001 in. with the use of this system.

CASE INSTITUTE OF TECHNOLOGY
10900 Euclid, Cleveland, Ohio

The purpose of this visit was to see the numerical punched-tape control system for machine tools developed by H. W. Mergler and exhibited at the Production Engineering Show, Chicago, in September 1955.

The system uses a binary-decimal code and needs about 120 electronic tubes in order to obtain two-dimensional control. The addition of a third control dimension would require about another 50 valves. This controller has been applied to a universal milling machine (Kearney and Trecker Corporation, Milwaukee, Wis.), and will traverse the workpiece in 'x' and 'y' directions through a maximum distance per instruction of 1 in. in steps of 0.001 in. Its main advantage is its low cost as compared with other electronic digital control systems.

An interview with Professor D. P. Eckman, who teaches instrumentation and automatic control, revealed that research in this field is rarely sponsored by the manufacturers because they are neither big enough nor rich enough to provide adequate funds. The sponsors are the very large firms in the electric and process industries. One group has recently sponsored research on the application of electronic computers to continuous optimizing product control in the batch production of margarine. This was one of two pilot-plant studies (one of a continuous process and one of a batch process) which were proposed in a brochure, published by the Case Institute of Technology, with a view to attracting industrial support.

Another activity in Professor Eckman's section has been the development of positional servo-mechanisms of very high performance. This work has been supported by Warner and Swazey, Cleveland, Ohio, makers of turret lathes.

Most of the sponsored research was initiated by the Institute, which then approached industry for financial backing; the forward thinking is being done at the Institute. In some industries, financial backing for research was readily forthcoming; in others it was not.

CINCINNATI MILLING MACHINE COMPANY
4701 Marburg Ave., Cincinnati 9, Ohio

This visit was made for two reasons: to see the foundry, which had been very well spoken of by the British Cast Iron Research Association, and to discuss the electronic control of machine tools.

The works employs about 4000 people and covers a very large area. It is one of the finest engineering plants the writers has seen, with wide gangways giving free movement along the well-illuminated shops.

The young engineer who acted as guide had graduated at Cincinnati University after serving with the U.S. Strategic Air Force in the U.K. during the war. His five-year course was a 'sandwich' course in which alternate periods of seven weeks were spent at the works and at the University. Young engineers taking courses of this kind work in pairs, so that one is at the works whilst the other is at the University. Such a system is far more attractive to the employer than the British part-time day-release scheme, for he is never short of a man. After his five-years course, the guide became a graduate engineer and was at the same time accepted as a fully qualified professional engineer by reason of his works experience.

The foundry was very well planned, all melting being carried out in four cupolas, all of which had basic linings and two of which were water cooled. The outer lining was 9 in. thick in acid brick and the inner lining 9 in. thick in basic brick with a hearth diameter of 50 in. The guide said that the firm experienced a great deal of trouble with the linings at first, but has developed excellent methods of repairing them, and they now give quite satisfactory performance. The castings produced in the foundry have considerably improved in quality since the basic-lined furnaces have been in operation. The percentage of scrap was around 2 per cent.

The Cincinnati Milling Machine Company, a licensee of the International Meehanite Corporation, makes the usual additions of calcium silicide and calcium carbide to the ladle. It makes nodular-graphite cast iron by a Meehanite process, adding magnesium-ferrosilicon to the ladle. It was said that no other patent royalties are involved. The iron produced is a low-manganese, high-silicon material which is partly ferritic as cast, but contains sufficient pearlite to enable the parts made of it to be flame-hardened where necessary.

A zirconia or fine silica core and mould wash was being widely used in the foundry. The moulders liked it because they preferred to get dirty with a white powder than with a black one.

No shell moulding was seen in the foundry, but the guide said that the company had been experimenting with it. He did not think it was suited to the company's type of product; according to costing studies he had so far completed, it would be about twice as expensive as conventional moulding processes.

The chief engineer took me for a short trip round the machine shop and the assembly hall. He said that a quarter of the machines produced (including milling machines, grinding machines and die-sinking machines) now embodied specialized loading and measuring facilities which would enable the machines to be included in highly automatic production lines. He pointed out that the company had been producing for several years a centreless grinder incorporating feed-back control of the size of workpiece.

One interesting device seen was an automatic wheel-balancing mechanism which the company patented some years ago and have fitted to their grinding machines ever since. This device functions by temporarily mounting the grinding-wheel spindle on extremely flexible supports so that it can be run for a short time above its critical speed. Three balls located in the hub of the wheel can be locked in their angular positions by axial movement of a conical inner member. When this lock is removed, with the wheel running above its critical speed, the three balls automatically position themselves in such a way that the centre of gravity of the assembly lies on the axis of rotation. The centre cone is locked so that the balls cannot move; the wheel is then in perfect balance.

A large area of the administrative offices was occupied by demonstration models of the various machines that are at present in production. A notable feature of all these machines was the centralized control box from which all the tool movements and gear changes could be operated without movement by the machinist.

There was a discussion of the analogue-controlled milling machine that Cincinnati were developing in conjunction with the British firm Electrical and Musical Industries. It was said that the E.M.I. variable-voltage analogue was to be used for the direct control of cutting-tool movement, and that it was expected that movements would be controllable to within one part in 100,000: this would enable the dimensions of workpieces up to 100 in. long to be held within 0.001 in. The main reason why Cincinnati liked the E.M.I. analogue control system was that it controlled rates of movement accurately and was not subject to cumulative errors such as can result from the miscounting of individual sets of control pulses in a digital control system.

The new canteen includes a museum where the original Cincinnati Milling Machine Works is reproduced complete with its machine tools.

CONCORD CONTROL
1282 Soldiers' Field Road, Boston, Massachusetts

This company has recently been formed by the Giddings and Lewis Machine Tool Company to exploit the electronic machine-tool directors originally developed at the Massachusetts Institute of Technology. The President is James O. McDonough who was project leader at the Institute. He was interviewed at the company's earlier address, but no electrical equipment could be seen there; pending the occupation of new premises, it was being made by Giddings and Lewis and the M.I.T. to the latter's designs.

The Concord Control system is that described in the Giddings and Lewis visit report. It has a big advantage over the M.I.T.'s original machine—a wider range of control and a decimal input, which relieves the engineer of any need to handle binary numbers. It is capable of controlling five simultaneous motions at speeds up to 120 in. per minute, compared with the three motions up to 15 in. per minute of the M.I.T. machine.

Mr. McDonough did not consider that there was any call for a very complicated director into which equations of tool paths could be fed direct, except perhaps for certain radar components. The great majority of engineering forms could be approximated with sufficient accuracy by a tool which cut along a multitude of tangent planes. He showed two circular plates which had been milled, one with tangent planes every 5°, and the other every 1°; the latter would be smooth enough for most applications in aircraft engineering. It was only necessary to work out once the sets of feed distances and times for a series of tangent planes to commonly used curves; they could then be recorded on a punched tape and fed into the programme automatically. Once a library of such master tapes had been produced, there would be no need to complicate the machine-tool controller by making provision for curve following from an equation.

Mr. McDonough was asked if he thought the demand for electronic digitally controlled machine tools would be small, in view of the fact that they could only replace tracer-controlled tools, which themselves only comprised a very small part of machine-tool production. He said that Concord are relying on the aircraft industry to provide the major demand for large precision machines of this type. They argue that congestion in and around airports, and the rapid growth of air transport, will force manufacturers to develop ever larger aircraft, which will only be required in very small numbers (less than 200). If the craft are not to be obsolete by the time they are produced, the last of each batch must not follow too long after the first. Rapid production will be essential, and complicated tooling is too slow and too expensive. In many other industries, too, he said, there is an as yet unsatisfied demand for smaller and simpler systems of machine-tool control.

There was mention of the British boot and shoe industry's requirement for a cheap electronic control for machines cutting and stitching uppers. It was considered that the General Electric Company's record-play-back control would do admirably, as templates were already available from which the records could be made. As designed for extreme accuracy for machine tools, the system is expensive (a figure of $60 000 was quoted), but it was thought that a very much simpler and cheaper system could be developed for the boot and shoe industry since it would require much lower accuracy.

Questioned about the reliability and serviceability of the Institute's milling machine, Mr. McDonough said that during three years' operation, the 'down-time' for maintenance was 7 per cent —half of it planned maintenance. Three out of 300 parts produced for the U.S. Air Force had been scrapped owing respectively to a jack plug falling out, a failure of power supply and a relay failure.

CROSS COMPANY
3250 Bellevue, Detroit 7, Michigan

This company was founded in 1898 as a sub-contractor to the earliest automobile builders, who had their original plants in the Bellevue area. It made transmissions and gears for these firms and later manufactured geartooth rounding machines.

During World War II, the Cross Company designed and manufactured its first automatic transfer machine: this was a special lathe for the manufacture of railway-truck axles. It purposely chose a component which was fairly simple to machine but heavy and awkward to handle, so that the automatic features of the machine would be very attractive. Later in the war it made a very large number of special shell lathes for the production of high-explosive shells in Government establishments. It has continued to specialize in automatic equipment and the manufacture of special machine tools, and has done its best to standardize transfer-machine components, of which it carries a large stock.

The Cross Company now has between five and six hundred employees, of whom 100 are engineering, 100 administrative, and 300-400 skilled craftsmen. All castings, forgings, sheet metal work and heavy fabricated machine frames are brought out—but to its own designs. It does all the machining on these components in a very well equipped general-purpose machine shop. It buys all electrical components out but builds its own control panels, such as the 'Toolometer.'

A striking feature of this Company is the youth of its management. The President and Vice-President are brothers in their middle forties. The Cross brothers decided after the war that the firm would have to expand if it was to remain competitive, because the big American machine-tool firms did not show as much reluctance as many of their British counterparts to enter the market for transfer machines. Its employees have more than doubled in number since the end of the war, and the premises are now extremely crowded. The floor space requirements for a firm manufacturing transfer machines are much greater than those needed for the manufacture of individual machine tools, because the complete transfer line has to be assembled and tested by the maker before it is shipped to the customer. A new plant is being built on the outskirts of Detroit, and this is expected to be ready late in 1956. It is noteworthy that the Cross Company has paid no dividends to the common stockholders for more than seven years, all the profits being ploughed back into the business to help finance this expansion. The turnover has increased from $3.5 million in 1951 to $12.1 million in 1955. The company is so busy that each of the two shifts has been working 58 hours a week over the past ten years—ten hours daily from Monday to Friday, and eight hours on Saturday. Even so, the delivery times were over twelve months at the time of writing.

Besides making machines for the home industries, the company has supplied machines to the Ford Motor Company and Vauxhall Motors in the U.K., and has recently supplied one to Adam Opel A/G in Germany. The last-named customer has arranged to make some of the parts in order to reduce dollar expenditure, but this is very unusual. The Cross Company considered setting up a branch in the U.K. two years ago, but decided to press ahead with home expansion first.

A special machine was being built for facing and boring the bosses for fuel burners on the steel diffuser of a turbo-jet aircraft engine. This machine was arranged with a T-slotted central pillar, around which the boring heads could be adjusted as necessary when the machine was being installed. A change of component design altering the number of burners would involve removing the machine from its foundations and respacing the heads.

GIDDINGS AND LEWIS MACHINE TOOL COMPANY
Fond Du Lac, Wisconsin

This visit was made to see the 'Numericord' system of machine-tool control developed by the Servomechanisms Laboratory of the Massachusetts Institute of Technology, under contract, from this company.

Fond du Lac is a small town of some 30 000 inhabitants, and Giddings and Lewis is its largest industrial firm. The firm has a very well laid out engineering shop which specializes in the production of horizontal boring, drilling, and milling machines, but also makes planing machines, planer-type millers and vertical boring mills. It does not engage in mass-production and most of the 1000 employees are highly skilled craftsmen. The average gross pay of the machinists is $3.19 an hour.

In a discussion about possible applications of electronic, digitally controlled, path-following machine tools, the Chief Engineer gave his opinion that the total market in the United States was possibly of the order of 50 machines. This matched the writer's own estimate for the much smaller British market. It seems doubtful whether there is sufficient work to keep even one machine-tool company in business making this specialized type of automatically controlled machine tool. There is, however, a much larger potential demand for electronic control systems which will provide the feed motions found on conventional (non-copying) milling machines and lathes, or the precise movements between fixed points required on drilling, spot welding, and riveting machines. Giddings and Lews hold that machine tools having this type of control system may one day supersede manually set machine tools.

There was a large spar and skin milling machine operated by the 'Numericord' control. It was milling the channels in the lower components of North American F 100 D Super Sabre horizontal stabilizers, each of which was about 8 ft. long × 2 ft. wide. A pair (port and starboard) were accommodated side by side on the machine table; they were held in position by a vacuum chuck, with a light oil to maintain the vacuum and strips of black plastics material to help seal the edges and any holes that passed through the plate. They were machined consecutively without re-setting the machine tool: the 'Numericord' system is so designed that magnetic tapes for components having mirror-image symmetry can be produced from a single punched-paper tape by merely throwing a switch. In this case, a set of three reels of magnetic tape controlled all machining operations on a pair of stabilizer lower surfaces; another set of reels would control the machining of a pair of upper surfaces.

The machine on which this skin milling is performed carries two cutting heads, each of which is driven by a 60-cycle water-cooled electric motor running at 3600 rev. min.: the left-hand head has a 50 h.p. motor; the right-hand head a 100 h.p. motor, which is needed for heavy channel milling. The chips produced by the cutters are removed by suction through a large duct, which fits closely around the cutting head. Removal is extremely efficient and very few chips fall on to the chuck or on to the machine-way covers. Automatic control from the 'Numericord' system governs:

(1) vertical movement of cutter head No. 1;
(2) vertical movement of cutter head No. 2;
(3) longitudinal movement of the machine table;
(4) transverse movement of either head No. 1 or head No. 2; and
(5) the angular setting of the vacuum chuck on the machine table.

All machine-ways are covered either by telescopic metal covers or by accordion-type plastics-impregnated canvas covers. All traversing lead-screws employ circulating ball nuts to reduce friction and backlash. Automatic path-following within limits of ± 0.000125 in. was said to be available at feed rates up to 120 in. a minute.

It was also said that the programming time for one horizontal stabilizer component was about 500 man-hours. It was not possible to compare this time in man hours with the time required to make steel profile templates for making the horizontal-stabilizer components on the Giddings and Lewis planer-type milling machine, equipped with the General Electric Company's electronic-tracer control. But one set of templates bought out for this job had cost $6000 and at this rate the programming cost would not exceed the template cost. Also, with the 'Numericord' system, horizontal stabilizers could be machined in less than half the time taken by the tracer-controlled machine because there was no down-time whilst the templates were being changed and the machine adjusted.

This is how the 'Numericord' system works. The information fed in by the programmer is of three kinds: first, a code which tells the machine which degree of freedom is being altered—A, B, C, D, or E; secondly, the distance over which it is required to move in that degree of freedom; and third, the time this movement is to occupy. This information is fed into a special electric typewriter, which records it on a foolscap sheet and at the same time punches the information on tape. The distances and times are fed into the machine on the keyboard in ordinary decimal units, the distances being in 10/000ths of an inch and the times in seconds.

The 'Numericord' system provides tool movements in steps of 0.005 in., so the last figure of the decimal input must be 0 or 5. The typed record gives a permanent account in everyday engineering units of the programme that has been fed in, and it is very useful for checking purposes.

Having fed in the desired motion in one direction, the programmer types the code for the next direction and follows it up with the distance and time concerned. The tape emerges from the electric typewriter with a whole series of consecutive machine movements, which must be taken in series of five and somehow fed to the machine simultaneously. This is done by passing the tape into an electronic digital computer having a magnetic-memory system. The series of five sets of information goes into the magnetic memory, where it is stored until complete and then transferred as phase-modulated electric signals to a magnetic-tape recorder equipped with 14 recording heads. One of these heads records a standard carrier frequency of 200 or 400 cycles per second. Five more record the information from the memory, leaving eight channels available for controlling other machine-tool actions, such as lifting one cutter head out of the way whilst another operates.

The computer was produced by the M.I.T. and contains about 350 valves mounted on a large number of panels, which are quickly detachable from the chassis. All the panels employ printed circuits. The tape used is a standard one-inch magnetic sound-recording tape; three 4800 ft. reels of this tape are needed for the complete programming of the pair of horizontal stabilizer lower components that were being machined. A thinner tape now available enables 7200 ft—enough for 1½ hours' continuous operation —to be carried on one reel.

The Giddings and Lewis machine incorporates the closed-loop 'feed-back' principle in all its motions. Racks are mounted on each moving table or tool head and drive pinions directly coupled to selsyns. Any backlash that may exist in the tool and table drives is thereby fully corrected. Like the original M.I.T. digitally controlled milling machine, the Giddings and Lewis machine can only move in straight lines between consecutive instructions: any curve-following has to be done by the programmer from tables or from previously calculated co-ordinates at a sufficient number of points to give an approximation to the curve required. However, this is all the engineer requires for most purposes.

From the above description, it will be evident that the 'Nu-

mericord' system is an improved and simplified digital electronic three-dimensional control combining the best features of the original M.I.T. Servomechanisms Laboratory programming with General Electric magnetic-tape direction. It was certainly the most highly developed system seen on this visit to the U.S.A. A number of *'Numericord'* directors are now being manufactured by Concord Control, a company associated with Giddings and Lewis, for the United States Air Force Air Materiel Command.

MASSACHUSETTS INSTITUTE OF TECHNOLOGY
Cambridge, Massachusetts

This visit was made in order to see recent progress with digitally controlled machine tools, notably the electronic digitally controlled milling machine which had been developed from a Cincinnati 28 in. *'Hydro-Tel'* for the United States Air Force to produce components of complicated shape for prototype aircraft. It is fundamentally a much simpler machine than some that have been produced in the U.K., for it makes no attempt to follow a curve from its mathematical equation. Its designers recognized that there was little real need in aircraft engineering to generate perfect curves of this kind, and that a machine would follow a number of tangent planes was all that was really necessary. It moves the workpiece in straight lines between a series of points by means of binary digital input from punched tape of distances and times of traverse along three orthogonal axes. The accuracy of curve-following is under the control of the programmer, who simply feeds in more tangent planes for a given length of curve if he wants a smoother finish.

Many airframe components, produced in the M.I.T. machine, were on show. A typical wing-root bracket had taken six weeks to programme, but the weeks could have been reduced to hours by using a large-scale electronic computer to determine the feed rates and times. Most circular arcs had been programmed at 1° tangent plane intervals, and the finish appeared to be quite acceptable. Several people who had worked on the development of the M.I.T. machine expressed great admiration for a later British machine which would follow curves automatically from simple mathematical equations; but they did not believe the extra complication and cost of this system was worth while in engineering, though it might be essential for radar work (wave guides).

Although unable to meet any of the professors concerned, the writer formed the impression that work on the digital control of machine tools at the M.I.T. had virtually ceased. The project leader who was in charge of this work during its final stages has left to form a manufacturing company, Concord Control.

NATIONAL ADVISORY COMMITTEE FOR AERONAUTICS
Lewis Flight Propulsion Laboratory, Hopkins Airport, Cleveland, Ohio

The purpose of the visit was to see the digital analogue system for the control of machine tools, which has been developed for the production of blades for experimental gas turbines and compressors.

Compressor and turbine blades for aricraft gas turbines are of very complicated shape. They are tapered and twisted from root to tip, and the blade section is usually precisely defined at only three reference sections, the drawing generally indicating that the surface must blend smoothly with these three profiles. To make such a blade on a digitally controlled machine tool, one might draw to a large scale the profiles at a very large number of radial stations and measure the co-ordinates of many points on each profile, so working out a programme for the machine to follow. The draughting time alone rules out this method, for it would be quicker to make a master blade by hand and produce copies in a tracer-controlled milling machine in the time-honoured way. What is required is some automatic method of interpolation which will produce a smooth curve through a reasonable number of points at each of the three reference sections, and which will also do the same for taper and allow for twist. The N.A.C.A. machine uses a spline interpolator for the first, mathematical interpolation for the second, and automatic co-ordinate phase shift for the last.

The blade profile at one of the three design radii is drawn to a large scale, and the drawing is mounted on a rotary table in such a way that its centre of rotation is at the point about which the reference section concerned will be turned during manufacture. The platform on which the rotary table turns carries a synchro, which signals the angular position of the profile relative to a fixed datum line. It also carries a potentiometer device, upon the slider of which can be mounted a disc representing (to the same scale) the cutter which is to be used. The axis of the slider is set up to cross the rotary table along a path similar to that traced by the cutter centre in the machine tool. Output from the angular synchro and the potentiometer are fed to an I.B.M. 604 punched-card machine.

When programming a given profile, the machine operator sets the profile in the reference position, then slides the potentiometer until the cutter disc touches the profile: he then presses a switch, which transmits the angular position of the profile and the cutter displacement to an I.B.M. punched card. After indexing the profile round 10°, this operation is repeated, and so on until 36 cards have been produced for that profile.

This procedure is repeated for the other two reference sections, but the large-scale drawings of these are mounted on the rotary table in the angular positions they would occupy if the blade to be machines had no twist from root to tip. The co-ordinates for each section are stored in the I.B.M. machine as before.

Several light-alloy compressor blades, which had been manufactured in the N.A.C.A. machine, were available for inspection. Even with 10° initial co-ordinate spacing and the temporary machine tool, the finish was quite good and little hand-smoothing would be required.

MEAT PROCESSING

CANADA PACKERS
St. Boniface, Manitoba

This firm, which claims to be the largest meat-packing firm in the British Commonwealth, employs 1500 people on all kinds of food processing and packaging, including packed and canned meats, bacon, sausage, margarine, lard and peanut butter. It has the usual deodorizing equipment for natural oils, which gives partial distillation under vacuum and hydrogenation. The margarine plant uses automatic mixing equipment with displacement pumps metering the various ingredients, which are fed into a common mixing chamber. Automatic machines are also used for all kinds of packaging and, of course, for sausage manufacture. The cellophane packages for sliced meat and bacon are automatically filled, evacuated and heat sealed.

Lard was aerated before measurement and was metered by volume under a standard controlled pressure before packaging. Gas for the hydrogenation of oils was produced by electrolysis of water, the oyxgen being passed to waste as the quantity involved was stated to be not worth recovering.

Usually, only one shift is worked. One automatic production line handles 100 steers an hour and another 450 hogs an hour. Only the first was seen; the other was said to be quite similar apart from the method of killing. Three steers are received into the line, where they enter a pen which closely confines them side by side. They are then stunned, and the bodies are hitched to an overhead monorail conveyor which grasps them by one leg and hangs them head downwards. Each body is struck and bled, and is then carried by conveyor to a pneumatically operated automatic hide-stripping machine, which removes the complete hide at one operation. This machine is assisted by only one man, who has to make a quck incision in the flesh at one point in the operation to prevent the hide from tearing.

After the hide has been stripped, the carcass is sawn in half from top to bottom by a hand-held reciprocating saw driven by an electric motor; the two halves then proceed, one after the other, along the overhead monorail. After complete dismemberment by many specially designed machine saws, the two half carcasses are steam-sterilized before passing over an ingenious delay mechanism, which weighs the pair of them.

The conventional method of handling steers is to remove the carcasses from the monorail and do the difficult hide-stripping operation on the floor. A very large floor area is needed and it is hard to keep the meat free from contamination.

It was said that the new method of handling the carcass had considerably reduced the dressing time and had greatly improved the dressing quality. The system is known as the 'Can-Pac' system of beef dressing and is fully patented: Globe Machinery of Chicago, Ill., have the manufacturing rights in the U.S.A.; the North British Lifting and Moving Appliance Company, London, in the U.K., France, Denmark, Norway and Sweden; and the Cincinnati Butchers Supply Company, Cincinnati, Ohio, elsewhere.

The whole of the equipment used in this slaughterhouse had been designed and developed by Canada Packers. Overhead conveyors facilitated the movement of by-products everywhere and even the pulleys and hooks running along the conveyors were automatically sterilized by being passed through a steam chamber before returning to the beginning of the line.

PIPE FITTINGS

WALWORTH COMPANY
789 E. First Street, Boston 8, Mass.

The Walworth Company has four plants in the U.S.A. producing valves and pipe fittings of all kinds. The show piece was the shell-moulding foundry, producing valves and fittings up to 2 in. bore in brass and steel. 1400 shell moulds are produced each day on a rotary-table machine developed by Walworth in conjunction with the Link Belt Company. The machine is entirely automatic; it is fed with mixed sand from a hopper and unloaded to a buffer store by one man, who also acts as inspector.

In the store, pairs of completed moulds are suspended by wire hooks from axially rotating and very slightly inclined bars; they feed down automatically to the end of the store, where they are removed as required. One half is placed on an open conveyor, where a synthetic resin cement is extruded on to the joint face through a plastics hose. Meanwhile, the other half is passing along a parallel conveyor which takes it through an oven electrically heated to 240°F. The ends of these two conveyors terminate opposite a merry-go-round which carries aluminium boxes with hinged lids, each box and lid containing an inflatable rubber bladder. The two half moulds are placed on top of one another in the box, the lid is closed by hand and a clasp fixed. Air pressure is then automatically applied to the bladders, and by the time the other side of the merry-go-round is reached, the cement has set.

Here, the cemented shell mould is removed from the box, and is passed to a second merry-go-round. This carries triangular section buckets about 3 ft. long, 2 ft. deep and 1 ft. wide, into which two moulds are placed. The bucket with its moulds passes under a station where steel shot ($\frac{1}{8}$ in.) is automatically poured around the moulds; the supply is cut off when the right shot depth is reached. The backed-up moulds then pass along a platform, which is synchronized to move at the same speed as the moulds.

Brass melted in an oil-fired furnace is poured from small ladles, which are supported from an overhead monorail and which move freely with the pourer along the conveyor. At a later station, the buckets are automatically tipped, discharging the castings, moulds, and shot into a vibrating separator, which shakes out the sand and shot. The shot is caught by a screen, and returned by air blast to an overhead hopper, from which it is passed back to the merry-go-round as required.

A similar but less highly automatic plant is used for steel valve bodies; here 3 per cent red iron oxide is added to the sand/resin mix to improve the surface finish of the castings. No sand is recovered from either plant.

In the machine shop there is an impressive special-purpose machine (Goss and de Leeuw Machine Company, Kensington, Conn.). It has four work stations and three mandrels, and it machines all three valve faces at the same setting, producing 240 1-in. valves or 180 2-in. valves an hour. Alongside it is a machine which finishes the valve seating. One operator runs both machines, so the production figures relate to the complete machining of brass-valve bodies by one operator.

Also in the machine shop were several ram-type turret lathes which had been converted to automatic reciprocation with hydraulic drives (The Lynn Manufacturing Company, Minneapolis, Minn.). The drive cost $3000, but with it one operator could handle two machines. The cost of converting two machines was $6000 and the annual saving on one operator $4000; the capital could thus be recovered in 18 months.

PLYWOOD AND PLYWOOD DOORS

MACMILLAN & BLOEDEL
Vancouver Plywood Division, Cromwell Street, Vancouver, B.C.

This plywood manufacturing division of an old established firm now has about 1500 employees. Douglas fir logs, from water-storage in and near the Fraser River, have their bark skimmed off in a special mill, which is loaded and operated by one man. The bark is passed straight into the boiler plant and burned without previous drying. Clean wood waste is chipped for the manufacture of sulphate pulp at the company's Harmac pulp mill near Nanaimo, B.C. Other wood waste is briquetted in eight machines (Wood Briquettes, Lewiston, I.) which automatically load waste into a multiplicity of cylindrical chambers of about 5 in. diameter incorporated in the rim of a large wheel. As the wheel is indexed round, the contents of each chamber are highly compressed into briquettes about a foot long, the resin in the wood providing the only bond. The briquettes are automatically unloaded on to a conveyor and taken to storage. They sell for 7¼ cents each and are used for domestic heating.

After peeling, the veneers are dried during their continuous travel through large ovens; the speed of travel is varied for different veneer thicknesses. The veneers are then passed along a roller conveyor, and under an electronic moisture meter which causes any sheet with excess water content to be marked with a red reject mark. The veneers then pass visual inspection before stacking. All gluing, press loading and unloading are automatic. Machines of German manufacture (B. Raimann, St. Georgen-Freibourg) punch elliptical holes at knots, and insert sound veneer at the same stroke. Pneumatic routers of the firm's own design cut thin channels out of completed ply at cracks, and these are then filled with sound veneer strip about 0.2 in. wide.

Plywood-covered doors, of which up to 2500 are produced per day, are assembled in a pneumatically clamped jig of the firm's own design and manufacture. Glue is applied simultaneously to both sides of the frames by passing them through a roller-type glue spreader; the frames are then faced with three-ply panels. After gluing, the doors are passed along a roller conveyor and through a machine which routs them to correct width and then chamfers the edges.

A large quantity of plywood is now produced with a pattern hot-rolled on one face to represent fine carving. This is used for wall panelling.

In the plant, many operations had become automatic with a consequent reduction in labour. In one operation, labour had been reduced by 67 per cent, but the average reduction over the past two years had been about 10 per cent. However, the resulting expansion of the business had allowed the people displaced to be absorbed in other jobs as well as providing employment for other workers. There was stated to be no trade union opposition to this automation. The relations between management and staff were obviously very friendly.

STEEL

ALLEGHENY LUDLAM STEEL CORPORATION
River Road, Brackenridge, Pittsburgh, Pennsylvania

This plant employs about 5000 people, and produces stainless steel and high silicon steels for electrical applications: the total monthly melting capacity is about 75,000 tons. All the stainless steel is melted in electric arc furnaces while the silicon steels are melted in open-hearth furnaces fired by natural gas. A large proportion of the furnace charge for the arc furnaces was made up of machining scrap, which emitted very large volumes of dense smoke and fumes; these polluted the atmosphere for miles around the works.

The only point of metallurgical interest noted during the visit was the use of nodular-graphite cast-iron for the charging doors of the electric arc furnaces and for the hot tops of ingot moulds. It was said that the nodular iron doors and mould tops were lasting two or three times as long in service as previous grey-iron components. Both doors and hot tops were made at the Brackenridge Plant by the nickel-magnesium process, but this was regarded as a sideline and visitors were not shown the plant used.

The writer was unable to see the initial stages of rolling the ingot, because the turning gear on the roller tables of the blooming mill had become unserviceable. The hot strip mill was still working correctly and seemed to follow the same general practice as in the United Kingdom. There was no evidence from a distance of automatic gauge control in use on the cold strip mill and the guide did not know whether it was available. Even from a distance what appeared to be a pair of distant-reading gauges could be seen mounted above the control panel for each mill; the operator watched these very carefully throughout the process, adjusting various controls one way and the other as the strip went through. It could be assumed that the gauges measured the thickness of the strip at two points across the width, and that the operator had to adjust the mill screws as necessary to keep both sides of the trip within the required limits.

An unusual feature of this plant is the large Lamination Stamping Department, which produces a great variety of laminations for electric motors and transformers. There are big advantages in performing this work at the steel mill in that shipping weights are reduced and the scrap can be fed back to the furnaces, with no external transport charges. The punch presses varied in capacity from 25 to 175 tons, whilst many of the dies were of tungsten carbide and were said to have a service life of more than 25 million stampings. As might be expected, nearly all the punch presses incorporated automatic feed from continuous strip and automatic unloading. Used in conjunction with the carbide dies, they could be left in continuous operation for very long periods without individual attention.

REPUBLIC STEEL CORPORATION
Harvard Avenue at East Seventeenth Street, Cleveland, Ohio

The 98-in. continuous strip mill here is claimed to be the widest in the world: it can produce hot and cold rolled sheet up to 91 in. wide, and it makes over 160,000 tons of sheet steel a month.

Time only permitted the rolling operations to be seen at this works. There were no unusual features in the hot rolling mill. There was no automatic positioning control for the slab, but apparently one was being developed in the company's research department.

There are three cold-rolling mills at this plant, the oldest being 54 in. wide, the next 98 in. wide, and the most recent addition 72 in. wide. The last-mentioned is a three-stand, four high reversing mill (United States Steel Corporation, U.S. Steel Supply Division, Chicago, Ill.) equipped with automatic gauge control (Sheffield Corporation, Dayton, Ohio) on the two end stands. Gauging is performed by β-ray absorption, the radioactive source being carried in a vizor located between 18 in. and 24 in. above the strip emerging from the work rolls. The operator's control panels had only two dials, graduated in thousands of an inch and tens of thousands of an inch: all he had to do was to set the gauge size desired to the nearest thousandth and the automatic gauge contol did the rest. No figures could be obtained comparing the percentages of off-range sheet rolled in the mills with and without automatic gauge control.

Selected Bibliography

BARTON, H. K. and BARTON, L. C. Low frequency induction heating in the die-casting shop. *Machinery,* Lond., 1954, LXXV (2184), 676-85.

BENNETT, F. C. A hot chamber die-casting machine for magnesium. *Modern Metals,* 1954, X (11), 76.

BRENNER, A. Electroless plating comes of age. *Metal Finishing,* Westwood, N. J., November 1954, 68: December 1954, 61.

ECKMAN, D. P., TAFT, C. K. and SCHUMAN, R. L. An electro-hydraulic servomechanism with ultra-high frequency response. Case Institute of Technology, November 1, 1955.

FARMER, P. J. Analogue control: application of the E.M.I. control system to a standard vertical milling machine. *Aircr. Prod. Lond.*, April, 1956, 126.

McKEE, K. M. Dynasert automatic component assembly machine for printed circuits. *Machinery, Lond.*, 1956 LXXXIX (2296), 1115-21.

MERGLER, H. W. A numerical punched tape, machine tool control system. *Control Engineering Lond.*, 1955, II (a), 132.

MERGLER, H. W. A numerical machine tool control system operated from coded punched paper tape. Case Institute of Technology, Cleveland, Ohio, Ph.D. Thesis, 1956.

MERGLER, H. W. A digital-analog machine tool control system. Western Computer Conference (sponsored by the American Institute of Electrical Engineers, the Institute of Radio Engineers, and the Association for Computing Machines), Los Angeles, Calif., February 11, 1954.

PROKOPOVICH, A. E. Automatic piston works. (Public lecture delivered at the All-Union Society for the Propagation of Political and Scientific Knowledge) (Zavod. Avtomat.), Moscow, 1951. (Translation available on loan from Department of Scientific and Industrial Research, Records Section, 5-11 Regent Street, London, S.W. 1.)

SIEGEL, A. Automatic programming of numerically-controlled machine tools. *Control Engineering, Lond.*, October, 1956, III, 65.

TERBORGH, G. Dynamic equipment policy. (New York: McGraw-Hill, 1949.)

Reports

A numerically-controlled milling machine: final report to the U.S. Airforce on construction and initial operation. (Servomechanisms Laboratory, Massachusetts Institute of Technology, Cambridge, Mass.; Part 1, July 30, 1952; Part 2, May 31, 1953.)

Airslip forming: a new vacuum technique for thermoplastics, *Engineering, Lond.*, 1956, CLXXXI, 566.

Aluminio nella industria automobilistica Americana. *Alluminio*, June, 1956, 294-6.

Aluminium engine block successfully die cast. *New York Times,* December 2, 1955. See also BAUER, A. F. The V-8 is next: diecast six cylinder block. *Modern Metals,* 1956, XII (5) , 72.

Automation: a report on the technical trends and their impact on management and labour. (London: H.M. Stationery Office, 1956.)

Computer control in industrial processes. Case Institute of Technology. August 19, 1953. (Copy available in Department of Scientific and Industrial Research.)

Graduate courses in instrumentation, servomechanisms and process control. Case Institute of Technology, 1956 (Copy available in Department of Scientific and Industrial Research) .

Reynolds to expand plant to fill big aluminum order for Ford. *New York Times,* January 12, 1956.

Zinc and aluminium die casting. Anglo-American Council on Productivity, London, 1952.